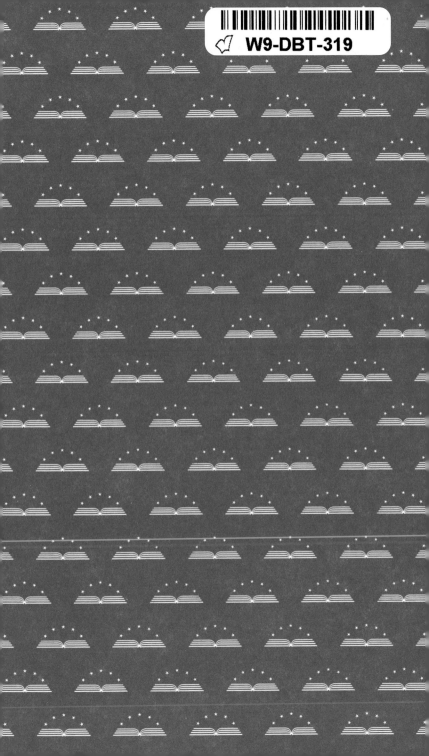

Library of America, a nonprofit organization,
champions our nation's cultural heritage
by publishing America's greatest writing in
authoritative new editions and providing resources
for readers to explore this rich, living legacy.

LOREN EISELEY
COLLECTED ESSAYS

VOLUME TWO

LOREN EISELEY

COLLECTED ESSAYS ON EVOLUTION, NATURE, AND THE COSMOS

———————————

VOLUME TWO

The Invisible Pyramid
The Night Country
Essays from *The Star Thrower*

———————————

William Cronon, *editor*

THE LIBRARY OF AMERICA

Visit our website at www.loa.org.

This paper meets the requirements of
ANSI/NISO Z39.48–1992 (Permanence of Paper).

Distributed to the trade in the United States
by Penguin Random House Inc.
and in Canada by Penguin Random House Canada Ltd.

Library of Congress Control Number: 2016935164
ISBN 978–1–59853–507–5

First Printing
The Library of America—286

Manufactured in the United States of America

Loren Eiseley:
Collected Essays on Evolution, Nature, and the Cosmos
is published with support from

THE GOULD FAMILY FOUNDATION

and will be kept in print by its gift to
the Guardians of American Letters Fund,
established by the Library of America
to ensure that every volume in the series
will be permanently available.

Contents

THE INVISIBLE PYRAMID

DEDICATED TO THE MEMORY OF

FRANK G. SPECK

to me,

the last magician

Preface

THE THEME of this book was developed through a series of lectures delivered under the auspices of the John Danz Fund at the University of Washington in Seattle in the Fall of 1969. It gives me pleasure to express to the members of the Danz family my appreciation of their interest and generosity, as well as to the administrative staff of the University, who were my hosts.

I should also like to express my thanks to my friend and former colleague, the astronomer Frank Bradshaw Wood, of the University of Florida, Gainesville, for information cheerfully supplied me upon the elliptic of Halley's comet. Similarly I am glad to seize this opportunity to mention the many provocative conversations which have taken place with my university colleagues, Froelich Rainey, Director of the University Museum, and Dale Coman, M.D., of the University Medical School. As men concerned with the growing problems of our environment, we share equal anxieties and hopes.

In this book I have chosen, for literary reasons extending into the seventeenth century, to use somewhat interchangeably the terms Halley's star and Halley's comet, since for the latter no satisfactory synonym exists. I do not think anyone will be confused by this interchange, which has stylistic advantages in a book of this nature.

Once in a cycle the comet
Doubles its lonesome track.
Enriched with the tears of a thousand years,
Aeschylus wanders back.
—JOHN G. NEIHARDT

Prologue

MAN WOULD NOT be man if his dreams did not exceed his grasp. If, in this book, I choose to act in the ambivalent character of pessimist and optimist, it is because mankind itself plays a similar contradictory role upon the stage of life. Like John Donne, man lies in a close prison, yet it is dear to him. Like Donne's, his thoughts at times overleap the sun and pace beyond the body. If I term humanity a slime mold organism it is because our present environment suggests it. If I remember the sunflower forest it is because from its hidden reaches man arose. The green world is his sacred center. In moments of sanity he must still seek refuge there.

If I dream by contrast of the eventual drift of the star voyagers through the dilated time of the universe, it is because I have seen thistledown off to new worlds and am at heart a voyager who, in this modern time, still yearns for the lost country of his birth. As an anthropologist I know that we exist in the morning twilight of humanity and pray that we may survive its noon. The travail of the men of my profession is to delve amid the fragments of civilizations irretrievably lost and, at the same time, to know man's enormous capacity to create.

But I dream, and because I dream, I severally condemn, fear, and salute the future. It is the salute of a gladiator ringed by the indifference of the watching stars. Man himself is the solitary arbiter of his own defeats and victories. I have mused on the dead of all epochs from flint to steel. They fought blindly and well against the future, or the cities and ourselves would not be here. Now all about us, unseen, the final desperate engagement continues.

If man goes down I do not believe that he will ever again have the resources or the strength to defend the sunflower forest and simultaneously to follow the beckoning road across the star fields. It is now or never for both, and the price is very high. It may be, as A. E. Housman said, that we breathe the air that kills both at home and afar. He did not speak of pollution; he spoke instead of the death that comes with memory. I have wondered how long the social memory of a great culture can be sustained without similarly growing lethal. This also our century may decide.

I confess that the air that kills has been breathed upon the pages of this book, but upon it also has shone the silver light of flying thistledown. In the heart of the city I have heard the wild geese crying on the pathways that lie over a vanished forest. Nature has not changed the force that drives them. Man, too, is a different expression of that natural force. He has fought his way from the sea's depths to Palomar Mountain. He has mastered the plague. Now, in some final Armageddon, he confronts himself.

As a boy I once rolled dice in an empty house, playing against myself. I suppose I was afraid. It was twilight, and I forget who won. I was too young to have known that the old abandoned house in which I played was the universe. I would play for man more fiercely if the years would take me back.

ONE

The Star Dragon

Already at the origin of the species man was equal to what he was destined to become.

—JEAN ROSTAND

IN THE YEAR 1910 Halley's comet—the comet that among many visitations had flared in 1066 over the Norman invasion of England—was again brightening the night skies of earth. "Menace of the Skies," shrieked the more lurid newspapers.

Like hundreds of other little boys of the new century, I was held up in my father's arms under the cottonwoods of a cold and leafless spring to see the hurtling emissary of the void. My father told me something then that is one of my earliest and most cherished memories.

"If you live to be an old man," he said carefully, fixing my eyes on the midnight spectacle, "you will see it again. It will come back in seventy-five years. Remember," he whispered in my ear, "I will be gone, but you will see it. All that time it will be traveling in the dark, but somewhere, far out there"—he swept a hand toward the blue horizon of the plains—"it will turn back. It is running glittering through millions of miles."

I tightened my hold on my father's neck and stared uncomprehendingly at the heavens. Once more he spoke against my ear and for us two alone. "Remember, all you have to do is to be careful and wait. You will be seventy-eight or seventy-nine years old. I think you will live to see it—for me," he whispered a little sadly with the foreknowledge that was part of his nature.

"Yes, Papa," I said dutifully, having little or no grasp of seventy-five years or millions of miles on the floorless pathways of space. Nevertheless I was destined to recall the incident all my life. It was out of love for a sad man who clung to me as I to him that, young though I was, I remembered. There are long years still to pass, and already I am breathing like a tired runner, but the voice still sounds in my ears and I know with

7

the sureness of maturity that the great wild satellite has reversed its course and is speeding on its homeward journey toward the sun.

At four I had been fixed with the compulsive vertigo of vast distance and even more endless time. I had received, through inherited temperament and inclination, a nostalgic admonition to tarry. Besides, I had given what amounted to a desperate promise. "Yes, Papa," I had said with the generosity of childhood, not knowing the chances that men faced in life. This year, after a visit to my doctor, I had written anxiously to an astronomer friend. "Brad," I had asked, "where is Halley's comet reported on the homeward track? I know it must have turned the elliptic, but where do you calculate it now, how far—and how long, how long—?"

I have his answer before me. "You're pushing things, old man," he writes. "Don't expect us to see it yet—you're too young. The orbit is roughly eighteen astronomical units or one billion six hundred and fifty million miles. It headed back this way probably in nineteen forty-eight."

Nineteen forty-eight. I grope wearily amidst memories of the Cold War, Korea, the Berlin blockade, spies, the impossible-to-be-kept secrets of the atom. All that time through the black void the tiny pinpoint of light has been hurrying, hurrying, running faster than I, thousands of miles faster as it curves toward home. Because of my father and the promise I had made, a kind of personal bond has been projected between me and the comet. I do not think of what it heralded over Hastings in 1066. I think it is racing sunward so that I can see it stretched once more across the heavens and momently restore the innocence of 1910.

But there is inner time, "personal, private chronometry," a brain surgeon once told me. There is also outer time that harries us ruthlessly to our deaths. Some nights in a dark room, staring at the ceiling, I can see the light like a mote in my eye, like a far-off train headlight glimpsed long ago as a child on the prairies of the West. The mournful howl of the train whistle echoes in my head and mingles with the night's black spaces. The voice is that of the comet as I hear it, climbing upward on the arc of space. At last in the dark I compose myself for sleep. I pull the blanket up to my chin and think of radar ceaselessly sweeping the horizon, and the intercontinental missiles resting in their blast-hardened pits.

But no, I dream deeper, slipping back like a sorcerer through the wood of time. Life was no better, not even as safe, proportionately, in the neolithic hill forts whose tiny trenches can be seen from the air over the British downs. A little band of men, with their families beside them, crouched sleepless with ill-made swords, awaiting an attack at dawn. And before that, the caves and the freezing cold, with the ice creeping ever southward autumn by autumn.

The dead we buried in red ochre under the fire pit, the red standing for blood, for we were quick in analogies and magic. The ochre was for life elsewhere and farewell. We tramped away in our furred garb and the leaves and snow washed over the place of our youth. We worked always toward the south across the tundra following the long trail of the mammoth. Someone saw a vast flame in the sky and pointed, but it was not called Halley's comet then. You could see it glinting through the green light and the falling snow.

Farther backward still across twin ice advances and two long interglacial summers. We were cruder now, our eyes wild and uncertain, less sure that we were men. We no longer had sewn garments, and our only weapon was a heavy pointed stone, unhafted and held in the hand. Even our faces had taken on the cavernous look of the places we inhabited. There were difficulties about making fire, and we could not always achieve it. The dead were left where they fell. Women wept less, and the bands were smaller. Our memories consisted of dim lights under heavy sockets of bone. We did not paint pictures, or increase, by magic, the slain beasts. We talked, but the words we needed were fewer. Often we went hungry. It was a sturdy child that survived. We meant well but we were terrifyingly ignorant and given to frustrated anger. There was too much locked up in us that we could not express.

We were being used, and perhaps it was against this that we unconsciously raged the most. We were neither beast nor man. We were only a bridge transmitting life. I say we were almost animals and knew little, but this we felt and raged against. There were no words to help us. No one could think of them. Sometimes we were stalked by the huge cats, but it was the inner stalking that was most terrible. I saw a star in the sky with a flaming tail and cowered, shaking, into a bush, making uncouth sounds. It is not laughable. Animals do not do this. They do not see the world as we do—even we.

I think we are now well across the last ice, toward the beginning. There is no fire of any sort but we do not miss it. We are far to the south and the climate is warm. We have no tools except an occasional bone club. We walk upright, but I think we are now animals. We are small—pygmies, in fact. We wear no clothes. We no longer stare at the stars or think of the unreal. The dead are dead. No one follows us at nightfall. Do not repeat this. I think we are animals. I think we have reached beyond the bridge. We are happy here. Tell no one.

I sigh in my sleep but I cannot hold to the other side of the bridge—the animal side. The comet turns blazing on its far run into space. Slowly I plod once more with the furred ones up the ladder of time. We cross one ice and then another. There is much weeping, too much of memory. It is all to do over again and go on. The white-robed men think well in Athens. I heard a man named Pindar acclaim something that implied we have a likeness to the immortals. "What course after nightfall," he questioned, "has destiny written that we must run to the end?"

What course after nightfall? I have followed the comet's track returning and returning while our minds and our bodies changed. The comet will appear once more. I will follow it that far. Then I will no longer be part of the bridge. Perhaps I will be released to go back. Time and space are my inheritance from my father and the star. I will climb no further up the ladder of fiery return. I will go forward only one more rung. What will await me there is not pleasant, but it is in the star's destiny as well as mine. I lie awake once more on the dark bed. I feel my heart beating, and wait for the hurrying light.

II

In 1804, well over a century and a half ago, Captain William Clark recorded in his diary far up the unknown Missouri that ahead of the little expedition that he shared with Meriwether

Lewis hung a formidable curtain of blowing dust through which they could not see.

"Tell us what is new," the few savants in the newborn American republic had advised the explorers when they departed westward. Men continued to have strange expectations of what lay hidden in the still uncharted wilds behind the screen of the great eastern forest. Some thought that the mammoth, whose bones had been found at Big Bone Lick, in Kentucky, might still wander alive and trumpeting in that vast hinterland. The "dreadful curtain" through which the youthful captains peered on that cold, forbidding day in January could have hidden anything. Indeed the cloud itself was symbolic. It represented time in inconceivable quantities—time, not safe, not contained in Christian quantity, but rather vast as the elemental dust storm itself.

The dust in those remote regions was the dust of ice ages, of mountains wearing away under the splintering of frost and sun. The Platte was slowly carrying a mountain range to the sea over giant fans of gravel. Frémont's men would later report the strange and grotesque sculptures of the wind in stone. It was true that a few years earlier the Scottish physician James Hutton had philosophically conceived such time as possible. His views had largely proved unwelcome and had been dismissed in Europe. On the far-western divide, however, amid the roar of waters falling toward an unknown western ocean, men, frontiersmen though they were, must have felt with an increasing tinge of awe the weight of ages unknown to man.

Huge bones bulked in the exposed strata and were measured with wonder. No man knew their names or their antiquity. New things the savants had sought surrounded the explorers, not in the sense of the living survival of great elephants but rather in the sense of a vaster novelty—the extension of time itself. It was as though man for the first time was intruding upon some gigantic stage not devised for him. Among these wastes one felt as though inhuman actors had departed, as though the drama of life had reached an unexpected climax.

One catches this same lost feeling in the remarks of another traveler, Alexis de Tocqueville, venturing into the virgin forest far from the pruned orchards of France. "Here," he said, "man seems to enter life furtively. Everything enters into a silence so profound, a stillness so complete that the soul feels penetrated by a sort of religious terror." Even in the untouched forest, time had taken on this same American quality: "Immense trees," de Tocqueville wrote in awe, "retained by the surrounding branches, hang suspended in the air and fall into dust without touching the earth."

It is perhaps a significant coincidence that man's full recognition of biological novelty, of the invisible transformations of the living substance itself, came close upon the heels of the discovery of the vast wilderness stage which still held the tumbled bones of the former actors. It was a domain which had remained largely unknown to Europeans. Sir Charles Lyell, who, in the 1830s, successfully revived Hutton's lost doctrines of geological antiquity, visited the United States in the 1840s and lectured here to enthralled thousands. Finally, it was Charles Darwin, the voyager-naturalist, who, as a convinced follower of Lyell, had gazed upon a comparable wilderness in South America and had succeeded, in his mind's eye, in peopling the abandoned stage with the creatures of former epochs. It was almost as though Europe, though rife with speculation since the time of the great voyagers, could not quite escape its man-centeredness or its preoccupation with civilized hedgerows and formal gardens. Its thinkers had still to breathe, like Darwin, the thin air of Andean highlands, or hear the falling of stones in mountain cataracts.

To see his role on the world stage, Western man had twice to revise his conception of time: once from the brevity of a few thousand years to eons of inconceivable antiquity, and, a second time, with far more difficulty, to perceive that this lengthened time-span was peopled with wraiths and changing cloud forms. Time was not just aged rocks and trees, alike since the beginning of creation; its living aspect did not consist merely of endless Oriental cycles of civilizations rising and declining. Instead, the living flesh itself was alterable. Our seeming stability of form was an illusion fostered by the few millennia of written history. Behind that history lay the vast and unrecorded gloom of ice ages inhabited by the great beasts which the explorers, at Thomas Jefferson's bidding, had sought through the blowing curtain of the dust.

Man, but not man in the garb we know, had cracked marrow

bones in those dim shadows before his footprints vanished amidst the grass of wild savannahs. For interminable ages winged reptiles had hovered over the shores of ancient seas; creatures still more strange had paddled in the silence of enormous swamps. Finally, in that long backward range of time, it was possible to emerge upon shores which no longer betrayed signs of life, because life had become mere potential.

At that point one could have seen life as the novelty it truly is. "Tell us what is new," reiterated the eager scientists to the explorers. Past mid-century, an answer could be made. It was life itself that was eternally, constantly new. Dust settled and blew the same from age to age; mountains were worn down to rise again. Only life, that furtive intruder drifting across marsh and field and mountain, altered its masks upon the age-old stage. And as the masks were discarded they did not come again as did the lava of the upthrust mountain cores. Species died as individuals died, or, if they did not perish, they were altered beyond recognition and recall. Man cannot restore the body that once shaped his mind. The bird upon the bough cannot, any more than a summer's yellow butterfly, again materialize the chrysalis from which it sprang.

Indeed, in the end, life can be seen not only as a novelty moving through time toward an endlessly diverging series of possible futures but also as a complete phantom. If we had only the scattered chemicals of the cast-off forms and no experience in ourselves of life's existence, we would not be able to identify its reality or its mutability by any chemical test known to us. The only thing which infuses a handful of dust with such uncanny potential is our empirical knowledge that the phenomenon called life exists, and that it constantly pursues an unseen arrow which is irreversible.

Through the anatomical effort and puzzle-fitting of many men, time, by the mid-nineteenth century, had become gigantic. When *On the Origin of Species* was published, the great stage was seen not alone to have been playing to remote, forgotten audiences; the actors themselves still went masked into a future no man could anticipate. Some straggled out and died in the wings. But still the play persisted. As one watched, one could see that the play had one very strange quality about it: the characters, most of them, began in a kind of generous latitude of living space and ended by being pinched out of existence in a grimy corner.

Once in a while, it is true, a prisoner escaped just when all seemed over for him. This happened when some oxygen-starved Devonian fish managed to stump ashore on their fins and become the first vertebrate invaders of the land. By and large, however, the evolutionary story had a certain unhappy quality.

The evolutionary hero became a victim of his success and then could not turn backward; he prospered and grew too large and was set upon by clever enemies evolving about him. Or he specialized in diet, and the plants upon which he fed became increasingly rare. Or he survived at the cost of shutting out the light and eating his way into living rock like some mollusks. Or he hid in deserts and survived through rarity and supersensitive ears. In cold climates he reduced his temperature with the season, dulled his heart to long-drawn spasmodic effort, and slept most of his life away. Or, parasitically, he slumbered in the warm intestinal darkness of the tapeworm's eyeless world.

Restricted and dark were many of these niches, and equally dark and malignant were some of the survivors. The oblique corner with no outlet had narrowed upon them all. Biological evolution could be defined as one long series of specializations—hoofs that prevented hands, wings that, while opening the wide reaches of the air, prevented the manipulation of tools. The list was endless. Each creature was a tiny fraction of the life force; the greater portion had died with the environments that created them. Others had continued to evolve, but always their transformations seemed to present a more skilled adaptation to an increasingly narrow corridor of existence. Success too frequently meant specialization, and specialization, ironically, was the beginning of the road to extinction. This was the essential theme that time had dramatized upon the giant stage.

III

It may now appear that I have been wandering mentally amidst irrelevant and strange events—time glimpsed through a blowing curtain of dust, and, among fallen stones and badland pinnacles, bones denoting not just the erosion of ages but the mysterious transformation of living bodies.

Man after man in the immediately post-Darwinian days would stare into his mirror at the bony contours of a skull that held some grinning secret beyond the simple fact of death. Anatomists at the dissecting table would turn up odd vestigial muscles and organs. Our bodies held outdated machinery as strange as that to be found in the attics of old houses. Into these anatomical depths few would care to probe. But there were scholars who were not averse to delving among fossils, and the skulls they found or diagnosed would multiply. These would be recognized at last for what they were, the dropped masks of the beginning of Nature's last great play—the play of man.

Strangely, it is a different play, though made partly of old ingredients. In three billion years of life upon the planet, this play had never been acted upon the great stage before. We come at a unique moment in geological history, and we ourselves are equally unique. We have brought with us out of the forest darkness a new unprophesiable world—a latent, lurking universe within our heads.

In the world of Charles Darwin, evolution was particulate; it contained and traced the history of fins, claws, wings, and teeth. The Darwinian circle was immersed in the study of the response of the individual organism to its environment, and the selective impact of the environment upon its creatures. By contrast, just as biological evolution had brought the magic of the endlessly new in organic form, so the evolving brain, through speech, had literally created a superorganic structure unimaginable until its emergence.

Alfred Russel Wallace, Darwin's contemporary, perceived that with the emergence of the human brain, man had, to a previously inconceivable degree, passed out of the domain of the particulate evolution of biological organs and had entered upon what we may call history. Human beings, in whom the power of communication had arisen, were leaving the realm of phylogeny for the realm of history, which was to contain, henceforth, our final destiny. After three billion years of biological effort, man alone had seemingly evaded the oblique trap of biological specialization. He had done so by the development of a specialized organ—the brain—whose essential purpose was to evade specialization.

The tongue and the hand, so disproportionately exaggerated in his motor cortex, were to be its primary instruments. With these he would elude channelized instinct and channelized organic development. The creature who had dropped from some long-ago tree into the grass had managed to totter upright and free the grasping forelimb. Brain, hand, and tongue would henceforth evolve together. Fin, fur, and paw would vanish into the mists of the past. Henceforth it would be the brain that clothed and unclothed man. Fire would warm him, flint would strike for him, vessels would carry him over dangerous waters.

In the end, with the naked body of an awkward and hastily readjusted climber, he would plumb the seas' depths and mount, with wings spun in his brain, the heights of air. Enormous computations upon the movements of far bodies in space would roll in seconds from his computers. His great machines would leap faster at his bidding than the slower speed of his own nerves.

Because of speech, drawn from an infinitesimal spark along a nerve end, the vague, ill-defined surroundings of the animal world would be transformed, named, and categorized. Mind would reach into a past before its becoming; the misty future experienced by dim animal instinct would leap into sudden, clear perspective. Language, whose constituents have come down the long traverse of millennia as rolled and pounded by circumstance as a flint ax churned in a river bed, leaves no direct traces of its dim beginnings. With the first hieroglyph, oral tradition would become history. Out of a spoken sound, man's first and last source of inexhaustible power, would emerge the phantom world which the anthropologist prosaically calls culture. Its bridges, its towers, and its lightnings lie potential in a little globe of gray matter that can fade and blow away on any wind. The novelty of evolutionary progression through time has begotten another novelty, the novelty of history, the evolutionary flow of ideas in the heads of men.

The role of the brain is analogous in a distant way to the action of mutation in generating improbabilities in the organic realm. Moreover, the human brain appears to be a remarkably solitary product of this same organic process which, in actuality, it has transcended. In this sense life has produced a newly emergent instrument capable of transmitting a greatly speeded-up social heredity based not upon the gene but instead upon communication. In its present technological phase it has brought the ends of the world into conflict and at the same time is reaching outward into space.

About ourselves there always lingers a penumbral rainbow—what A. L. Kroeber termed the superorganic—that cloud of ideas, visions, institutions which hover about, indeed constitute human society, but which can be dissected from no single brain. This rainbow, which exists in all heads and dies with none, is the essential part of man. Through it he becomes what we call human, and not otherwise.

Man is not a creature to be contained in a solitary skull vault, nor is he measurable as, say, a saber-toothed cat or a bison is measurable. Something, the rainbow dancing before his eyes, the word uttered by the cave fire at evening, eludes us and runs onward. It is gone when we come with our spades upon the cold ashes of the campfire four hundred thousand years removed.

Paradoxically, the purpose of the human brain is to escape physical specialization by the projections of thought. There is no parallel organism with which to compare ourselves. The creature from which we arose has perished. On the direct hominid line there is no twilight world of living fossils which we can subject to examination. At best we are forced to make inferences from less closely related primates whose activities lie below the threshold of speech.

The nineteenth century, in the efforts of men like Hughlings Jackson, came to see the brain as an organ whose primary parts had been laid down successively in evolutionary time, a little like the fossil strata in the earth itself. The centers of conscious thought were the last superficial deposit on the surface of a more ancient and instinctive brain. As the roots of our phylogenetic tree pierce deep into earth's past, so our human consciousness is similarly embedded in, and in part constructed of, pathways which were laid down before man in his present form existed. To acknowledge this fact is still to comprehend as little of the brain's true secrets as an individual might understand of the dawning of his own consciousness from a single egg cell.

The long, slow turn of world-time as the geologist has known it, and the invisibly moving hour hand of evolution perceived only yesterday by the biologist, have given way in the human realm to a fantastically accelerated social evolution induced by industrial technology. So fast does this change progress that a growing child strives to master the institutional customs of a society which, compared with the pace of past history, compresses centuries of change into his lifetime. I myself, like others of my generation, was born in an age which has already perished. At my death I will look my last upon a nation which, save for some linguistic continuity, will seem increasingly alien and remote. It will be as though I peered upon my youth through misty centuries. I will not be merely old; I will be a genuine fossil embedded in onrushing man-made time before my actual death.

IV

"There never was a first man or a first primate," Dr. Glenn Jepsen of Princeton once remarked iconoclastically. The distinguished paleontologist then added that the "billions of genetic filaments in our ancestral phyletic cord are of many lengths, no two precisely the same. We have not had our oversized brain very long but the pentadactyl pattern of our extremities originated deep in . . . the Paleozoic." Moreover, we have, of late, discovered that our bipedal, man-ape ancestors seem to have flourished for a surprisingly long time without any increase in their cranial content whatever—some four or five million years, in fact.

It used to be thought that the brain of proto-man would have had to develop very early to enable him to survive upright upon the ground at all. Oddly, it now appears that man survived so well under these circumstances that it is difficult to say why, in the end, he became man at all. His bipedal pre-man phase lasted much longer—five or six times at least—than his whole archaeological history down to this very moment. What makes the whole story so mystifying is that the expansion of his neurocranium took place relatively rapidly during the million years or so of Ice Age time, and has not been traced below this point. The supposed weak-bodied creature whom Darwin nervously tried to fit into his conception of the war of nature on the continents is thought to have romped through a longer geological time period than his large-brained descendants may ever see.

We know that at least two million years ago the creature could make some simple use of stones and bones and may possibly have fashioned crude windbreaks. He was still small-brained in human terms, however, and if his linguistic potentialities were increasing there remains no satisfactory evidence of the fact. Thus we are confronted with the question why man, as we know him, arose, and why, having arisen, he found his way out of the green confines of his original world. Not all the human beings even of our existing species did. Though their brains are comparable to our own, they have lingered on, something less than one per cent of today's populations, at the edge of a morning twilight that we have forgotten. There can thus be no ready assertion that man's departure from his first world, the world of chameleon-like shifts and forest changes, was either ordained or inevitable. Neither can it be said that visible tools created brains. Some of the forest peoples—though clever to adapt—survive with a paucity of technical equipment.

As to why our pygmoid ancestors, or, more accurately, some group of them, took the road to larger brains we do not know. Most of the suggestions made would just as readily fit a number of non-human primate forms which did not develop large brains. Our line is gone, and while the behavior of our existing relatives is worth examination we cannot unravel out of another genetic strand the complete story of our own.

Not without interest is the fact that much of this development is correlated with the advances and recessions of the continental ice fields. It is conceivable, at least, that some part of the human stock was being exposed during this time to relentless genetic pressures and then, in interglacial times, to renewed relaxation of barriers and consequent genetic mixture. A few scattered finds from remote portions of the Euro-Asiatic land mass will never clarify this suspicion. For hundreds of thousands of years of crucial human history we have not a single bone as a document.

There is another curious thing about the Ice Age. Except for the emergence of genuinely modern man toward the close of its icy winter, it is an age of death, not a birth time of species. Extinction has always followed life relentlessly through the long eras of earth's history. The Pleistocene above all else was a time of great extinctions. Many big animals perished, and though man's hunting technology was improving, his numbers were still modest. He did not then possess the capacity to ravage continents in the way he was later to do.

The dinosaurs vanished before man appeared on earth, and their disappearance has caused much debate. They died out over a period many millions of years in extent and at a time when the low warm continents lapped by inland seas were giving way to bleaker highlands. The events of the Ice Age are markedly different. First of all, many big mammals—mammoth, mastodon, sloth, long-horned bison—survived the great ice sheets only to die at their close. It is true that man, by then dispersing over the continents, may have had something to do with their final extermination, but there perished also certain creatures like the dire wolves, in which man could have taken little direct interest.

We are thus presented, in contrast to the situation at the close of the age of reptiles, with a narrowly demarcated line of a few thousand years in which a great variety of earth's northern fauna died out while man survived. Along with the growing desiccation in Southwest Asia, these extinctions gave man, the hunter, a mighty push outside his original game-filled Eden. He had to turn to plant domestication to survive, and plants, it just happens, are the primary road to a settled life and the basic supplies from which cities and civilizations arise. A half-dying green kingdom, one might say, forced man out of a relationship which might otherwise have continued down to the present.

But, the question persists, why did so many creatures die in so little time after marching back and forth with the advancing or retreating ice through so many thousand years? Just recently the moon voyage has hinted at a possible clue, though it must be ventured very tentatively when man's observational stay upon the moon has been so short.

The Apollo 11 astronauts observed and succeeded in photographing melted or glazed droplets concentrated on points and edges of moon rock. Dr. Thomas Gold, director of Cornell

University's Center for Radio Physics, has suggested that these glasslike concretions are evidence of melting, produced by a giant solar flare activated for only a few moments, but of an unexpected intensity. Giant storms are known to lick outward from the sun's surface, but a solar disturbance of the magnitude required to account for such a melting—if it was indeed sun-produced—would have seemed from earth like the flame of a dragon's breath. Most of the ultraviolet of the sun-storm, generated perhaps by a comet hurtling into the sun's surface, would have been absorbed by the earth's atmosphere. A temperature effect on earth need not have been pronounced so long as the flare was momentary. The unprotected surface of the moon, however, would have received the full impact of the dragon's tongue.

Dr. Gold has calculated by various means that the event, if actually produced by a solar flare, lies somewhere close to thirty thousand years from us in time and is therefore unrecorded in the annals of man. But here is the curious thing. The period involved lies in the closing Ice Age, in the narrow time zone of vast extinctions in the northern hemisphere. Was the giant flare, an unheard-of phenomenon, in some way involved with the long dying of certain of the great mammals that followed? Seemingly the earth escaped visible damage because of its enveloping blanket of air. No living man knows what the flicking tongue of a dragon star might do, however, or what radiation impact or atmospheric change might have been precipitated upon earth. Some scholars are loath to accept the solar-flare version of the moon glaze because of the stupendous energy which would have to be expended, and the general known stability of the sun. But men are short-lived, and solar catastrophes like the sunward disintegration of a comet would be exceedingly rare. Until more satisfactory evidence is at hand, most scientists will probably prefer to regard the glazed rock as splashed by the heat of meteoritic impact.

Nevertheless, the turbulent outpouring of even ordinary solar flares is on so gigantic a scale as to be terrifying in a close-up view. Until there is further evidence that ours is not a sleepy dragon star, one may wonder just what happened thirty thousand years ago, and why, among so many deaths, it was man who survived. Whatever occurred, whether by ice withdrawal or the momentary penetration of the ultraviolet into our atmosphere, man's world was changed. Perhaps there is something after all to the story of his eviction from the green Garden.

When I lie in bed now and await the hastening of Halley's comet, I would like to dream my way back just once to that single, precise instant when the star dragon thrust out its tongue. Perhaps the story of all dragons since comes from that moment. Men have long memories when the memories are clothed in myth. But I drowse, and the train whistle mingles and howls with the heaven-sweeping light in my dream. It is 1910. I am going back once more.

TWO

The Cosmic Prison

Not till we are lost . . . do we begin to understand ourselves.
—HENRY DAVID THOREAU

"A NAME IS A PRISON, God is free," once observed the Greek poet Nikos Kazantzakis. He meant, I think, that valuable though language is to man, it is by very necessity limiting, and creates for man an invisible prison. Language implies boundaries. A word spoken creates a dog, a rabbit, a man. It fixes their nature before our eyes; henceforth their shapes are, in a sense, our own creation. They are no longer part of the unnamed shifting architecture of the universe. They have been transfixed as if by sorcery, frozen into a concept, a word. Powerful though the spell of human language has proven itself to be, it has laid boundaries upon the cosmos.

No matter how far-ranging some of the mental probes that man has philosophically devised, by his own created nature he is forced to hold the specious and emerging present and transform it into words. The words are startling in their immediate effectiveness, but at the same time they are always finally imprisoning because man has constituted himself a prison keeper. He does so out of no conscious intention, but because for immediate purposes he has created an unnatural world of his own, which he calls the cultural world, and in which he feels at home. It defines his needs and allows him to lay a small immobilizing spell upon the nearer portions of his universe. Nevertheless, it transforms that universe into a cosmic prison house which is no sooner mapped than man feels its inadequacy and his own.

He seeks then to escape, and the theory of escape involves bodily flight. Scarcely had the first moon landing been achieved before one U.S. senator boldly announced: "We are the masters of the universe. We can go anywhere we choose." This statement was widely and editorially acclaimed. It is a striking example of the comfort of words, also of the covert substitutions and mental projections to which they are subject. The cosmic prison is not made less so by a successful journey of some two hundred and forty thousand miles in a cramped and primitive vehicle.

To escape the cosmic prison man is poorly equipped. He has to drag portions of his environment with him, and his life span is that of a mayfly in terms of the distances he seeks to penetrate. There is no possible way to master such a universe by flight alone. Indeed such a dream is a dangerous illusion. This may seem a heretical statement, but its truth is self-evident if we try seriously to comprehend the nature of time and space that I sought to grasp when held up to view the fiery messenger that flared across the zenith in 1910. "Seventy-five years," my father had whispered in my ear, "seventy-five years and it will be racing homeward. Perhaps you will live to see it again. Try to remember."

And so I remembered. I had gained a faint glimpse of the size of our prison house. Somewhere out there beyond a billion miles in space, an entity known as a comet had rounded on its track in the black darkness of the void. It was surging homeward toward the sun because it was an eccentric satellite of this solar system. If I lived to see it, it would be but barely, and with the dimmed eyes of age. Yet it, too, in its long traverse, was but a flitting mayfly in terms of the universe the night sky revealed.

So relative is the cosmos we inhabit that, as we gaze upon the outer galaxies available to the reach of our telescopes, we are placed in about the position that a single white blood cell in our bodies would occupy, if it were intelligently capable of seeking to understand the nature of its own universe, the body it inhabits. The cell would encounter rivers ramifying into miles of distance seemingly leading nowhere. It would pass through gigantic structures whose meaning it could never grasp—the brain, for example. It could never know there was an outside, a vast being on a scale it could not conceive of and of which it formed an infinitesimal part. It would know only the pouring tumult of the creation it inhabited, but of the nature of that great beast, or even indeed that it was a beast, it could have no conception whatever. It might examine the liquid in which it floated and decide, as in the case of the fall of Lucretius's atoms, that the pouring of obscure torrents had created its world.

It might discover that creatures other than itself swam in the torrent. But that its universe was alive, had been born and was destined to perish, its own ephemeral existence would never allow it to perceive. It would never know the sun; it would explore only through dim tactile sensations and react to chemical stimuli that were borne to it along the mysterious conduits of the arteries and veins. Its universe would be centered upon a great arborescent tree of spouting blood. This, at best, generations of white blood cells by enormous labor and continuity might succeed, like astronomers, in charting.

They could never, by any conceivable stretch of the imagination, be aware that their so-called universe was, in actuality, the prowling body of a cat or the more time-enduring body of a philosopher, himself engaged upon the same quest in a more gigantic world and perhaps deceived proportionately by greater vistas. What if, for example, the far galaxies man observes make up, across void spaces of which even we are atomically composed, some kind of enormous creature or cosmic snowflake whose exterior we will never see? We will know more than the phagocyte in our bodies, but no more than that limited creature can we climb out of our universe, or successfully enhance our size or longevity sufficiently to thrust our heads through the confines of the universe that terminates our vision.

Some further "outside" will hover elusively in our thought, but upon its nature, or even its reality, we can do no more than speculate. The phagocyte might observe the salty turbulence of an eternal river system, Lucretius the fall of atoms creating momentary living shapes. We suspiciously sense, in the concept of the expanding universe derived from the primordial atom—the monobloc—some kind of oscillating universal heart. At the instant of its contraction we will vanish. It is not given us, nor can our science recapture, the state beyond the monobloc, nor whether we exist in the diastole of some inconceivable being. We know only a little more extended reality than the hypothetical creature below us. Above us may lie realms it is beyond our power to grasp.

This, then, is the secret nature of the universe over which the ebullient senator so recklessly proclaimed our absolute mastery. Time in that universe is in excess of ten billion years. It recedes backward into a narrowing funnel where, at some inconceivable point of concentration, the monobloc containing all the matter that composes the galaxies exploded in the one gigantic instant of creation.

Along with that explosion space itself is rushing outward. Stars and the great island galaxies in which they cluster are more numerous than the blades of grass upon a plain. To speak of man as "mastering" such a cosmos is about the equivalent of installing a grasshopper as Secretary General of the United Nations. Worse, in fact, for no matter what system of propulsion man may invent in the future, the galaxies on the outer rim of visibility are fleeing faster than he can approach them. Moreover, the light that he is receiving from them left its source in the early history of the planet earth. There is no possible way of even establishing their present existence. As the British astronomer Sir Bernard Lovell has so appropriately remarked, "At the limit of present-day observations our information is a few billion years out of date."

Light travels at a little over one hundred and eighty-six thousand miles a second, far beyond the conceivable speed of any spaceship devised by man, yet it takes light something like one hundred thousand years just to travel across the star field of our own galaxy, the Milky Way. It has been estimated that to reach the nearest star to our own, four light-years away, would require, at the present speed of our spaceships, a time equivalent to more than the whole of written history, indeed one hundred thousand earthly years would be a closer estimate —a time as long, perhaps, as the whole existence of *Homo sapiens* upon earth. And the return, needless to state, would consume just as long a period.

Even if our present rocket speeds were stepped up by a factor of one hundred, human generations would pass on the voyage. An unmanned probe into the nearer galactic realms would be gone so long that its intended mission, in fact the country which sent it forth, might both have vanished into the mists of history before its messages began to be received. All this, be it noted, does not begin to involve us in those intergalactic distances across which a radio message from a cruising spaceship might take hundreds of thousands of years to be received and a wait of other hundreds of thousands before a reply would filter back.

We are, in other words, truly in the position of the blood cell exploring our body. We are limited in time, by analogy a miniature replica of the cosmos, since we too individually ascend from a primordial atom, exist, and grow in space, only to fall back in dissolution. We cannot, in terms of the time dimension as we presently know it, either travel or survive the interstellar distances.

Two years ago I chanced to wander with a group of visiting scholars into a small planetarium in a nearby city. In the dark in a remote back seat, I grew tired and fell asleep while a lecture was progressing. My eyes had closed upon a present-day starry night as represented in the northern latitudes. After what seemed in my uneasy slumber the passage of a long period of time, I started awake in the dark, my eyes fixed in amazement upon the star vault overhead. All was quiet in the neighboring high-backed seats. I could see no one. Suddenly I seemed adrift under a vast and unfamiliar sky. Constellations with which I was familiar had shifted, grown minute, or vanished. I rubbed my eyes. This was not the universe in which I had fallen asleep. It seemed more still, more remote, more enormous, and inconceivably more solitary. A queer sense of panic struck me, as though I had been transported out of time.

Only after some attempt to orient myself by a diminished pole star did the answer come to me by murmurs from without. I was not the last man on the planet, far in the dying future. My companions had arisen and left, while the lecturer had terminated his address by setting the planetarium lights forward to show the conformation of the heavens as they might exist in the remote future of the expanding universe. Distances had lengthened. All was poised, chill, and alone.

I sat for a moment experiencing the sensation all the more intensely because of the slumber which left me feeling as though ages had elapsed. The sky gave little sign of movement. It seemed drifting in a slow indeterminate swirl, as though the forces of expansion were equaled at last by some monstrous tug of gravity at the heart of things. In this remote night sky of the far future I felt myself waiting upon the inevitable, the great drama and surrender of the inward fall, the heart contraction of the cosmos.

I was still sitting when, like the slightest leaf movement on a flooding stream, I saw the first faint galaxy of a billion suns race like a silverfish across the night and vanish. It was enough: the fall was equal to the flash of creation. I had sensed it waiting there under the star vault of the planetarium. Now it was cascading like a torrent through the ages in my head. I had experienced, by chance, the farthest reach of the star prison. I had similarly lived to see the beginning descent into the maelstrom.

III

There are other confinements, however, than that imposed by the enormous distances of the cosmos. One could almost list them. There is, for example, the prison of smells. I happen to know a big black hunting poodle named Beau. Beau loves to go for walks in the woods, and at such times as I visit his owners this task of seeing Beau safely through his morning adventures is happily turned over to me.

Beau has eyes, of course, and I do not doubt that he uses them when he greets his human friends by proffering a little gift such as his food dish. After this formality, which dates to his puppyhood, is completed, Beau immediately reverts to the world of snuffles. As a long-time trusted friend, I have frequently tried to get Beau to thrust his head out of the world of smells and actually to see the universe. I have led him before the mirror in my bedroom and tried to persuade him to see himself, his own visible identity. The results, it turns out, are totally unsatisfactory if not ludicrous. Beau peers out from his black ringlets as suspiciously as an ape hiding in a bush. He drops his head immediately and pretends to examine the floor. It is evident that he detests this apparition and has no intention of being cajoled into some dangerous and undoggy wisdom by my voice.

He promptly brings his collar and makes appropriate throaty conversation. To appease his wounded feelings, I set out for a walk in the woods. It is necessary to do this with a long chain and a very tight grasp upon it. Beau is a big, powerful animal, and ringlets or no he has come from an active and carnivorous past. Once in the woods all this past suddenly emerges. One is dragged willy-nilly through leaf, thorn, and thicket on intangible trails that Beau's swinging muzzle senses upon the wind.

His deep, wet nose has entered a world denied to me—a mad world whose contours and direction change with every gust of air. I leap and bound with a chafed wrist through a smell universe I cannot even sense. Occasionally something squawks or bounds from under our feet and I am flung against trees or wrapped around by a flying chain.

Finally, on one memorable occasion, after a rain, Beau paused, sniffing suspiciously between two rocks on a hillside. Another rabbit? I groaned mentally, taking a tighter hold on the chain. Beau then began some careful digging, curving and patting the soil aside in a way I had never before witnessed. A small basin shaped by Beau's forepaws presently appeared, and up from the bottom of it welled a spring-fed pool in which Beau promptly buried his snout and lapped long and lustily of water that I am sure carried the living tastes and delicate nuances disseminated from an unseen watershed.

Beau had had a proper drink of tap water before we started from home, but this drink was different. I could tell from the varied, eager, slurping sounds that emanated from Beau. He was intoxicated by living water which dim primordial memories had instructed him how to secure. I looked on, interested and sympathetic, but aware that the big black animal lived in a smell prison as I, in my way, lived in a sight prison. Our universes intersected sufficiently for us to recognize each other in a friendly fashion, but Beau would never admit the mirror image of himself into his mind, and, try as I would, the passing breeze would never inform me about the shadowy creatures that passed unglimpsed in the forest.

IV

Other prisons exist besides those dominated by the senses of smell or sight or temperature. Some involve the length of a creature's lifetime, as in the case of five-year-old Beau, who gambols happily about his master, knowing him to be one of the everlasting immortals of his universe.

It is my belief that there has never been a culture that represented man any more than there has been a man who represented men. Our prisons, both societal and cultural, are far too complex for this. In one age religion drives the scientist-philosopher into hiding in narrow corners or castigates him as a public enemy. Such was the fate of Nicolaus Copernicus, Galileo Galilei, and others of equal importance. The pharaohs, by contrast, dreamed of traversing the sky after death in solar boats which they prepared after the fashion of Mediterranean sea craft. The Old Kingdom pharaohs, however, were entranced by a pole-star conception of their final voyage. Only later did the solar journey take precedence. I mention these cultural prisons only to indicate that man's cosmic yearnings are very old but subjected to the vicissitudes of history.

The dream of men elsewhere in the universe alleviating the final prison of loneliness dies very hard. Nevertheless a wise remark of George Santayana's made many years ago should discourage facile and optimistic thinking upon this very point. "An infinite number of solar systems," the philosopher meditated, "must have begun as ours began, but each of them must have deviated at one point from ours in its evolution, all the previous incidents being followed in each case by a different sequel." In voicing this view Santayana betrays a clearer concept of the chance-filled course of genetics and its unreturning pathways than some astronomers. The Mendelian pathways are prisons of no return. Advances are made but always a door swings shut behind the evolving organism. It can no longer mate with its one-time progenitors. It can only press forward along roads that increasingly will fix its irrevocable destiny.

Ours is a man-centered age. Not long ago I was perusing a work on space when I came across this statement by a professional astronomer: "Other stars, other planets, other life, and other races of men are evolving all along, so that the net effect is changeless." Implied in this remark was an utter confidence that the evolutionary process was everywhere the same, ran through the same succession of forms, and emerged always with men at the helm of life, men presumably so close to ourselves that they might interbreed—a supposition fostered by our comic strips.

In the light of this naive concept, for such it is, let us consider just two worlds we know about, not worlds in space, but continents on our own planet. These continents exist under the same sun and are surrounded by the same waters as our own; their life bears a distant relationship to ours but has long been isolated. Man never arose in the remote regions of South America and Australia. He only reached them by migration from outside. They are laboratories of agelong evolution which tell us much about the unique quality of the human experience.

The southern continents of our earth do not maintain the intimacy of faunal exchange that marks the land masses encircling the basin of the polar sea. Instead, they are lost in the southern latitudes of the oceans and for long intervals their faunas have evolved in isolation. These lands have been in truth "other worlds."

The most isolated of these worlds is Australia, and it is a marsupial world. With the insignificant exception of a few late drifters from outside, ground life, originally represented by a few marsupial forms, has, since the Mesozoic, evolved untroubled by invading placental mammals from without. Every possible ecological niche from forest tree to that of underground burrower has been occupied by the evolutionary radiation of a slower-brained mammal whose young are born in a far more embryonic condition than are those of the true Placentalia.

This world remained unknown to Western science until the great exploratory voyages. Somewhere in the past, life had taken another turn. Chance mutation, "total contingency" in the words of the paleontologist William King Gregory, had led to another universe. The "world" of Australia contained no primates at all, nor any hint of their emergence. Upon that "planet" lost in the great waters, they were one of an infinite number of random potentialities that had remained as unrealized as the whole group of placental mammals, of which the Primate order is a minor part.

If we now turn to South America, we encounter still another isolated evolutionary center—but one not totally unrelated to that of Eurasia. Here, so the biogeographers inform us, an attenuated land bridge, at intervals completely severed, has both stimulated local evolutionary development and at times interrupted it by migrations from North America. Our concern is with just one group of animals, the South American monkeys. They are anatomically distinct from the catarrhine forms of the Old World and constitute an apparent parallel emergence from the prosimians of the early Tertiary.

Once more, however, even though the same basic Primate stock is involved, things have gone differently. There are no great apes in the New World, no evidence of ground-dwelling experiments of any kind. Though fewer carnivores are to be found on the South American grasslands than in Africa, the rain-forest monkeys, effectively equipped with prehensile tails, still cling to their archaic pathways. One can only observe that South America's vast rivers flow through frequently flooded lowlands, and that by contrast much of Africa is high, with open savannah and parkland. The South American primates appear confined to areas where descent to the ground proved less inviting. Here ended another experiment which did not lead to man, even though it originated within the same order from which he sprang. Another world has gone astray from the human direction.

If, as some thinkers occasionally extrapolate, man was so ubiquitous, so easy to produce, why had two great continental laboratories, Australia and South America—"worlds," indeed —failed to produce him? They had failed, we may assume, simply because the great movements of life are irreversible, the same mutations do not occur, circumstances differ in infinite particulars, opportunities fail to be grasped, and so what once happened is no more. The random element is always present, but it is selected on the basis of what has preceded its appearance.

There is no trend demanding man's constant reappearance, either on the separate "worlds" of this earth or elsewhere. There can be no more random duplication of man than there is a random duplication of such a complex genetic phenomenon as fingerprints. The situation is not one that is comparable to a single identical cast of dice, but rather it is an endless addition of new genes building on what has previously been incorporated into a living creature through long ages. Nature gambles but she gambles with constantly new and altering dice. It is this well-established fact which enables us to call long-range evolution irreversible.

Finally, there are even meteorological prisons. The constant circulation of moisture in our atmosphere has actually played an important role in creating the first vertebrates and, indirectly, man. If early rivers had not poured from the continents into the sea, the first sea vertebrates to penetrate streams above sea level would not have evolved a rigid muscular support, the spine, to enable them to wriggle against down-rushing currents. And, if man, in his early history, had not become a tree climber in tropical rain forests, he would never have further tilted that same spine upright or replaced the smell prison of the horizontal mammal with the stereoscopic, far-ranging "eye brain" of the higher primates, including man. Such final dice throws, in which leaf and grass, wave and water, are inextricably commingled with the chemistry of the body, could be multiplied. The cosmic prison is subdivided into an infinite number of unduplicable smaller prisons, the prisons of form.

V

We are now in a position to grasp, after an examination of the many prisons which encompass life, that the cosmic prison which many men, in the excitement of the first moon landing, believed we had escaped still extends immeasurably beyond us. The present lack of any conceivable means of star travel and the shortness of our individual lives appear to prevent the crossing of such distances. Even if we confined ourselves to unmanned space probes of far greater sophistication than any we now possess, their homing messages through the void could be expected to descend upon the ruined radio scanners of a civilization long vanished, or upon a world whose scholars would have long since forgotten what naive dreams had been programmed into such instruments. We have, in other words, detected that we exist in a prison of numbers, otherwise known as light-years. We are also locked in a body which responds to biological rather than sidereal time. That body, in turn, sees the universe through its own senses and no others.

At every turn of thought a lock snaps shut upon us. As societal men we bow to a given frame of culture—a world view we have received from the past. Biologically each of us is unique, and the tight spiral of the DNA molecules conspires to doom us to mediocrity or grandeur. We dream vast dreams of Utopias and live to learn the meaning of the two-thousand-year-old judgment of a Greek philosopher: "The flaw is in the vessel itself"—the flaw that defeats all governments.

By what means, then, can we seek escape from groveling in mean corners of despair? Not, certainly, by the rush to depart upon the night's black pathways, nor by attention to the swerving wind vane of the senses. We are men, and despite all our follies there have been great ones among us who have counseled us in wisdom, men who have also sought keys to our prison. Strangely, these men have never spoken of space; they have spoken, instead, as though the farthest spaces lay within the mind itself—as though we still carried a memory of some light of long ago and the way we had come. Perhaps for this reason alone we have scanned the skies and the waters with what Henry Vaughan so well labeled the "Ecclips'd Eye," that eye incapable of quite assembling the true meaning of the universe but striving to do so "with Hyeroglyphicks quite dismembered."

These are the words of a seventeenth-century mystic who has mentally dispatched inward vision through all the creatures until he comes to man who "shines a little" and whose depths he finds it impossible to plumb. Thomas Traherne, another man of that century of the Ecclips'd Eye, when religion was groping amidst the revelations of science, stated well the matter of the keys to the prison.

"Infinite love," he ventured, "cannot be expressed in finite room. Yet it must be infinitely expressed in the smallest moment, . . . Only so is it in both ways infinite."

Can this insight be seen to justify itself in modern evolutionary terms? I think it can.

Close to a hundred years ago the great French medical scientist Claude Bernard observed that the stability of the inside environment of complex organisms must be maintained before an outer freedom can be achieved from their immediate surroundings. What Bernard meant was profound but simple to illustrate.

He meant that for life to obtain relative security from its fickle and dangerous outside surroundings the animal must be able to sustain stable, unchanging conditions within the body. Warm-blooded mammals and birds can continue to move about in winter; insects cannot. Warm-blooded animals such as man, with his stable body temperature, can continue to think and reason in outside temperatures that would put a frog to sleep in a muddy pond or roll a snake into a ball in a crevice. In winter latitudes many of the lower creatures are forced to sleep part of their lives away.

Many millions of years of evolutionary effort were required before life was successful in defending its internal world from the intrusion of the heat or cold of the outside world of nature. Yet only so can life avoid running down like a clock in winter or perishing from exposure in the midday sun. Even the desert rattlesnake is forced to coil in the shade of a bush at midday. Of course our tolerance is limited to a few degrees of temperature when measured against the great thermometer of the stars, but this hard-won victory is what creates the ever-active brain of the mammal as against the retarded sluggishness of the reptile.

A steady metabolism has enabled the mammals and also the birds to experience life more fully and rapidly than cold-blooded creatures. One of the great feats of evolution, perhaps the greatest, has been this triumph of the interior environment over exterior nature. Inside, we might say, has fought invading outside, and inside, since the beginning of life, by slow degrees has won the battle of life. If it had not, man, frail man with his even more fragile brain, would not exist.

Unless fever or some other disorder disrupts this internal island of safety, we rarely think of it. Body controls are normally automatic, but let them once go wrong and outside destroys inside. This is the simplest expression of the war of nature—the endless conflict that engages the microcosm against the macrocosm.

Since the first cell created a film about itself and elected to carry on the carefully insulated processes known as life, the creative spark has not been generalized. Whatever its principle may be it hides magically within individual skins. To the day of our deaths we exist in an inner solitude that is linked to the nature of life itself. Even as we project love and affection upon others we endure a loneliness which is the price of all individual consciousness—the price of living.

It is, though overlooked, the discontinuity beyond all others: the separation both of the living creature from the inanimate and of the individual from his kind. These are star distances. In man, moreover, consciousness looks out isolated from its own body. The body is the true cosmic prison, yet it contains, in the creative individual, a magnificent if sometimes helpless giant. John Donne, speaking for that giant in each of us, said: "Our creatures are our thoughts, creatures that are borne Gyants. . . . My thoughts reach all, comprehend all. Inexplicable mystery; I their Creator am in a close prison, in a sick bed, anywhere, and any one of my Creatures, my thoughts, is with the Sunne and beyond the Sunne, overtakes the Sunne, and overgoes the Sunne in one pace, one steppe, everywhere."

This thought, expressed so movingly by Donne, represents the final triumph of Claude Bernard's interior microcosm in its war with the macrocosm. Inside has conquered outside. The giant confined in the body's prison roams at will among the stars. More rarely and more beautifully, perhaps, the profound mind in the close prison projects infinite love in a finite room. This is a crossing beside which light-years are meaningless. It is the solitary key to the prison that is man.

THREE

The World Eaters

Really we create nothing. We merely plagiarize nature.
—JEAN BATAILLON

IT CAME TO ME in the night, in the midst of a bad dream, that perhaps man, like the blight descending on a fruit, is by nature a parasite, a spore bearer, a world eater. The slime molds are the only creatures on the planet that share the ways of man from his individual pioneer phase to his final immersion in great cities. Under the microscope one can see the mold amoebas streaming to their meeting places, and no one would call them human. Nevertheless, magnified many thousand times and observed from the air, their habits would appear close to our own. This is because, when their microscopic frontier is gone, as it quickly is, the single amoeboid frontiersmen swarm into concentrated aggregations. At the last they thrust up overtoppling spore palaces, like city skyscrapers. The rupture of these vesicles may disseminate the living spores as far away proportionately as man's journey to the moon.

It is conceivable that in principle man's motor throughways resemble the slime trails along which are drawn the gathering mucors that erect the spore palaces, that man's cities are only the ephemeral moment of his spawning—that he must descend upon the orchard of far worlds or die. Human beings are a strange variant in nature and a very recent one. Perhaps man has evolved as a creature whose centrifugal tendencies are intended to drive it as a blight is lifted and driven, outward across the night.

I do not believe, for reasons I will venture upon later, that this necessity is written in the genes of men, but it would be foolish not to consider the possibility, for man as an interplanetary spore bearer may be only at the first moment of maturation. After all, *mucoroides* and its relatives must once have performed their act of dissemination for the very first time. In

man, even if the feat is cultural, it is still possible that some incalculable and ancient urge lies hidden beneath the seeming
rationality of institutionalized science. For example, a young
space engineer once passionately exclaimed to me, "We must
give all we have . . ." It was as though he were hypnotically
compelled by that obscure chemical, acrasin, that draws the
slime molds to their destiny. And is it not true also that observers like myself are occasionally bitterly castigated for daring to
examine the motivation of our efforts toward space? In the
intellectual climate of today one should strive to remember
the words that Herman Melville accorded his proud, fate-
confronting Captain Ahab, "All my means are sane, my motive
and my object mad."

The cycles of parasites are often diabolically ingenious. It is
to the unwilling host that their ends appear mad. Has earth
hosted a new disease—that of the world eaters? Then inevitably the spores must fly. Short-lived as they are, they must fly.
Somewhere far outward in the dark, instinct may intuitively
inform them, lies the garden of the worlds. We must consider
the possibility that we do not know the real nature of our kind.
Perhaps *Homo sapiens*, the wise, is himself only a mechanism in
a parasitic cycle, an instrument for the transference, ultimately,
of a more invulnerable and heartless version of himself.

Or, again, the dark may bring him wisdom.

I stand in doubt as my forebears must once have done at the
edge of the shrinking forest. I am a man of the rocket century;
my knowledge, such as it is, concerns our past, our dubious
present, and our possible future. I doubt our motives, but I
grant us the benefit of the doubt and, like Arthur Clarke, call
it, for want of a better term, "childhood's end." I also know, as
did Plato, that one who has spent his life in the shadow of
great wars may, like a captive, blink and be momentarily
blinded when released into the light.

There are aspects of the world and its inhabitants that are
eternal, like the ripples marked in stone on fossil beaches.
There is a biological preordination that no one can change.
These are seriatim events that only the complete reversal of
time could undo. An example would be the moment when the
bats dropped into the air and fluttered away from the insectivore line that gave rise to ourselves. What fragment of man,

perhaps a useful fragment, departed with them? Something, shall we say, that had it lingered, might have made a small, brave, twilight difference in the mind of man.

There is a part of human destiny that is not fixed irrevocably but is subject to the flying shuttles of chance and will. Everyone imagines that he knows what is possible and what is impossible, but the whole of time and history attest our ignorance. One evening, in a drab and heartless area of the metropolis, a windborne milkweed seed circled my head. On impulse I seized the delicate aerial orphan which otherwise would have perished. Its long midwinter voyage seemed too favorable an augury to ignore. Placing the seed in my glove, I took it home into the suburbs and found a field in which to plant it. Of a million seeds blown on a vagrant wind into the city, it alone may survive.

Why did I bother? I suppose, in retrospect, for the sake of the future and the memory of the bats whirling like departing thoughts from the tree of ancestral man. Perhaps, after all, there lingered in my reflexes enough of a voyager to help all travelers on the great highway of the winds. Or perhaps I am not yet totally a planet eater and wished that something green might survive. A single impulse, a hand outstretched to an alighting seed, suggests that something is still undetermined in the human psyche, that the time trap has not yet closed inexorably. Some aspect of man that has come with him from the sunlit grasses is still instinctively alive and being fought for. The future, formidable as a thundercloud, is still inchoate and unfixed upon the horizon.

II

Man is "a tinkerer playing with ideas and mechanisms," comments a recent and very able writer upon technology, R. J. Forbes. He goes on to state that, if those impulses were to disappear, man would cease to be a human being in the sense we know. It is necessary to concede to Forbes that for Western man, *Homo faber*, the tool user, the definition is indeed appropriate. Nevertheless, when we turn to the people of the wilderness we must place certain limitations upon the words. That man has utilized tools throughout his history is true, but that man has been particularly inventive or a tinkerer in the sense of seeking constant innovation is open to question.

Students of living primitives in backward areas such as Australia have found them addicted to immemorial usage in both ideas and tools. There is frequently a prejudice against the kind of change to which our own society has only recently adjusted. Such behavior is viewed askance as disruptive. The society is in marked ecological balance with its surroundings, and any drastic innovation from within the group is apt to be rejected as interfering with the will of the divine ancestors.

Not many years ago I fell to chatting with a naturalist who had had a long experience among the Cree of the northern forests. What had struck him about these Indians, even after considerable exposure to white men, was their remarkable and yet, in our terms, "indifferent" adjustment to their woodland environment. By indifference my informant meant that while totally skilled in the use of everything in their surroundings, they had little interest in experiment in a scientific sense, or in carrying objects about with them. Indeed they were frequently very careless with equipment or clothing given or loaned to them. Such things might be discarded after use or left hanging casually on a branch. One was left with the impression that these woodsmen were, by our standards, casual and feckless. Their reliance upon their own powers was great, but it was based on long traditional accommodation and a psychology by no means attuned to the civilized innovators' world. Plant fibers had their uses, wood had its uses, but everything from birch bark to porcupine quills was simply "given." Raw materials were always at hand, to be ignored or picked up as occasion demanded.

One carried little, one survived on little, and little in the way of an acquisitive society existed. One lived amidst all one had use for. If one shifted position in space the same materials were again present to be used. These people were ignorant of what Forbes would regard as the technological necessity to advance. Until the intrusion of whites, their technology had been long frozen upon a barely adequate subsistence level. "Progress" in Western terms was an unknown word.

Similarly I have heard the late Frank Speck discuss the failure of the Montagnais-Naskapi of the Labrador peninsula to take advantage, in their winter forest, either of Eskimo snow goggles, for which they substituted a mere sprig of balsam thrust under the cap, or of the snow house, which is far more comfortable than their cold and draft-exposed wigwams. The same indifference toward technological improvement or the acceptance of innovations from outside thus extended even to their racial brothers, the Eskimo. Man is a tool user, certainly, whether of the stone knife or less visible hunting magic. But that he is an obsessive innovator is far less clear. Tradition rules in the world's outlands. Man is not on this level driven to be inventive. Instead he is using the sum total of his environment almost as a single tool.

There is a very subtle point at issue here. H. C. Conklin, for example, writes about one Philippine people, the Hanunóo, that their "activities require an intimate familiarity with local plants. . . . Contrary to the assumption that subsistence level groups never use but a small segment of the local flora, ninety-three percent of the total number of native plant types are recognized . . . as culturally significant." As Claude Lévi-Strauss has been at pains to point out, this state of affairs has been observed in many simple cultures.

Scores of terms exist for objects in the natural environment for which civilized man has no equivalents at all. The latter is engaged primarily with his deepening shell of technology which either exploits the natural world or thrusts it aside. By contrast, man in the earlier cultures was so oriented that the total natural environment occupied his exclusive attention. If parts of it did not really help him practically, they were often inserted into magical patterns that did. Thus man existed primarily in a carefully reorganized nature—one that was watched, brooded over, and managed by magico-religious as well as practical means.

Nature was actually as well read as an alphabet; it was the real "tool" by which man survived with a paucity of practical equipment. Or one could say that the tool had largely been forged in the human imagination. It consisted of the way man had come to organize and relate himself to the sum total of his environment. This world view was comparatively static. Nature was sacred and contained powers which demanded careful propitiation. Modern man, on the other hand, has come to look upon nature as a thing outside himself—an object to be manipulated or discarded at will. It is his technology and its vocabulary that makes his primary world. If, like the primitive, he has a sacred center, it is here. Whatever is potential must be unrolled, brought into being at any cost. No other course is conceived as possible. The economic system demands it.

Two ways of life are thus arrayed in final opposition. One way reads deep, if sometimes mistaken, analogies into nature and maintains toward change a reluctant conservatism. The other is fiercely analytical. Having consciously discovered sequence and novelty, man comes to transfer the operation of the world machine to human hands and to install change itself as progress. A reconciliation of the two views would seem to be necessary if humanity is to survive. The obstacles, however, are great.

Nowhere are they better illustrated than in the decades-old story of an anthropologist who sought to contact a wild and untouched group of aborigines in the red desert of central Australia. Traveling in a truck loaded with water and simple gifts, the scientist finally located his people some five hundred miles from the nearest white settlement. The anthropologist lived with the bush folk for a few weeks and won their confidence. They trusted him. The time came to leave. Straight over the desert ran the tracks of his car, and the aborigines are magnificent trackers.

Things were not the same when their friend had left; something had been transposed or altered in their landscape. The gifts had come so innocently. The little band set out one morning to follow the receding track of their friend. They were many days drifting on the march, drawn on perhaps by that dim impulse to which the slime molds yield. Eventually they came to the white man's frontier town. Their friend was gone, but there were other and less kindly white men. There were also drink, prostitution, and disease, about which they were destined to learn. They would never go back to the dunes and the secret places. In five years a begging and degraded remnant would stray through the outskirts of the settlers' town.

They had learned to their cost that it is possible to wander out of the world of the ancestors, only to become an object of scorn in a world directed to a different set of principles for which the aborigines had no guiding precedent. By leaving the timeless land they had descended into hell. Not all the tiny beings of the slime mold escape to new pastures; some wander, some are sacrificed to make the spore cities, and but a modicum of the original colony mounts the winds of space. It is so in the cities of men.

III

Over a century ago Samuel Taylor Coleridge ruminated in one of his own encounters with the universe that "A Fall of some sort or other—the creation as it were of the non–absolute— is the fundamental postulate of the moral history of man. Without this hypothesis, man is unintelligible; with it every phenomenon is explicable. The mystery itself is too profound for human insight."

In making this observation Coleridge had come very close upon the flaw that was to create, out of a comparatively simple creature, the world eaters of the twentieth century. How, is a mystery to be explored, because every man on the planet belongs to the same species, and every man communicates. A span of three centuries has been enough to produce a planetary virus, while on that same planet a few lost tribesmen with brains the biological equal of our own peer in astonishment from the edges of the last wilderness.

One of the scholars of the scientific twilight, Joseph Glanvill, was quick to intimate that to posterity a voyage to the moon "will not be more strange than one to America." Even in 1665 the ambitions of the present century had entered human consciousness. The paradox is already present. There is the man *Homo sapiens* who in various obscure places around the world would rarely think of altering the simple tools of his forefathers, and, again, there is this same *Homo sapiens* in a wild flurry of modern thought patterns reversing everything by which he had previously lived. Such an episode parallels the rise of a biological mutation as potentially virulent in its effects as a new bacterial strain. The fact that its nature appears to be cultural merely enables the disease to be spread with fantastic rapidity. There is no comparable episode in history.

There are two things which are basic to an understanding of

the way in which the primordial people differ from the world eaters, ourselves. Coleridge was quite right that man no more than any other living creature represented the absolute. He was finite and limited, and thus his ability to wreak his will upon the world was limited. He might dream of omniscient power, he might practice magic to obtain it, but such power remained beyond his grasp.

As a primitive, man could never do more than linger at the threshold of the energy that flickered in his campfire, nor could he hurl himself beyond Pluto's realm of frost. He was still within nature. True, he had restructured the green world in his mind so that it lay slightly ensorceled under the noonday sun. Nevertheless the lightning still roved and struck at will; the cobra could raise its deathly hood in the peasant's hut at midnight. The dark was thronged with spirits.

Man's powerful, undisciplined imagination had created a region beyond the visible spectrum which would sometimes aid and sometimes destroy him. Its propitiation and control would occupy and bemuse his mind for long millennia. To climb the fiery ladder that the spore bearers have used one must consume the resources of a world. Since such resources are not to be tapped without the drastic reordering of man's mental world, his final feat has as its first preliminary the invention of a way to pass knowledge through the doorway of the tomb—namely, the achievement of the written word.

Only so can knowledge be made sufficiently cumulative to challenge the stars. Our brothers of the forest have not lived in the world we have entered. They do not possess the tiny figures by which the dead can be made to speak from those great cemeteries of thought known as libraries. Man's first giant step for mankind was not through space. Instead it lay through time. Once more in the words of Glanvill, "That men should speak after their tongues were ashes, or communicate with each other in differing Hemisphears, before the Invention of Letters could not but have been thought a fiction."

In the first of the world's cities man had begun to live against the enormous backdrop of the theatre. He had become self-conscious, a man enacting his destiny before posterity. As ruler, conqueror, or thinker he lived, as Lewis Mumford has put it, by and for the record. In such a life both evil and good come to cast long shadows into the future. Evil leads to evil, good to good, but frequently the former is the most easy for the cruel to emulate. Moreover, when invention lends itself to centralized control, the individualism of the early frontiers easily gives way to routinized conformity. If life is made easier it is also made more dependent. If artificial demands are stimulated, resources must be consumed at an ever-increasing pace.

As in the microscopic instance of the slime molds, the movement into the urban aggregations is intensified. The most technically advanced peoples will naturally consume the lion's share of the earth's resources. Thus the United States at present, representing some six percent of the world's population, consumes over thirty-four percent of its energy and twenty-nine percent of its steel. Over a billion pounds of trash are spewed over the landscape in a single year. In these few elementary facts, which are capable of endless multiplication, one can see the shape of the future growing—the future of a planet virus *Homo sapiens* as he assumes in his technological phase what threatens to be his final role.

Experts have been at pains to point out that the accessible crust of the earth is finite, while the demand for minerals steadily increases as more and more societies seek for themselves a higher, Westernized standard of living. Unfortunately many of these sought-after minerals are not renewable, yet a viable industrial economy demands their steady output. A rising world population requiring an improved standard of living clashes with the oncoming realities of a planet of impoverished resources.

"We live in an epoch of localized affluence," asserts Thomas Lovering, an expert on mineral resources. A few shifts and subterfuges may, with increasing effort and expense, prolong this affluence, but no feat of scientific legerdemain can prevent the eventual exhaustion of the world's mineral resources at a time not very distant. It is thus apparent that to apply to Western industrial man the term "world eater" is to do so neither in derision nor contempt. We are facing, instead, a simple reality to which, up until recently, the only response has been flight— the flight outward from what appears unsolvable and which threatens, in the end, to leave an impoverished human remnant clinging to an equally impoverished globe.

So quick and so insidious has been the rise of the world virus that its impact is just beginning to be felt and its history to be studied. Basically man's planetary virulence can be ascribed to just one thing: a rapid ascent, particularly in the last three centuries, of an energy ladder so great that the line on the chart repesenting it would be almost vertical. The event, in the beginning, involved only Western European civilization. Today it increasingly characterizes most of the planet.

The earliest phase of the human acquisition of energy beyond the needs of survival revolves, as observed earlier, around the rise of the first agricultural civilizations shortly after the close of the Ice Age. Only with the appearance of wealth in the shape of storable grains can the differentiation of labor and of social classes, as well as an increase in population, lay the basis for the expansion of the urban world. With this event the expansion of trade and trade routes was sure to follow. The domestication of plants and animals, however, was still an event of the green world and the sheepfold. Nevertheless it opened a doorway in nature that had lain concealed from man.

Like all earth's other creatures, he had previously existed in a precarious balance with nature. In spite of his adaptability, man, the hunter, had spread across the continents like thin fire burning over a meadow. It was impossible for his numbers to grow in any one place, because man, multiplying, quickly consumes the wild things upon which he feeds and then himself faces starvation. Only with plant domestication is the storage granary made possible and through it three primary changes in the life of man: a spectacular increase in human numbers; diversification of labor; the ability to feed from the countryside the spore cities into which man would presently stream.

After some four million years of lingering in nature's shadow, man would appear to have initiated a drastic change in the world of the animal gods and the magic that had seen him through millennial seasons. Such a change did not happen overnight, and we may never learn the details of its incipient beginnings. As we have already noted, at the close of the Ice Age, and particularly throughout the northern hemisphere, the big game, the hairy mammoth and mastodon, the giant long-horned bison, had streamed away with the melting glaciers. Sand was blowing over the fertile plains of North Africa and the Middle East. Gloomy forests were springing up in the Europe of the tundra hunters. The reindeer and musk ox had withdrawn far into the north on the heels of the retreating ice.

Man must have turned, in something approaching agony and humiliation, to the women with their digging sticks and arcane knowledge of seeds. Slowly, with greater ceremonial, the spring and harvest festivals began to replace the memory of the "gods with the wet nose," the bison gods of the earlier free-roving years. Whether for good or ill the world would never be the same. The stars would no longer be the stars of the wandering hunters. Halley's comet returning would no longer gleam on the tossing antlers and snowy backs of the moving game herds. Instead it would glimmer from the desolate tarns left by the ice in dark forests or startle shepherds watching flocks on the stony hills of Judea. Perhaps it was the fleeting star seen by the three wise men of legend, for a time of human transcendence was approaching.

To comprehend the rise of the world eaters one must leap centuries and millennia. To account for the rise of high-energy civilization is as difficult as to explain the circumstances that have gone into the creation of man himself. Certainly the old sun-plant civilizations provided leisure for meditation, mathematics, and transport energy through the use of sails. Writing, which arose among them, created a kind of stored thought-energy, an enhanced social brain.

All this the seed-and-sun world contributed, but no more. Not all of these civilizations left the traditional religious round of the seasons or the worship of the sun-kings installed on earth. Only far on in time, in west Europe, did a new culture and a new world emerge. Perhaps it would be best to limit our exposition to one spokesman who immediately anticipated its appearance. "If we must select some one philosopher as the hero of the revolution in scientific method," maintained William Whewell, the nineteenth-century historian, "beyond all doubt Francis Bacon occupies the place of honor." This view is based upon four simple precepts, the first of which, from *The Advancement of Learning*, I will give in Bacon's own words. "As the foundation," he wrote, "we are not to imagine or suppose, but to *discover* what nature does or may be made to do." Today this sounds like a truism. In Bacon's time it was a novel, analytical, and unheard-of way to explore nature. Bacon was thus the herald of what has been called "the invention of inventions"—the scientific method itself.

He believed also that the thinker could join with the skilled worker—what we today would call the technologist—to conduct experiments more ably than by simple and untested meditation in the cloister. Bacon, in other words, was groping toward the idea of the laboratory, of a whole new way of schooling. Within such schools, aided by government support, he hoped for the solution of problems not to be dealt with "in the hourglass of one man's life." In expressing this hope he had recognized that great achievement in science must not wait on the unaided and rare genius, but that research should be institutionalized and supported over the human generations.

Fourth and last of Bacon's insights was his vision of the future to be created by science. Here there clearly emerges that orientation toward the future which has since preoccupied the world of science and the West. Bacon was preeminently the spokesman of *anticipatory* man. The long reign of the custombound scholastics was at an end. Anticipatory analytical man, enraptured by novelty, was about to walk an increasingly dangerous pathway.

He would triumph over disease and his numbers would mount; steam and, later, air transport would link the ends of the earth. Agriculture would fall under scientific management, and fewer men on the land would easily support the febrile millions in the gathering cities. As Glanvill had foreseen, thought would fly upon the air. Man's telescopic eye would rove through the galaxy and beyond. No longer would men be burned or tortured for dreaming of life on far-off worlds.

There came, too, in the twentieth century to mock the dream of progress the most ruthless and cruel wars of history. They were the first wars fought with total scientific detachment. Cities were fire-bombed, submarines turned the night waters into a flaming horror, the air was black with opposing air fleets.

The laboratories of Bacon's vision produced the atom bomb and toyed prospectively with deadly nerve gas. "Overkill" became a professional word. Iron, steel, Plexiglas, and the deadly mathematics of missile and anti-missile occupied the finest constructive minds. Even before space was entered, man began to assume the fixed mask of the robot. His courage was unbreakable, but in society there was mounting evidence of strain. Billions of dollars were being devoured in the space effort, while at the same time an affluent civilization was consuming its resources at an ever-increasing rate. Air and water and the land itself were being polluted by the activities of a creature grown used to the careless ravage of a continent.

Francis Bacon had spoken one further word on the subject of science, but by the time that science came its prophet had been forgotten. He had spoken of learning as intended to bring an enlightened life. Western man's ethic is not directed toward the preservation of the earth that fathered him. A devouring frenzy is mounting as his numbers mount. It is like the final movement in the spore palaces of the slime molds. Man is now only a creature of anticipation feeding upon events.

"When evil comes it is because two gods have disagreed," runs the proverb of an elder people. Perhaps it is another way of saying that the past and the future are at war in the heart of man. On March 7, 1970, as I sit at my desk the eastern seaboard is swept by the shadow of the greatest eclipse since 1900. Beyond my window I can see a strangely darkened sky, as though the light of the sun were going out forever. For an instant, lost in the dim gray light, I experience an equally gray clarity of vision.

IV

There is a tradition among the little Bushmen of the Kalahari desert that eclipses of the moon are caused by Kingsfoot, the lion who covers the moon's face with his paw to make the night dark for hunting. Since our most modern science informs us we have come from animals, and since almost all primitives have tended to draw their creator gods from the animal world with which they were familiar, modern man and his bush contemporaries have arrived at the same conclusion by very different routes. Both know that they are shape shifters and changelings. They know their relationship to animals by different ways of logic and different measures of time.

Modern man, the world eater, respects no space and no thing green or furred as sacred. The march of the machines has entered his blood. They are his seed boxes, his potential wings and guidance systems on the far roads of the universe. The fruition time of the planet virus is at hand. It is high autumn, the autumn before winter topples the spore cities. "The living memory of the city disappears," writes Mumford of this phase of urban life; "its inhabitants live in a self-annihilating moment to moment continuum." The ancestral center exists no longer. Anonymous millions roam the streets.

On the African veldt the lion, the last of the great carnivores, is addressed by the Bushmen over a kill whose ownership is contested. They speak softly the age-old ritual words for the occasion, "Great Lions, Old Lions, we know that you are brave." Nevertheless the little, almost weaponless people steadily advance. The beginning and the end are dying in unison and the one is braver than the other. Dreaming on by the eclipse-darkened window, I know with a sudden sure premonition that Kingsfoot has put his paw once more against the moon. The animal gods will come out for one last hunt.

Beginning on some winter night the snow will fall steadily for a thousand years and hush in its falling the spore cities whose seed has flown. The delicate traceries of the frost will slowly dim the glass in the observatories and all will be as it had been before the virus wakened. The long trail of Halley's comet, once more returning, will pass like a ghostly match-flame over the unwatched grave of the cities. This has always been their end, whether in the snow or in the sand.

FOUR

The Spore Bearers

Either the machine has a meaning to life that we have not yet been able to interpret in a rational manner, or it is itself a manifestation of life and therefore mysterious.

—GARET GARRETT

I T IS A remarkable fact that much of what man has achieved through the use of his intellect, nature had invented before him. *Pilobolus*, another fungus which prepares, sights, and fires its spore capsule, constitutes a curious anticipation of human rocketry. The fungus is one that grows upon the dung of cattle. To fulfill its life cycle, its spores must be driven up and outward to land upon vegetation several feet away, where they may be eaten by grazing cows or horses.

The spore tower that discharges the *Pilobolus* missile is one of the most fascinating objects in nature. A swollen cell beneath the black capsule that contains the spores is a genuinely light-sensitive "eye." This pigmented eye controls the direction of growth of the spore cannon and aims it very carefully at the region of greatest light in order that no intervening obstacle may block the flight of the spore capsule.

When a pressure of several atmospheres has been built up chemically within the cell underlying the spore container, the cell explodes, blasting the capsule several feet into the air. Since firing takes place in the morning hours, the stalks point to the sun at an angle sure to carry the tiny "rocket" several feet away as well as up. Tests in which the light has been reduced to a small spot indicate that the firing eye aims with remarkable accuracy. The spore vessel itself is so equipped with a quick-drying glue as to adhere to vegetation always in the proper position. Rain will not wash it off, and there it waits an opportunity to be taken up by munching cattle in order that *Pilobolus* can continue its travels through the digestive tract of the herbivores.

The tiny black capsule that bears the living spores through space is strangely reminiscent, in miniature, of man's latest adventure. Man, too, is a spore bearer. The labor of millions and the consumption of vast stores of energy are necessary to hurl just a few individuals, perhaps eventually people of both sexes, on the road toward another planet. Similarly, for every spore city that arises in the fungus world, only a few survivors find their way into the future.

It is useless to talk of transporting the excess population of our planet elsewhere, even if a world of sparkling water and green trees were available. In nature it is a law that the spore cities die, but the spores fly on to find their destiny. Perhaps this will prove to be the rule of the newborn planet virus. Somehow in the *mysterium* behind genetics, the tiny pigmented eye and the rocket capsule were evolved together.

In an equal mystery that we only pretend to understand, man, in the words of Garet Garrett, "reached with his mind into emptiness and seized the machine." Deathly though some of its effects have proved, robber of the earth's crust though it may appear at this human stage to be, perhaps there are written within the machine two ultimate possibilities. The first, already, if primitively, demonstrated, is that of being a genuine spore bearer of the first complex organism to cross the barrier of the void. The second is that of providing the means by which man may someday be able to program his personality, or its better aspects, into the deathless machine itself, and thus escape, or nearly escape, the mortality of the body.

This may well prove to be an illusory experiment, but we who stand so close under the green primeval shade may still be as incapable of evaluating the human future as the first ape-man would have been to chart the course of *Homo sapiens*. There are over one hundred thousand spores packed in a single capsule of *Pilobolus*, and but few such capsules will ever reach their destiny. This is the way of the spore cities, in the infinite prodigality of nature. It may well be the dictum that controls the fate of man. Perhaps Rome drove blindly toward it and failed in the marches of the West. In the dreaming Buddhist cities that slowly ebbed away beneath the jungle, something was said that lingers, not entirely forgotten—namely, "Thou canst not travel on the Path before thou hast become the Path itself."

Perhaps written deep in ourselves is a simulacrum of the Way and the mind's deep spaces to travel. If so, our goal is light-years distant, even though year by year the gantries lengthen over the giant rockets.

Man possesses the potential power to reach all the planets in this system. None, so far as can be presently determined, offers the prospects of extended colonization. The journey, however, will be undertaken as President Nixon has announced. It will be pursued because the technological and psychological commitment to space is too great for Western culture to abandon. In spite of the breadth of the universe we have previously surveyed, a nagging hope persists that someday, by means unavailable at present, we might achieve the creation of a rocket ship operating near the speed of light. At this point we would enter upon unknown territory, for it has been argued on the basis of relativity theory that men in such a mechanism might exist on a different time scale and age less rapidly than man upon earth. Assuming that such were the case, a question arises whether such a ship, coasting around the galaxy or beyond, might return to find life on our planet long departed. The disparities and the problems are great, and the conflicts of authorities have not made them less so.

It has been pointed out that so great a physicist as Sir Ernest Rutherford, as late as 1936, had pronounced the use of nuclear energy to be Utopian, at least within this century. Similar speculations on the part of others suggest that a great scientist's attempts to extrapolate his knowledge into the future may occasionally prove as inaccurate as the guesses of laymen. Scientific training is apt to produce a restraint, laudable enough in itself, that can readily degenerate into a kind of institutional conservatism. Darwin saw and commented upon this in his time. History has a way of outguessing all of us, but she does it in retrospect.

Nevertheless, because man is small and growing ever lonelier in his expanding universe, there remains a question he is unlikely ever to be able to answer. It involves the discovery of other civilizations in the cosmos. In some three billion years of life on this planet, man, who occupies a very small part of the geological time scale, is the one creature of earth who has achieved the ability to reason on a high abstract level. He has only grasped the nature of the stars within the last few generations. The number of such stars in the universe cannot be counted. Some may possess planets. Judging by our own solar system, of those planets few will possess life. Fewer still, infinitely fewer, will possess what could be called "civilizations" developed by other rational creatures.

On the basis of pure statistical chance, the likelihood that such civilizations are located in our portion of the galaxy is very small. Man's end may well come upon him long before he has had time to locate or even to establish the presence of other intelligent creatures in the universe. There are far more stars in the heavens than there are men upon the earth. The waste to be searched is too great for the powers we possess. In gambling terms, the percentage lies all with the house, or rather with the universe. Lonely though we may feel ourselves to be, we must steel ourselves to the fact that man, even far future man, may pass from the scene without possessing either negative or positive evidence of the existence of other civilized beings in this or other galaxies.

This is said with all due allowance for the fact that we may learn to make at least some satellites or planets within our own solar system artificially capable, in a small way, of sustaining life. For man to spread widely on the dubious and desert worlds of this sun system is unlikely. Much more unlikely is the chance that coursing at near the speed of light over a single arm of this galaxy would ever reveal intelligence, even if it were there. The speed would be too high and the planetary body too small. The size of our near neighbors, Mars and Venus, is proportionately tiny beside the sun's diameter or even that of the huge outer planets, Jupiter and Saturn.

I have suggested that man-machines and finally pure intelligent machines—the product of a biology and a computerized machine technology beyond anything this century will possess —might be launched by man and dispersed as his final spore flight through the galaxies. Such machines would not need to trouble themselves with the time problem and, as the capsules of *Pilobolus* carry spores, might even be able to carry refrigerated human egg cells held in suspended animation and prepared to be activated, educated, and to grow up alone under the care of the machines.

The idea is fantastically wasteful, but so is life. It would be sufficient if the proper planetary conditions were discovered once in a thousand times. These human-machine combinations are much spoken of nowadays under the term "cyborg"—a shorthand term for "cybernetic organism." The machine structures would be intimately controlled by the human brain but built in such a manner as to amplify and extend the powers of the human personality. Other machines might be controlled by human beings deliberately modified by man's increasing knowledge of micro-surgery and genetics. Science has speculated that man has reached an evolutionary plateau. To advance beyond that plateau he must either intimately associate himself with machines in a new way or give way to "exosomatic evolution" and, in some fashion, transfer himself and his personality to the machine.

These are matters of the shadowy future and must be considered only as remote possibilities. More likely is the stricture that, even if we do not destroy ourselves as a planet virus, we will exhaust the primary resources of earth before we can produce the kind of spore carriers of which we dream. Again, the conception may lie forever beyond us. There is a certain grandeur, however, in the thought of man in some far future hour battling against oblivion by launching a final spore flight of cyborgs through the galaxy—a haven-seeking flight projected by those doomed never to know its success or failure, a flight such as life itself has always engaged in since it arose from the primeval waters.

One must repeat that nature is extravagant in the expenditure of individuals and germ cells. Our remote half-human ancestors gave themselves and never expected, or got, an answer as to the destiny their descendants might serve or if, indeed, they would survive. This is still the road we tread in the twentieth century. Sight of the future is denied us, and life was never given to be bearable. To what far creature, whether of metal or of flesh, we may be the bridge, no word informs us. If such a being is destined to come, there can be no assurance that it will spare a thought for the men who, in the human dawn, prepared its way. Man is a part of that torrential living river, which, since the beginning, has instinctively known the value of dispersion. He will yearn therefore to spread beyond the planet he now threatens to devour. This thought persists and is growing. It is rooted in the psychology of man.

A story has been told of the founder of one of the world's great religions—a religion which seeks constantly, in its higher manifestations, to wipe clean the mirror of the mind. Buddha is reported to have said to his sorrowing disciples as he lay dying, "Walk on." He wanted his people to be free of earthly entanglement or desire. That is how one should go in dignity to the true harvest of the worlds. It is a philosophy transferred from the old sun civilizations of earth. It implies that one cannot proceed upon the path of human transcendence until one has made interiorly in one's soul a road into the future. This is the warning of one who knew that the spaces within stretch as far as those without. Cyborgs and exosomatic evolution, however far they are carried, partake of the planet virus. They will never bring peace to man, but they will harry him onward through the circle of the worlds.

II

A scientific civilization in the full sense is an anomaly in world history. The civilizations of the sun never developed it. Only one culture, that of the West, has, through technology, reduced the religious mystique so long attached to agriculture. Never before have such large masses of people been so totally divorced from the land or the direct processing of their own foodstuffs.

This phenomenon has undoubtedly contributed to the alienation of man from nature, as more and more acres go under cement for parking lots, shopping centers, and superhighways. A steadily mounting population threatens increasing damage to the natural environment from which food and breathable air are drawn. All kinds of sidelong, not very visible or dramatic dangers lurk about the edges of such an unstable situation. Any one of them could at some point become lethal, and an obscure and ignored problem turn into a disaster.

The tragedy of a single man in the New York blackout in 1965 could easily become the symbol for an entire civilization. This man, as it happened, was trapped by the darkness on an upper floor of a skyscraper. A Negrito or any one of the bush folk would have known better than to go prowling in a spirit-haunted, leopard-infested jungle after nightfall. The forest dwellers would have remained in their huts until daybreak.

In this case civilized man was troubled by no such inhibitions. Seizing a candle from a desk in his office, he made his way out into the corridor. Since the elevators were not running, he cast about for a stairway. Sighting what in the candlelight appeared to be a small service doorway near the bank of elevators, he opened it and, holding his small candle at eye level, stepped in. He was found the next day at the bottom of an elevator shaft, the extinguished candle still clutched in his hand.

I have said that this episode is symbolic. Man, frail, anticipatory man, no longer possessed the caution to find his way through a disturbance in his nightly routine. Instead he had seized a candle, the little flickering light of human knowledge, with which to confront one of his own giant creations in the dark. A janitor had left a door unlocked that should have been secured. Urban man, used to walking on smooth surfaces, had never glanced below his feet. He and his inadequate candle had plunged recklessly forward and been swallowed up as neatly by a machine in its tunnel as by a leopard on a dark path.

I have seen similar errors made at the onset of floods by men who no longer had the wit or conditioning to harken to the whispered warnings of wild nature. They had grown too confident of the powers of their own world, from which nature, so they thought, had been excluded. In the wider context of civilization, our candle flame may illuminate the next few moments but scarcely more. The old precarious world from which we came lurks always behind the door. It will find a way to be present, even if we should force it to retreat to the nearest star. Moreover, if, after the crust of the earth has been rifled and its resources consumed, civilization were to come upon evil times, man would have to start over with incredibly less than lay potentially before the flint users of the Ice Age.

But there emerges to haunt us the question of why this peculiar civilization arose. In the first part of the twentieth century appeared a man destined to be widely read, criticized, and contended against, even to be called wicked. He was destined to influence the philosophical historians who followed him in the attempt to observe some kind of discernible pattern in the events of history. Our concern with this man, Oswald Spengler, and his book, *The Decline of the West*, relates to just one aspect of Spengler's thought: the rise of our scientific civilization. That Spengler is periodically declared outmoded or resurrected need not involve us. What does affect us is that the man is basically a German poet-philosopher who glimpsed the leitmotif of the era we have been discussing and who pictured it well. It is the world in which we of the West find ourselves. Spengler is difficult, but in this aspect of his work he pictures the idea forms, the *zeitgeist*, lurking within the culture from which the rocket was to emerge.

Perhaps what he terms the Faustian culture—our own—began as early as the eleventh century with the growing addiction to great unfillable cathedrals with huge naves and misty recesses where space seemed to hover without limits. In the words of one architect, the Gothic arch is "a bow always tending to expand." Hidden within its tensions is the upward surge of the space rocket.

Again, infinite solitude tormented the individual soul. A too guilty hunger for forbidden knowledge beset the introverted heroes of this culture. The legend of Faust to this day epitomizes the West; the Quest of the Holy Grail is another of its Christian symbols. The bell towers of Western Europe have rung of time and death and burial in a way characteristic of no other culture. The bells were hung high and intended to reach far across space.

Faustian man is never at rest in the world. He is never the contemplative beneath the sacred Bô tree of the Buddha. He is, instead, a spokesman of the will. He is the embodiment of a restless, exploratory, and anticipating ego. In that last word we have the human head spun round to confront its future—the future it has created. It well may be that the new world, which began amidst time-tolling bells and the stained glass and dim interiors of Gothic cathedrals, laid an enchantment upon the people of Western Europe that provided at least a portion of the seedbed for the later rise of science—just as guilt has also haunted us. In its highest moments, science could also be said, not irreverently, to be a search for the Holy Grail. There the analogy lies—a poet's vision perhaps, but a powerful one. I would merely add one observation: that the owl, Minerva's symbol of wisdom, is able to turn its head through an angle of one hundred eighty degrees. It can be not visually anticipatory alone, it can look backward. Perhaps it is the lack of this ability that gives modern man and his children a slightly inhuman cast of countenance.

III

Giordano Bruno was burned at Rome in 1600. His body perished, but the ideas for which he died—the heretical concepts of the great depths of the universe and of life on other worlds —ran on with similar dreams across the centuries to enlighten our own time. Space travel unconsciously began when the first hunters took their bearings on the North Star or saw the rising of the Southern Cross. It grew incipiently with the mathematics and the magnetic needle of the mariners. Man, in retrospect, seems almost predestined for space. To master the dream in its entirety, however, man had to invent in two categories: inventions of power and inventions of understanding. The invention of the scientific method itself began as an adventure in understanding. Inventions of power without understanding have been the bane of human history.

The word "invention" can denote ideas far removed from the machines to which the people of our mechanically inclined era seek constantly to limit the word. Let us take one refined example. The zero, invented twice in the mists of prehistory, once by the Hindus and once by the Maya, lies at the root of all complicated mathematics, yet it is not a "thing." Rather, it is a "no thing," a "nothing," without which Roman mathematics was a heavy, lumbering affair. In our time that necessary zero leaps instantaneously through the circuits of computers, helping to guide a rocket on the long pathway to Mars. One might say that an unknown mathematical genius seeking pure abstract understanding was a necessary prehistoric prelude to the success of the computer. He was also, and tragically, the possible indirect creator of world disaster in the shape of atomic war.

"Traveling long journeys is costly, at all times troublesome, at some times dangerous," warned a seventeenth-century writer.

These were true words spoken of great seas and unmapped continents. They can also be spoken of the scientific journey itself. Today, magnified beyond the comprehension of that ancient wayfarer, we contemplate roads across the planetary orbits, the penetration of unknown atmospheres, and the defiance of solar flares. This effort has become the primary obsession of the great continental powers. Into the organization of this endeavor has gone an outpouring of wealth and inventive genius so vast that it constitutes a public sacrifice equivalent in terms of relative wealth to the building of the Great Pyramid at Giza almost five thousand years ago. Indeed, there is a sense in which modern science is involved in the construction of just such a pyramid, though an invisible one.

Science, too, demands great sacrifice, persistence of purpose across the generations, and an almost religious devotion. Whether its creations will loom to future ages as strangely antiquated as the sepulchres of the divine pharaohs, time alone will tell. Perhaps, in the final reckoning, only understanding will enable man to look back upon his pathway. For if inventions of power outrun understanding, as they now threaten to do, man may well sink into a night more abysmal than any he has yet experienced. Understanding increasingly begets power, but, as perceptive statesmen have long observed, power in the wrong hands has a way of corrupting understanding.

There is an eye atop Palomar Mountain that peers at fleeing galaxies so remote that eons have elapsed since the light which reaches that great lens began its journey. There is another eye, that of the electron microscope, which peers deep into our own being. Both eyes are important. They are eyes of understanding. They balance and steady each other. They give our world perspective; they place man where he belongs. Such eyes, however, are subject to their human makers. Men may devise or acquire, and use beautiful or deathly machines and yet have no true time sense, no tolerance, no genuine awareness of their own history. By contrast, the balanced eye, the rare true eye of understanding, can explore the gulfs of history in a night or sense with uncanny accuracy the subtle moment when a civilization in all its panoply of power turns deathward. There are such troubled seers among us today—men who fear that the ramifications of the huge industrial complex centering upon

space are draining us of energy and wealth for other enterprises
—that it has about it a threatening, insensitive, and cataclysmic
quality.

A term in military parlance, "the objective," may be pertinent here. It is intended to secure the mind against the diffuse
and sometimes inept opportunism of the politician, or the
waves of uninformed emotion to which the general public is so
frequently subjected. An objective is delimited with precision
and care. Its intention is to set a clearly defined goal. Armies,
or for that matter sciences, do not advance on tides of words.
Instead, they must be supplied logistically. Schedules must
equate with a realistic appraisal of resources.

There will never be enough men or material for a multitudinous advance on all fronts—even for a wealthy nation. Thus,
as our technological feats grow more costly, the objectives of
our society must be assessed with care. From conservation to
hospitals, from defense to space, we are forced by circumstance
to live more constantly in the future. Random "tinkering,"
random response to the unexpected, become extraordinarily
costly in the industrial world which Western society has created. Yet, paradoxically, the unexpected comes with increasing
rapidity upon future-oriented societies such as ours. Psychological stresses appear. The current generation feels increasingly alienated from its predecessors. There is a quickening of
vibrations running throughout the society. One might, in
physiological terms, say that its metabolism has been feverishly
accelerated. For this, a certain price in stability has been ex-
acted, the effects of which may not be apparent until long
afterward.

The attempt to conquer space has seized the public imagination. To many of this generation, the sight of rockets roaring
upward has brought home the feats of science so spectacularly
that we sometimes forget the medical researcher brooding in
his laboratory, or the archaeologist striving amidst broken
shards and undeciphered hieroglyphs to understand what doom
destroyed a city lying beneath the sands of centuries. The estimated cost of placing the unmanned Surveyor 3 upon the
moon amounted to more than eighty million dollars. Just one
unmanned space probe, in other words, equaled or exceeded
the entire endowment of many a good college or university;

the manned flight of Apollo 12 cost two hundred and fifty millions. The total space program is inconceivably costly, yet the taxpayer, up until recently, accepted it with little question. By contrast, his elected officials frequently boggle over the trifling sums necessary to save a redwood forest or to clear a river of pollution.

What then, we are forced to ask, is our objective? Is it scientific? Is it purely military? Or is it these and more? Is there some unconscious symbolism at work? At heart, does each one of us, when a rocket hurtles into space, yearn once more for some lost green continent under other skies? Is humanity, like some ripening giant puffball, feeling the mounting pressure of the spores within? Are we, remote though we may be from habitable planets, driven by the same irresistible migrating impulse that descends upon an overpopulated hive? Are we each unconsciously escaping from the mechanized routine and urban troubles which increasingly surround us? Beneath our conscious rationalizations does this play a role in our willingness to sustain the growing burden?

Any answers to these questions would be complex and would vary from individual to individual. They are worth asking because they are part of the venture in understanding that is necessary to human survival. Three successful moon landings, it goes without saying, are an enormous intellectual achievement. But what we must try to understand is more difficult than the mathematics of a moon shot—namely, the nature of the scientific civilization we are in the process of creating. Science has risen in a very brief interval into a giant social institution of enormous prestige and governmentally supported power. To many, it replaces primitive magic as the solution for all human problems.

IV

In the coastal jungles of eastern Mexico the archaeologist comes at intervals upon giant stone heads of many tons weight carved in a strikingly distinct style far different from that of the Maya. They mark the remains of the lost Olmec culture of the first millennium before the Christian era. Around the globe, more than one such society of clever artisans has arisen and placed its stamp, the order of its style, upon surrounding objects, only to lapse again into the night of time. Each was self-contained. Each, with the limited amount of wealth and energy at its disposal, placed its greatest emphasis upon some human dream, some lost philosophy, some inner drive beyond the satisfaction of the needs of the body. Each, in turn, vanished.

Western man, with the triumph of the experimental method, has turned upon the world about him an intellectual instrument of enormous power never fully exploited by any previous society. Its feats of understanding include the discovery of evolutionary change as revealed in the stratified rocks. It has looked far down the scale of life to reveal man magically shrunken to a tiny tree shrew on a forest branch. Science has solved the mysteries of microbial disease and through the spectroscope has determined the chemical composition of distant stars. It has groped its way into a knowledge of the gigantic distances of the cosmos—distances too remote for short-lived man ever to penetrate. It has learned why the sun endures and at what pace light leaps across the universe.

Man can speak into infinite spaces, but in this time in which I write violence and contention rage, not alone on opposite sides of the world, but here at home. How far are all these voices traveling, I wonder? Out beyond earth's farthest shadow, on and on into the depths of the universe? And suppose that there were, out yonder, some hidden listening ear, would it be able to discern any difference between the sounds man made when he was a chittering tree shrew contesting for a beetle and those produced at his appearance before a parliament of nations?

It is a thing to consider, because with understanding arise instruments of power, which always spread faster than the inventions of calm understanding. The tools of violence appeal to the fanatic, the illiterate, the blindly venomous. The inventions of power have grown monstrous in our time. Man's newfound ingenuity has given him health, wealth, and increase, but there is added now the ingredient of an ever-growing terror. Man is only beginning dimly to discern that the ultimate menace, the final interior zero, may lie in his own nature. It is said in an old tale that to understand life man must learn to shudder. This century seems doomed to master the lesson.

Science, in spite of its awe-inspiring magnitude, contains one flaw that partakes of the nature of the universe itself. It can solve problems, but it also creates them in a genuinely confusing ratio. They escape unseen out of the laboratory into the body politic, whether they be germs inured to antibiotics, the waiting death in rocket silos, or the unloosed multiplying power of life. There are just so many masterful and inventive brains in the human population. Even with the growth of teamwork and the attempted solution of future problems now coming to be known as systems analysis, man is our most recalcitrant material. He does not yield cherished beliefs with rapidity; he will not take pills at the decree of some distant, well-intentioned savior.

No one knows surely what was the purpose of Olmec art. We do know something of the seemingly endless political expansion and ethnic dilution that precipitated the fall of Rome. We know also that the pace of technological innovation in the modern world has multiplied throughout our lifetimes. The skills expended now upon space may in the end alter our philosophies and rewrite our dreams, even our very concepts of the nature of life—if there is life—beyond us in the void. Moreover the whole invisible pyramid is itself the incidental product of a primitive seed capsule, the human brain, whose motivations alter with time and circumstance.

In summary, we come round again to the human objective. In the first four million years of man's existence, or, even more pointedly, in the scant second's tick during which he has inhabited cities and devoted himself to an advanced technology, is it not premature to pronounce either upon his intentions or his destiny? Perhaps it is—as the first man-ape could not have foreseen the book-lined room in which I write. Yet something of that creature remains in me as he does in all men. I compose, or I make clever objects with what were originally a tree dweller's hands. Fragments of his fears, his angers, his desires, still stream like midnight shadows through the circuits of my brain. His unthinking jungle violence, inconceivably magnified, may determine our ending. Still, by contrast, the indefinable potentialities of a heavy-browed creature capable of pouring his scant wealth into the grave in a gesture of grief and self-abnegation may lead us at last to some triumph beyond the realm of technics. Who is to say?

Not long ago, seated upon a trembling ladder leading to a cliff-house ruin that has not heard the voice of man for centuries, I watched, in a puff of wind, a little swirl of silvery thistle-down rise out of the canyon gorge beneath my feet. One or two seeds fell among stony crevices about me, but another, rising higher and higher upon the light air, ascended into the blinding sunshine beyond my vision. It is like man, I thought briefly, as I resumed my climb. It is like man, inside or out, off to new worlds where the chances, the stairways, are infinite. But like the seed, he has to grow. That impulse, too, we bring with us from the ancestral dark.

Another explosion of shimmering gossamer circled about my head. I held to the rickety ladder and followed the erratic, windborne flight of seeds until it mounted beyond the constricting canyon walls and vanished. Perhaps the eruption of our giant rockets into space had no more significance than this, I saw finally, as in a long geological perspective. It was only life engaged once more on an old journey. Here, perhaps, was our supreme objective, hidden by secretive nature even from ourselves.

Almost four centuries ago, Francis Bacon, in the years of the voyagers, had spoken of the new world of science as "something touching upon hope." In such hope do all launched seeds participate. And so did I, did unstable man upon his ladder or his star. It was no more than that. Within, without, the climb was many-dimensioned and over imponderable abysses. I placed my foot more carefully and edged one step farther up the face of the cliff.

FIVE

The Time Effacers

The savage mind deepens its knowledge with the help of imagines
mundi.

—CLAUDE LÉVI-STRAUSS

THERE ARE TWO diametrically opposed forces forever at war
in the heart of man: one is memory; the other is forgetful-
ness. No one knows completely the nature of the inner turmoil
which creates this struggle. Some rare individuals possess almost total recall; others find certain events in life so painful that
they are made to sink beneath the surface of consciousness. Sigmund Freud himself learned, as he practiced the arts of healing,
to dip his hands into the dark waters which contain our lurking
but suppressed memories. There are those among us who wish,
even in death, not a name or a memory to survive.

Once I sat in the office of a county coroner, having come
there at his request. We had previously had many discussions
involving cases of human identification that had come his way.
Some of his problems demanded the specialties of my own
field, and though I am not an expert in forensic medicine I had
been glad to listen to his experiences, as well as occasionally to
offer advice on some anatomical point.

When I was ushered into his office on that particular afternoon, a carefully prepared skull gleamed upon his desk. My
friend looked up at me with a grin of satisfaction. "You have
told me something about what the archaeologist is able to
infer concerning the habits of our remote ancestors," he said.
"Now I would like you to look at this specimen. The body
from which it came was discovered by accident in a drained
pond. It had been there for some time and was almost totally
decomposed."

I picked up the skull and slowly turned it over. A glitter of
platinum wire immediately caught my eye. I drew the mandi-
ble aside. "Look," I said in surprise, "this is one of the most

expensive and elaborate pieces of dental work I have ever seen."

"Precisely," said my friend. "The job was obviously done by a gifted specialist, and it could easily have cost a thousand dollars. So we know what?"

"That the individual had means and took care of himself," I sparred. "Surely an identification can be made on this basis."

Slowly the coroner shook his head. "We have tried," he said, "tried hard. The man did not come from around here. He came most probably from a far-off big city. It is in such places that this kind of work is done, but which place?"—he shrugged—"One could spend years on such a task and come up with nothing. Our office has neither the time, the staff, nor the money for such investigations, particularly if no evidence of a crime exists."

"You mean—?" I asked.

"Yes," he said, "it was very likely suicide by drowning."

"Then shouldn't there be some identification remaining—a wallet, a ring, something?"

My friend eyed me quizzically. "I wanted you to see this," he said, "not because skulls are new to you, but because you have always worked in the past—with another set of problems. What you see here in this individual specimen we encounter as a single category." He tapped the magnificent bridgework with a pencil to emphasize his point. "We find a certain persistent number of suicides—people like this one, very likely a man of wealth—who, when they have decided to depart this life, do so with the determination at the same time to obliterate their identity.

"Sometimes they travel far before the final act is carried out. At the last, every conceivable trace of identity is abandoned. Wallets with their cards may be hurled away, jewelry similarly disposed of; it is as though the individual were not satisfied to destroy himself, he must, as this man apparently did, bury his name so thoroughly that no one will be heard to pronounce it again."

"But murder," I interjected.

"Of course, of course, we have such cases and such conceal-ment." He turned once more to the skull. "I tell you now, how-ever, that their number is minute compared to these." He elevated the face and looked into it as if for an answer, but the skull stared beyond him unheeding and stubbornly triumphant.

The coroner sighed once more and eased the skull back upon the desk, a certain gentleness evident in his manner. "Well," he said, "I rather think this one will have his wish to be forgotten." He fingered one of the fine wires of the bridge-work. "Strange," he added. "He took care of himself—up to the last, that is. You can see it here. But then this thing—this shadow, whatever it was—came on him until he was forced to flee out of the body itself. But no one, if there was by then anyone, was to witness his final defeat. He saw to it well; he had given it thought, he left us a blank wall. Except for a new drainage ditch we would not have found even this."

My friend gestured politely. "A kind of gentleman's end, don't you think?" he said. "Perhaps there was an intent to spare someone, somewhere; who knows? It would appear he came a long way for this and went to some trouble. I won't bother you with any more details. I just wanted you to know what can lurk in these little boxes you and your colleagues handle with such scientific precision. Here in this office we are forced to build a different world with the same bones." He gently touched the skull again. "They are individuals to me, not phenotypes."

"You mistake us," I countered, "if you think we are not aware of the darkness in the human mind. Have you never heard of the *damnatio memoriae*?"

The coroner's eyes twinkled.

"Of couse not," he said. "That is what I got you in here for, to stir me up." He leaned back expectantly.

"It is a different matter from the case of your anonymous client here," I explained slowly, "and is frequently done for obscure or depraved reasons. Do you know that history is full of evidence of hatred for the past, of a desire on the part of some men to destroy even the memory of their predecessors? Public monuments are effaced, names destroyed, histories re-written. Sometimes to achieve these ends a whole intellectual elite may be massacred in order that the peasantry can be de-liberately caused to forget its past. The erasure of history plays a formidable role in human experience. It extends from the smashing of the first commemorative monuments right down to the creation of the communist 'non-person' of today. Carthage was a victim of that animus. So was the pharaoh Akhenaton, who introduced solar monotheism into Egypt."

I paused, but my friend the coroner only nodded. "Go on," he said.

"The French revolutionists sought from 1792 to 1805 permanently to eliminate the Christian calendar. Today's youth revolt is partly aimed at the destruction of the past and the humiliation of the previous generation. Just as the individual mind thrusts unwelcome thoughts below the level of consciousness, so there are times, when, in revulsion against painful or uncontainable thoughts and symbols, the social memory similarly reacts against itself. Or, again, it may reclothe old myths and traditions in new and more pleasing garments."

I pointed at the skull upon which my friend's hand rested. "Men get tired, you see. This man in the end wanted complete oblivion—not alone for his physical body—he wished to make sure he lived in no man's mind.

"The masses," I continued, "can be stirred by the same impulse. There are times of social disruption when they grow tired of history. If they cannot remake the past they intend at least to destroy it—efface the dark memory from their minds and so, in a sense, pretend that history has never been. There are plenty of examples—the assault of Cromwell's Puritans upon the statuary in the English cathedrals, or earlier, in Henry the Eighth's time, the breaking up of the great abbeys and the reckless dispersal of their ancient documents and treasures. Even worse was the total overthrow of Inca and Aztec civilization at the hands of the Spaniards. An entire writing system perished on the verge of the modern era."

The coroner's office seemed to grow darker from an impending storm gathering outside. For a moment I had a feeling of inexplicable terror, as though both of us crouched in some cranny beside a torrent that was sweeping everything to destruction.

"What you are saying"—the coroner's voice came from somewhere beyond the skull—"is that to know time is to fear it, and to know civilized time is to be terror-stricken."

I nodded. The room grew oppressively dark. I felt an impulse, somewhat against my better judgment, to speak further.

The skull had taken on a faintly watchful expression, as though it had in reality projected my thought. Beyond it, all seemed slipping into shadows.

"I am speaking as a gravedigger only," I said, my eyes fixed blindly forward. "But there is a paradox to all digging that only an archaeologist would understand. The best way to be resurrected is to be forgotten. Consider the case of Tutankhamen."

The coroner opened his window. The rain had begun to fall and its scent stole into the room along with a fresh breath of air.

"I know what you mean," he said, as the skull with its gleaming denture was deposited in a drawer. "Sometimes an individual, perhaps a great artist, or a civilization, has to be held off stage for a millennium or so until they can be understood. Like the art of Lascaux, fifteen thousand years forgotten in a sealed cave. In a case like that, even time has to be rediscovered. Not even discovered, but interpreted. It consists of more than the marks on a dial."

I arose and stood beside my friend, looking down on the wet pavement beneath us, where the rain was pushing fallen leaves along the gutter.

"Look," he said, waving a hand toward the street, "every culture in the world has a built-in clock, but in what other culture than ours has time been discovered to contain novelty? In what other culture would leaves, these yellow falling leaves, be said to be emergent and not eternal?"

"Evolutionary time," I added, "the time of the world eaters —ourselves."

We both stood silent, watching below the window the serrated shapes of the leaves as they spun past in the gathering dark.

II

"Every man," Thoreau once recorded in his journal, "tracks himself through life." Thoreau meant that the individual in all his reading, his traveling, his observations, would follow only his own footprints through the snows of this world. He would see what his temperament dictated, hear what voices his ears allowed him to hear, and not one whit more. This is the fate of every man. What is less well known is that civilizations, which are the products of men, are in their way equally obtuse. They follow their own tracks through a time measurable in centuries or millennia, but they approach the final twilight with much the same set of postulates with which they began. In Ruth Benedict's words, they resemble a human personality thrown large upon the screen, given gigantic features and a long time span.

Of these personalities the most intensely aggressive has been that of the West, particularly in the last three centuries which have seen the rise of modern science. When I say "aggressive," I mean an increasingly time-conscious, future-oriented society of great technical skill, which has fallen out of balance with the natural world about it. First of all, it is a consumer society which draws into itself raw materials from remote regions of the globe. These it processes into a wide variety of goods which a high standard of living enables it to consume. This vast industrial activity, in turn, enables the scientist and technologist to take command of business.

Scientists are not necessarily rich or the owners of business. The process is more subtle. With the passage of time and the growth of the urban structure, funds for research and development take up a far greater proportion of the budget of a particular industry. So long as the industry is in competition with others, it cannot afford to cling for long to a particular industrial process because of the fear that rival technicians will develop something more attractive or cheaper. The drive for miniaturization in the computer industry is a case in point. Thus the laboratory and its priesthood take an increasing share of the profits as they become a necessity for business survival. They also intensify the rate of social change which contributes both to human expectations and the alienation between the generations. Advertising becomes similarly important in order to encourage the acceptance of the new products as they are made available to the public. National defense is swept into the same expensive pattern in the technological war for survival.

In simple terms, the rise of a scientific society means a society of constant expectations directed toward the oncoming future. What we have is always second best, what we expect to have is "progress." What we seek, in the end, is Utopia. In the endless pursuit of the future we have ended by engaging to destroy the present. We are the greatest producers of non-degradable garbage on the planet. In the cities a winter snow-fall quickly turns black from the pollutants we have loosed in the atmosphere.

This is not to denigrate the many achievements and benefits of modern science. On a huge industrial scale, however, we have unconsciously introduced a mechanism which threatens to run out of control. We are tracking ourselves into the future—a future whose "progress" is as dubious as that which we experience today. Once the juggernaut is set in motion, to slow it down or divert its course is extremely difficult because it involves the livelihood and social prestige of millions of workers. The future becomes a shibboleth which chokes our lungs, threatens our ears with sonic booms, and sets up a pop-ulation mobility which is destructive in its impact on social institutions.

In the extravagant pursuit of a future projected by science, we have left the present to shift for itself. We have regarded science as a kind of twentieth-century substitute for magic, instead of as a new and burgeoning social institution whose ways are just as worthy of objective study as our political or economic structures. In short, the future has become our pri-mary obsession. We constantly treat our scientists as soothsay-ers and project upon them questions involving the destiny of man over prospective millions of years.

As evidence of our insecurity, these questions multiply with our technology. We are titillated and reassured by articles in the popular press sketching the ways in which the new biology will promote our health and longevity, while, at the other end of the spectrum, hovers the growing shadow of a locust swarm of human beings engendered by our successful elimination of famine and plague. To meet this threat to our standard of liv-ing we are immediately encouraged to believe in a "green rev-olution" brought about by ingenious plant scientists. That the green revolution, even if highly successful, would not long re-store the balance between nature and man, goes unremarked.

Thus science, as it leads men further and further from the first world they inhabited, the world we call natural, is beguil-ing them into a new and unguessed domain. In a world where contingencies multiply at a fantastic rate and nations react like fevered patients whose metabolism is seriously disturbed, the scientist is forced into a new and hitherto unsought role in society. From the seclusion of the laboratory he is being drawn into the role of an Eastern seer, with all the dangers and exacerbations this entails. To shepherd the recalcitrant masses, or indeed to guide himself safely through a world of his own unconscious creation, is a well-nigh impossible task which has come upon him by insidious degrees. He does not possess marked political power, yet he has transformed the world in which power operates.

The scientist is now in the process of learning that the social world is stubbornly indifferent to the elegant solutions of the lecture hall, and that to guide a future-oriented world along the winding path to Utopia demands an omniscience that no human being possesses. We have long passed the simple point at which science presented to us beneficent medicines and where, in the words of José Ortega y Gasset, science and the civilization shaped by it could be regarded as the self-objectivation of human reason. It is one thing successfully to plan a moon voyage; it is quite another to solve the moral problems of a distraught, unenlightened, and confused humanity.

Men wandering in the infinitude of space and time which their science has revealed are trapped in a world of darkening shadows, like those depicted in Giambattista Piranesi's eighteenth-century etchings, the *Carceri*. The pictures reveal giant buildings in which the human figure wanders lost amidst huge beams and winding stairways ascending or descending into vacancy. Thick ropes hang from spiked machines of unspeakable intent. This world of the prisons is the world of man; the vast maze offers no exit.

Unlike the cyclic time of the classical world, the time of the Christian era is novel. It proceeds to an end and it has arisen through a creative act. Though science has enormously extended the cosmic calendar, it has never succeeded in eliminating that foreknowledge of the non-existence of life and of individual genera and species which the Christian creation introduced. We bear in our actual bodies traces of our formation out of the animal remnants of the past. Thus causality plays a significant role in our thinking because we have been stitched together from the bones and tissues of creatures which are now extinct.

As our knowledge of the evolutionary past has increased, we have unconsciously transferred the observed complexity of forms leading up through the geological strata to our cultural behavior. We speak of "progress" in a rather ill-defined way, and from surveying time past we have become devotees of time future. We maintain "think tanks" in which experts are employed to play all possible games, to create models, military or otherwise, that the future might produce. We are handicapped in just one way: the future may be guessed at, but only as a series of unknown alternatives.

Some decades ago Henry Phillips of the Massachusetts Institute of Technology expressed this dilemma succinctly. "What will happen five minutes from now is pretty well determined," he wrote, "but as that period is gradually lengthened a larger and larger number of purely accidental occurrences are included. Ultimately a point is reached beyond which events are more than half determined by accidents which have not yet happened. Present planning loses significance when that point is reached. . . . Here is the fundamental dilemma of civilization . . . there is serious doubt whether the way forward is known."

Mathematical statistics arose as a technique for achieving some insight into the contingencies of life—that is, variable phenomena. This technique does not, however, aid us in discerning those events which Dr. Phillips describes as not yet having happened. It can, at best, inform us that the urban mass is reaching fantastic numbers and that the birthrate must be reduced. Of the end result of these phenomena, as of pollution, it can tell us little. It can only inform us that the trends which are extrapolated into the future spell disaster. Infinitesimal man is beginning to draw the macrocosm into himself, or, rather, it might be stated that through his evolutionary advancement the cosmos is beginning to reach into him. He has yet to prove that he can master the powers he has summoned up.

Man, oriented completely toward the future, suppresses his past as dissatisfying. He unconsciously resents continuity and causality; he is *event*-oriented. This is frequently reflected in the new motion pictures made to appeal to a youthful audience.

Plot gives way to episode. The existential world of the hippy provokes sensate experience but does not demand dramatic continuity. The result is the onset of that chaos in which societal order threatens to disappear.

Thus, with the near destruction of emotional continuity between the generations, time past is vilified or extinguished in favor of oncoming uncoordinated activist time. "Make the revolution," exhorts one youngster. "Afterwards we will decide what to do about it." Another pauses merely to exclaim, "Scrap the system." With those words, we have reached the final culmination of the Faustian hunger for experience. Creative time has been obliterated in order to welcome something new in human history—the pure and disconnected "event" that has replaced reality. The LSD trip has reached a level with the experience of the classroom. It is not a coincidence that the memory effacers have emerged in the swarming time of the spore cities. Man has followed his own tracks in a circle as great as that of the cometary visitant from space that had ushered in my childhood.

III

As an anthropologist I once came upon time wrapped in a small leathern bag—a bag such as that in which Odysseus might once have carried the four winds. The way of the matter was this. It was a small Pawnee medicine bundle. The bundle, among other objects, contained some feathers, a mineralized fossil tooth, an archaic, square-headed iron nail, and a beautifully flaked Ice Age spearpoint of agate. The date of the latter was easy to identify because of its shape, which related it immediately to a long-gone mammoth-hunting people.

The warrior to whom this precious bundle had once belonged must have had an alert eye for the things his guardian spirit had advised him to seek. There was no way of telling from this cracked receptacle what powers had been given its possessor or what had been his dreams. They had come swirling, presumably upon demand, from that dark region which contains the past. The square-headed nail represented the man's own time, but it had obviously been regarded as too sacred to hammer into an arrowpoint.

Most probably all this contained past in the little bundle was filled with streaming darkness and sudden emergences. For, as Joseph Campbell has so aptly pointed out, "where there is magic there is no death." I was not so insensitive in undoing the crumbling hide that I failed to feel the shadow of an extinct animal, or touch with longing the intricately flaked weapon of a vanished day. The bundle held in my hand had been a sacred object among a people who believed implicitly in its powers and who understood the prayers and fastings through which the owner had been instructed. I was an outsider to whom the nail could never denote more than a nail, or the flaked weapon stand for more than a bygone historical moment. I was afflicted by causality, by technological time rather than the magic of genuine earth time.

I have spoken of the time effacers in Western culture as those who would destroy all memory of the dead or, turning from the ancient institutions which have sustained our society, would engage in an orgiastic and undiscriminating embrace of the episodic moment—the statistical happening without significance. The deliberate effacement of defeated men and broken cultures is, as I have said, an ancient act in world history. The attempt to leap forward into the future or grudgingly to accept the fleeting moment as the only abode of man is particularly a phenomenon of our turbulent era. The causes I have to some degree explored. I have not, however, paused to examine the nature of time in those simple cultures which are without causal or novel time, and where the veil between life and death wavers fitfully at best. Borders are undefined, and animals or men, rather easily exchanging shapes, pass to and fro in ways unknown to the sophisticated world.

Certainly this world of the primitive is a novel one by civilized standards, but not in terms of primitive thought. There are several reasons why this is so. The distinctions between animals and men that have been established by biological science do not obtain in the primitive mind. Animals talk, they carry messages, they may be supernaturals. Time itself may exhibit highly eccentric behavior. It may stand still, as it does to those who intrude into fairyland, or again, two kinds of time, the time of human beings and the time of the supernaturals, may exist side by side. This last is a phenomenon on which

it would be advisable to dwell for a moment. It is highly characteristic of many peoples outside the urban swarming phase.
It tells us something of the psychology of man while he still
clung to the savage environment from which he had arisen.

With the Australian aborigines, for example, there are two
separate time scales: that of the immediate present, the common day of ordinary existence, and, in addition, that of the
period of "dreamtime" or dawn beings which precedes the
workaday world. This latter epoch is a sacred, mythological era
in which man was first created and in which the supernatural
beings laid down the laws that have since governed him.

The past of the dreamtime, however, is not really past at all.
The world, it is true, perhaps no longer visibly responds to the
forces that were exerted in the time of the ancestors. Nevertheless the atmosphere of the dreamtime still persists, like an
autumnal light, across the landscape of the aborigines. It is
elusive, it is immaterial, but it is there. The divine beings still
exist, even though they may have shifted form or altered their
abodes. Man survives by their aid and sufferance. He did not
come into existence merely once at their behest. They are still
his preceptors and guides. They continue to order his ways in
the difficult environment that surrounds him—a countryside
that has been appropriately described as his living age-old
family tree.

Totemic rituals establish for each generation the living reality of man's relationship to the plant and animal world which
sustains him. His occupations came from the totemic ancestors. Thus man, in his human time, subsists also in a kind of
surviving dreamtime which is eternal and unchanging. Both
men and animals come and go through the generations a little
like actors slipping behind the curtain, in order to reappear
later, drawn through the totemic center to precisely similar
renewed roles in society. Sacred time is of another and higher
dimension than secular time. It is, in reality, timeless; past and
future are contained within it. All of primitive man's meaningful relationship to his world is thus not history, not causality in
a scientific sense, but a mythical ordering of life which has not
deviated and will not in future deviate from the traditions of
eternity.

The emergence of novel events has no meaning to the timeless people. Sacred time enables man to escape from or, to a degree, to ignore the profane time in which he actually exists. Among the world's most simple people, we find remarkably effective efforts to erase or ignore all that is not involved with the transcendent search for timelessness, the happy land of no change. Perhaps this was what Plato sought in his doctrine of the forms— the world beyond reality so poetically expressed by Margaret Mead as the "world of the first rose, and the first lark song."

Perhaps at this stage of human culture man has sought psychic protection for himself by buttressing the stability of his environment. In his own fashion, he has remolded nature in mytho-poetic terms. He lives his life amidst talking animals and the marks of the going and coming of the dream divinities who are both his creators and the guardians of his days. As closely as a mortal can manage, he exists in eternity.

In this stage, in spite of numerous variations in religious practice over the world, man is basically not a consciously malicious time effacer. He is, instead, a creature to whom secular time has no meaning and no value; he is living in the perpetual light of a past dreamtime which still enfolds him. He is at peace with the seasons and, through decreed ritual, even with the animals he hunts. Frequently he is terrified by the unusual when it is thrust too prominently before his eyes or cannot be fitted into his accepted cosmology.

This world, a world that man has inhabited for a far longer period than high civilization has existed upon earth, contrasts spectacularly with the secular domain of science. If it is illusory, we must admit that it is at the same time relatively stable. Man lived safely within the confines of nature. A few stones from the riverbed, a bit of shaped clay, some wild seeds, and a receptacle made of bark perhaps sufficed him. On a subsistence level of economic activity, the primitives had actually arrived at an ecological balance with nature. They had created another world of reentry into that nature upon a psychical level. One might say that, like a turtle, man had thrust out his curious head just far enough to glimpse the harshness of the profane landscape and had quickly chosen to mythologize and thereby make peace with it. On a simple basis he had achieved what modern man in his thickening shell of technology is only now seeking unsuccessfully to accomplish.

Man, in other words, is not by innate psychology a world eater. He possesses, in his far-ranging mind, only the latent potentiality. The rise of Western urbanism, accompanied by science, produced the world eaters just as surely as those other less sophisticated primitives reentered nature by means of the sacred, never-to-be-disturbed time that was not, except as it developed in their heads.

So vast is the gap that now yawns between the degraded remnants of the hunting folk and their brothers, the world eaters, that even a perceptive anthropologist of a generation ago, Paul Radin, was not prepared in the atmosphere of his time to recognize the unconscious irony in the conclusion of his book *The World of Primitive Man.* After narrating many things about the Eskimo and their philosophy, including their abhorrence of overweening pride, Radin quotes the statement of one Eskimo who had been taken to New York. After gazing down into the great canyons of the streets, the wondering native finally remarked, according to Radin's informant, "Nature is great but man is greater still." One can understand the confusion of an Eskimo brought from far ice fields to peer down upon the greatest city of the world eaters. More appalling is the discovery that an anthropologist could have quoted this simple remark with approbation as a profound message to modern man. The irony is deepened when we learn that nature sends no messages to man when all is well.

If the message which Radin interpreted was sent, perhaps, like so many messages from the dark powers, it should have been read with a different emphasis. When man becomes greater than nature, nature, which gave him birth, will respond. She has dealt with the locust swarm and she has led the lemmings down to the sea. Even the world eaters will not be beyond her capacities. Sila, as the Eskimo call nature, remains apart from mankind "*just as long as men do not abuse life.*" This is the message that a more able shaman might have found a raven to carry—a raven who could still wing with an undefaced warning from the country of primeval time.

SIX

Man in the Autumn Light

From a . . . dream . . . to touch the edges of Change where all numbers twist and break . . .

—JACK LINDSAY

THE FRENCH DRAMATIST Jean Cocteau has argued persuasively about the magic light of the theatre. People must remember, he contended, that "the theatre is a trick factory where truth has no currency, where anything natural has no value, where the only things that convince us are card tricks and sleights of hand of a difficulty unsuspected by the audience."

The cosmos itself gives evidence, on an infinitely greater scale, of being just such a trick factory, a set of lights forever changing, and the actors themselves shape shifters, elongated shadows of something above or without. Perhaps in the sense men use the word natural, there is really nothing at all natural in the universe or, at best, that the world is natural only in being unnatural, like some variegated, color-shifting chameleon.

In Brazilian rivers there exists a fish, one of the cyprinodonts, which sees with a two-lensed eye, a kind of bifocal adjustment that permits the creature to examine the upper world of sunlight and air, while with the lower half of the lens he can survey the watery depths in which he lives. In this quality the fish resembles Blake, the English poet who asserted he saw with a double vision into a farther world than the natural. Now the fish, we might say, looks simultaneously into two worlds of reality, though what he makes of this divided knowledge we do not know. In the case of man, although there are degrees of seeing, we can observe that the individual has always possessed the ability to escape beyond naked reality into some other dimension, some place outside the realm of what might be called "facts."

Man is no more natural than the world. In reality he is, as we have seen, the creator of a phantom universe, the universe we call culture—a formidable realm of cloud shapes, ideas, potentialities, gods, and cities, which with man's death will collapse into dust and vanish back into "expected" nature.

Some landscapes, one learns, refuse history; some efface it so completely it is never found; in others the thronging memories of the past subdue the living. In my time I have experienced all these regions, but only in one place has the looming future overwhelmed my sense of the present. This happened in a man-made crater on the planet earth, but to reach that point it is necessary to take the long way round and to begin where time had lost its meaning. As near as I can pinpoint the place, it was somewhere at the edge of the Absaroka range along the headwaters of the Bighorn years ago in Wyoming. I had come down across a fierce land of crags and upland short-grass meadows, past aspens shivering in the mountain autumn. It was the season of the golden light. I was younger then and a hardened foot traveler. But youth had little to do with what I felt. In that country time did not exist. There was only the sound of water hurrying over pebbles to an unknown destination—water that made a tumult drowning the sound of human voices.

Somewhere along a creek bank I stumbled on an old archaeological site whose beautifully flaked spearpoints of jasper represented a time level remote from me by something like ten thousand years. Yet, I repeat, this was a country in which time had no power because the sky did not know it, the aspens had not heard it passing, the river had been talking to itself since before man arose, and in that country it would talk on after man had departed.

I was alone with the silvery aspens in the mountain light, looking upon time thousands of years remote, yet so meaningless that at any moment flame might spurt from the ash of a dead campfire and the hunters come slipping through the trees. My own race had no role in these mountains and would never have.

I felt the light again, the light that was falling across the void on other worlds. No bird sang, no beast stirred. To the west the high ranges with their snows rose pure and cold. It was a place to meet the future quite as readily as the past. The fluttering aspens expressed no choice, and I, a youngster with but few memories, chose to leave them there. The place was of no true season, any more than the indifferent torrent that poured among the boulders through summer and deep snow alike.

I camped in a little grove as though waiting, filled with a sense of incompleteness, alert for some intangible message that was never uttered. The philosopher Jacques Maritain once remarked that there is no future thing for God. I had come upon what seemed to be a hidden fragment of the days before creation. Because I was mortal and not an omniscient creature, I lingered beside the stream with a growing restlessness. I had brought time in my perishable body into a place where, to all intents, it could not exist. I was moving in a realm outside of time and yet dragging time with me in an increasingly excruciating effort. If man was a creature obliged to choose, then choice was here denied me. I was forced to wait because a message from the future could not enter this domain. Here was pure, timeless nature—sequences as incomprehensible as pebbles—dropped like the shaped stones of the red men who had no history. The world eaters, by contrast, with their insatiable hunger for energy, quickly ran through nature; they felt it was exhaustible. They had, like all the spore-bearing organisms, an instinctive hunger for flight. They wanted more from the dark storehouse of a single planet than a panther's skin or a buffalo robe could offer. They wanted a greater novelty, only to be found far off in the orchard of the worlds.

Eventually, because the message never came, I went on. I could, I suppose, have been safe there. I could have continued to hesitate among the stones and been forgotten, or, because one came to know it was possible, I could simply have dissolved in the light that was of no season but eternity. In the end, I pursued my way downstream and out into the sagebrush of ordinary lands. Time reasserted its hold upon me but not quite in the usual way. Sometimes I could almost hear the thing for which I had waited in vain, or almost remember it. It was as though I carried the scar of some unusual psychical encounter.

A physician once described to me in detail the body's need to rectify its injuries, to restore, in so far as possible, mangled bones and tissues. A precisely arranged veil of skin is drawn over ancient wounds. Similarly the injured mind struggles, even in a delusory way, to reassemble and make sense of its shattered world. Whatever I had been exposed to among the snow crests and the autumn light still penetrated my being. Mine was the wound of a finite creature seeking to establish its own reality against eternity.

I am all that I have striven to describe of the strangest organism on the planet. I am one of the world eaters in the time when that species has despoiled the earth and is about to loose its spores into space. As an archaeologist I also know that our planet-effacing qualities extend to time itself. When the swarming phase of our existence commences, we struggle both against the remembered enchantment of childhood and the desire to extinguish it under layers of concrete and giant stones. Like some few persons in the days of the final urban concentrations, I am an anachronism, a child of the dying light. By those destined to create the future, my voice may not, perhaps, be trusted. I know only that I speak from the timeless country revisited, from the cold of vast tundras and the original dispersals, not from the indrawings of men.

II

The nineteenth-century novelist Thomas Love Peacock once remarked critically that "a poet in our times is a semi-barbarian in a civilized community. . . . The march of his intellect is like that of a crab, backward." It is my suspicion that though many moderns would applaud what Peacock probably meant only ironically, there is a certain virtue in the sidelong retreat of the crab. He never runs, he never ceases to face what menaces him, and he always keeps his pincers well to the fore. He is a creature adapted by nature for rearguard action and withdrawal, but never rout.

The true poet is just such a fortunate creation as the elusive crab. He is born wary and is frequently in retreat because he is a protector of the human spirit. In the fruition time of the world eaters he is threatened, not with obsolescence, but with being hunted to extinction. I rather fancy such creatures— poets, I mean—as lurking about the edge of all our activities, testing with a probing eye, if not claw, our thoughts as well as our machines. Blake was right about the double vision of poets. There is no substitute, in a future-oriented society, for eyes on stalks, or the ability to move suddenly at right angles from some dimly imminent catastrophe. The spore bearers, once they have reached the departure stage, are impatient of any but acceptable prophets—prophets, that is, of the swarming time. These are the men who uncritically proclaim our powers over the cosmic prison and who dangle before us ill-assorted keys to the gate.

By contrast, one of the most perceptive minds in American literature, Ralph Waldo Emerson, once maintained stubbornly: "The soul is no traveller." Emerson spoke in an era when it was a passion with American writers to go abroad, just as today many people yearn for the experience of space. He was not engaged in deriding the benefits of travel. The wary poet merely persisted in the recognition that the soul in its creative expression is genuinely *not* a traveler, that the great writer is peculiarly a product of his native environment. As an untraveled traveler, he picks up selectively from his surroundings a fiery train of dissimilar memory particles—"unlike things" which are woven at last into the likeness of truth.

Man's urge toward transcendence manifests itself even in his outward inventions. However crudely conceived, his rockets, his cyborgs, are intended to leap some void, some recently discovered chasm before him, even as long ago he cunningly devised language to reach across the light-year distances between individual minds. The spore bearers of thought have a longer flight history than today's astronauts. They found, fantastic though it now seems, the keys to what originally appeared to be the impregnable prison of selfhood.

But these ancient word-flight specialists the poets have another skill that enhances their power beyond even the contemporary ability they have always had to sway minds. They have, in addition, a preternatural sensitivity to the backward and forward reaches of time. They probe into life as far as, if not farther than, the molecular biologist does, because they touch life itself and not its particulate structure. The latter is a recent scientific disclosure, and hence we acclaim the individual discoverers. The poets, on the other hand, have been talking across the ages until we have come to take their art for granted. It is useless to characterize them as dealers in the obsolete, because this venerable, word-loving trait in man is what enables him to transmit his eternal hunger—his yearning for the country of the unchanging autumn light. Words are man's domain, from his beginning to his fall.

Many years ago I chanced to read a story by Don Stuart entitled "Twilight." It is an account of the further history of humanity many millions of years in the future. The story is told by a man a few centuries beyond our time, who in the course of an experiment had been accidentally projected forward into the evening of the race. He had then escaped through his own powers, but in doing so had overleaped his own era and reached our particular century. The time voyager sings an unbearably sad song learned in that remote future—a song that called and sought and searched in hopelessness. It was the song of man in his own twilight, a song of the final autumn when hope had gone and man's fertility with it, though he continued to linger on in the shadow of the perfect machines which he had created and which would long outlast him.

What lifts this story beyond ordinary science fiction is its compassionate insight into the basic nature of the race—the hunger that had accompanied man to his final intellectual triumph in earth's garden cities. There he had lost the will and curiosity to seek any further to transcend himself. Instead, that passion had been lavished upon his great machines. But the songs wept and searched for something that had been forgotten—something that could never be found again. The man from the open noonday of the human triumph, the scientist of the thirty-first century, before seeking his own return down the time channel, carries out one final act that is symbolic of man's yearning and sense of inadequacy before the universe, even though he had wished, like Emerson, "to climb the steps of paradise." Man had failed in the end to sanctify his own being.

The indomitable time voyager standing before the deathless machines performs the last great act of the human twilight. He programs the instruments to work toward the creation of a machine which would incorporate what man by then had lost: curiosity and hope. In dying, man had transmitted his hunger to the devices which had contributed to his death. Is this act to be labeled triumph or defeat? We do not learn; we are too far down in time's dim morning. The poet speaks with man's own Delphic ambiguity. We are left wondering whether the time voyager had produced the only possible solution to the final decay of humanity—that is, the transference of human values to the world of imperishable machines—or, on the other hand, whether less reliance upon the machine might have prevented the decay of the race. These are questions that only the long future will answer, but "Twilight" is a magnificent evocation of man's ending that only a poet of this century could have adequately foreseen—that man in the end forgets the message that started him upon his journey.

III

On a planet where snow falls, the light changes, and when the light changes all is changed, including life. I am not speaking now of daytime things but of the first snows of winter that always leave an intimation in each drifting flake of a thousand-year turn toward a world in which summer may sometime forget to come back. The world has known such episodes: it has not always been the world it is. Ice like a vast white amoeba has descended at intervals from the mountains and crept over the hills and valleys of the continents, ingesting forests and spewing forth boulders.

Something still touches me from that vanished world as remote from us in years as an earth rocket would be from Alpha Centauri. Certainly Cocteau spoke the truth: to add to all the cosmic prisons that surround us there is the prison of the golden light that changes in the head of man—the light that cries to memory out of vanished worlds, the leaf-fall light of the earth's eternally changing theatre. And then comes the night snow that in some late hour transports us into that other, that vanished but unvanquished, world of the frost.

Near my house in the suburbs is a remaining fragment of woodland. It once formed part of a wealthy man's estate, and in one corner of the wood a huge castle created by imported workmen still looms among the trees that have long outlasted their original owner. A path runs through these woods and the people of multiplying suburbia hurry past upon it. For a long time I had feared for the trees.

One night it snowed and then a drop in temperature brought on the clear night sky. Dressed in a heavy sheepskin coat and galoshes, I had ventured out toward midnight upon the path through the wood. Out of old habit I studied the tracks upon the snow. People had crunched by on their way to the train station, but no human trace ran into the woods. Many little animals had ventured about the margin of the trees, perhaps timidly watching; none had descended to the path. On impulse, for I had never done so before in this spot, I swung aside into the world of no human tracks. At first it pleased me that the domain of the wood had remained so far untouched and undesecrated. Did man still, after all his ravages, possess some fear of the midnight forest or some unconscious reverence toward the source of his origins? It seemed hardly likely in so accessible a spot, but I trudged on, watching the pole star through the naked branches. Here, I tried to convince myself, was a fragment of the older world, something that had momently escaped the eye of the world eaters.

After a time I came to a snow-shrouded clearing, and because my blood is, after all, that of the spore bearers, I sheltered my back against an enormous oak and continued to watch for a long time the circling of remote constellations above my head. Perhaps somewhere across the void another plotting eye on a similar midnight errand might be searching this arm of the galaxy. Would our eyes meet? I smiled a little uncomfortably and let my eyes drop, still unseeing and lost in contemplation, to the snow about me. The cold continued to deepen.

We were a very young race, I meditated, and of civilizations that had yet reached the swarming stage there had been but few. They had all been lacking in some aspect of the necessary technology, and their doom had come swift upon them before they had grasped the nature of the cosmos toward which they unconsciously yearned.

Egypt, which had planted in the pyramids man's mightiest challenge to effacing time, had conceived long millennia ago the dream of a sky-traveling boat that might reach the pole star. The Maya of the New World rain forests had also watched the drift of the constellations from their temples situated above the crawling vegetational sea about them. But of what their dreamers thought, the remaining hieroglyphs tell us little. We know only that the Maya were able to grope with mathematical accuracy through unlimited millions of years of which Christian Europe had no contemporary comprehension. The lost culture had remarkably accurate eclipse tables and precise time-commemorating monuments.

Ironically the fragments of those great stelae with all their learned calculations were, in the end, to be dragged about and worshiped upside down by a surviving peasantry who had forgotten their true significance. I, in this wintry clime under the shifting of the Bear, would no more be able to enter the mythology of that world of vertical time than to confront whatever eye might roam the dust clouds at this obscure corner of the galaxy. So it was, in turning, that I gazed in full consciousness at last upon the starlit clearing that surrounded me.

Except for the snow, I might as well have been standing upon the ruins that had thronged my mind. The clearing was artificial, a swath slashed by instruments of war through the center of the wood. Shorn trees toppled by bulldozers lay beneath the snow. Piles of rusted machinery were cast indiscriminately among the fallen trees. I came forward, groping like the last man out of a shell hole in some giant, unseen conflict. Iron, rust, timbers—the place was like the graveyard of an unseen, incessant war.

In the starlight my eye caught a last glimpse of living green. I waded toward the object but it lay upon its side. I rolled it over. It was a still-living Christmas tree hurled out with everything dispensable from an apartment house at the corner of the wood. I stroked it in wordless apology. Like others, I had taken the thin screen remaining from the original wood for reality. Only the snow, only the tiny footprints of the last surviving wood creatures, had led me to this unmasking. Behind this little stand of trees the world eaters had all the time been assiduously at work.

Well, and why not, countered my deviant slime-mold mind? The sooner men finished the planet, the sooner the spores would have to fly. I kicked vaguely at some geared piece of mechanism under its cover of snow. I thought of the last Mayan peasants worshiping the upended mathematical tablets of their forerunners. The supposition persisted in the best scientific circles that the astronomer priests had in the end proved too great a burden, they and their temples and observatories too expensive a luxury for their society to maintain—that revolt had cut them down.

But what if, a voice whispered at the back of my mind, as though the indistinct cosmic figure I had earlier conjured up had just spoken, what if during all that thousand years of computing among heavy unnatural numbers, they had found a way to clamber through some hidden galactic doorway? Would it not have been necessary to abandon these monumental cities and leave their illiterate worshipers behind?

I turned over a snow-covered cogwheel. Who, after all, among such ruins could be sure that we were the first of the planet viruses to depart? Perhaps in the numbers and the hieroglyphs of long ago there had been hidden some other formula than that based upon the mathematics of rocket travel —some key to a doorway of air, leaving behind only the empty seedpods of the fallen cities. Slowly my mind continued to circle that dead crater under the winter sky.

Suppose, my thought persisted, there is still another answer to the ruins in the rain forests of Yucatan, or to the incised brick tablets baking under the Mesopotamian sun. Suppose that greater than all these, vaster and more impressive, an invisible pyramid lies at the heart of every civilization man has created, that for every visible brick or corbeled vault or upthrust skyscraper or giant rocket we bear a burden in the mind to excess, that we have a biological urge to complete what is actually uncompletable.

Every civilization, born like an animal body, has just so much energy to expend. In its birth throes it chooses a path, the pathway perhaps of a great religion as in the time when Christianity arose. Or an empire of thought is built among the Greeks, or a great power extends its roads, and governs as did the Romans. Or again, its wealth is poured out upon science, and science endows the culture with great energy, so that far goals seem attainable and yet grow illusory. Space and time widen to weariness. In the midst of triumph disenchantment sets in among the young. It is as though with the growth of cities an implosion took place, a final unseen structure, a spore-bearing structure towering upward toward its final release.

Men talked much of progress and enlightenment on the path behind the thin screen of trees. I myself had walked there in the cool mornings awaiting my train to the city. All the time this concealed gash in the naked earth had been growing. I was wrong in just one thing in my estimate of civilization. I have said it is born like an animal and so, in a sense, it is. But an animal is whole. The secret tides of its body balance and sustain it until death. They draw it to its destiny. The great cultures, by contrast, have no final homeostatic feedback like that of the organism. They appear to have no destiny unless it is that of the slime mold's destiny to spore and depart. Too often they grow like a malignancy, in one direction only. The Maya had calculated the drifting eons like gods but they did not devise a single wheeled vehicle. So distinguished an authority as Eric Thompson has compared them to an overspecialized Jurassic monster.

A monster? My eyes swept slowly over the midnight clearing and its hidden refuse of fallen trunks and cogwheels. This was the pyramid that our particular culture was in the process of creating. It represented energy beyond anything the world of man had previously known. Our first spore flight had burst against the moon and reached, even now, toward Mars, but its base was a slime-mold base—the spore base of the world eaters. They fed upon the world, and the resources they consumed would never be duplicable again because their base was finite. Neither would the planet long sustain this tottering pyramid thrust upward from what had once been the soil of a consumed forest.

A rising wind began to volley snow across the clearing, burying deeper the rusted wheels and shrieking over the cast-off tree of Christmas. There was a hint in the chill air of a growing implacable winter, like that which finally descends upon an outworn planet—a planet from which life and oxygen are long since gone.

I returned to the shelter of the oak, my gaze sweeping as I did so the night sky of earth, now dark and overcast. It came to me then, in a lonely surge of feeling, that I was childless and my destiny not bound to my kind. With the tough oak at my back I remembered the feel of my father's face against my own on the night I had seen Halley's star. The comet had marked me. I was a citizen and a scientist of that nation which had first reached the moon. There in the ruined wood, remembering the shrunken seedpods of dead cities, I yearned silently toward those who would come after me if the race survived.

Four hundred years ago a young poet and potential rival of Shakespeare had written of the knowledge-hungry Faust of legend:

> Thou art still but Faustus
> and a man.

In that phrase Christopher Marlowe had epitomized the human tragedy: We were world eaters and knowledge seekers but we were also men. It was a well-nigh fatal flaw. Whether we, like the makers of stone spearpoints in Wyoming, are a fleeting illusion of the autumn light depends upon whether any remain to decipher Marlowe's words one thousand years in the future.

The events of my century had placed the next millennium as far off as a star. All the elaborate mechanisms of communication we have devised have not ennobled, nor brought closer, individual men to men. The means exist. It is Faustus who remains a man. Beyond this dark, I, who was also a man, could not penetrate.

In the deepening snow I made a final obeissance to the living world. I took the still green, everlasting tree home to my living room for Christmas rites that had not been properly accorded it. I suppose the act was blindly compulsive. It was the sort of thing that Peacock in his time would have termed the barbarism of poets.

SEVEN

The Last Magician

> The human heart is local and finite, it has roots, and if the intel-
> lect radiates from it, according to its strength, to greater and
> greater distances, the reports, if they are to be gathered up at all,
> must be gathered at the center.
>
> —GEORGE SANTAYANA

EVERY MAN in his youth—and who is to say when youth is ended?—meets for the last time a magician, the man who made him what he is finally to be. In the mass, man now confronts a similar magician in the shape of his own collective brain, that unique and spreading force which in its manipulations will precipitate the last miracle, or, like the sorcerer's apprentice, wreak the last disaster. The possible nature of the last disaster the world of today has made all too evident: man has become a spreading blight which threatens to efface the green world that created him.

It is of the last miracle, however, that I would write. To do so I have to describe my closing encounter with the personal magician of my youth, the man who set his final seal upon my character. To tell the tale is symbolically to establish the nature of the human predicament: how nature is to be reentered; how man, the relatively unthinking and proud creator of the second world—the world of culture—may revivify and restore the first world which cherished and brought him into being.

I was fifty years old when my youth ended, and it was, of all unlikely places, within that great unwieldy structure built to last forever and then hastily to be torn down—the Pennsylvania Station in New York. I had come in through a side doorway and was slowly descending a great staircase in a slanting shaft of afternoon sunlight. Distantly I became aware of a man loitering at the bottom of the steps, as though awaiting me there. As I descended he swung about and began climbing toward me.

At the instant I saw his upturned face my feet faltered and I almost fell. I was walking to meet a man ten years dead and buried, a man who had been my teacher and confidant. He had not only spread before me as a student the wild background of the forgotten past but had brought alive for me the spruce-forest primitives of today. With him I had absorbed their superstitions, handled their sacred objects, accepted their prophetic dreams. He had been a man of unusual mental powers and formidable personality. In all my experience no dead man but he could have so wrenched time as to walk through its cleft of darkness unharmed into the light of day.

The massive brows and forehead looked up at me as if to demand an accounting of that elapsed decade during which I had held his post and discharged his duties. Unwilling step by step I descended rigidly before the baleful eyes. We met, and as my dry mouth strove to utter his name, I was aware that he was passing me as a stranger, that his gaze was directed beyond me, and that he was hastening elsewhere. The blind eye turned sidewise was not, in truth, fixed upon me; I beheld the image but not the reality of a long dead man. Phantom or genetic twin, he passed on, and the crowds of New York closed inscrutably about him.

I groped for the marble railing and braced my continued descent. Around me travelers moved like shadows. I was a similar shadow, made so by the figure I had passed. But what was my affliction? That dead man and myself had been friends, not enemies. What terror save the terror of the living toward the dead could so powerfully have enveloped me?

On the slow train running homeward the answer came. I had been away for ten years from the forest. I had had no messages from its depths, such as that dead savant had hoarded even in his disordered office where box turtles wandered over the littered floor. I had been immersed in the postwar administrative life of a growing university. But all the time some accusing spirit, the familiar of the last wood-struck magician, had lingered in my brain. Finally exteriorized, he had stridden up the stair to confront me in the autumn light. Whether he had been imposed in some fashion upon a convenient facsimile or was a genuine illusion was of little importance compared to the message he had brought. I had starved and betrayed myself. It was this that had brought the terror. For the first

time in years I left my office in midafternoon and sought the
sleeping silence of a nearby cemetery. I was as pale and drained
as the Indian pipe plants without chlorophyll that rise after
rains on the forest floor. It was time for a change. I wrote a
letter and studied time tables. I was returning to the land that
bore me.

Collective man is now about to enter upon a similar though
more difficult adventure. At the climactic moment of his journey into space he has met himself at the doorway of the stars.
And the looming shadow before him has pointed backward
into the entangled gloom of a forest from which it has been his
purpose to escape. Man has crossed, in his history, two worlds.
He must now enter another and forgotten one, but with the
knowledge gained on the pathway to the moon. He must learn
that, whatever his powers as a magician, he lies under the spell
of a greater and a green enchantment which, try as he will, he
can never avoid, however far he travels. The spell has been laid
on him since the beginning of time—the spell of the natural
world from which he sprang.

II

Long ago Plato told the story of the cave and the chained
prisoners whose knowledge consisted only of what they could
learn from flickering shadows on the wall before them. Then
he revealed their astonishment upon being allowed to see the
full source of the light. He concluded that the mind's eye may
be bewildered in two ways, either from advancing suddenly
into the light of higher things or descending once more from
the light into the shadows. Perhaps more than Plato realized in
the spinning of his myth, man has truly emerged from a cave
of shadows, or from comparable leaf-shadowed dells. He has
read his way into the future by firelight and by moonlight, for,
in man's early history, night was the time for thinking, and for
the observation of the stars. The stars traveled, men noted, and
therefore they were given hunters' names. All things moved
and circled. It was the way of the hunters' world and of the
seasons.

In spite of much learned discourse upon the ways of our
animal kin, and of how purely instinctive cries slowly gave way

to variable and muddled meanings in the head of proto-man, I like to think that the crossing into man's second realm of received wisdom was truly a magical experience.

I once journeyed for several days along a solitary stretch of coast. By the end of that time, from the oddly fractured shells on the beach, little distorted faces began to peer up at me with meaning. I had held no converse with a living thing for many hours. As a result I was beginning, in the silence, to read again, to read like an illiterate. The reading had nothing to do with sound. The faces in the cracked shells were somehow assuming a human significance.

Once again, in the night, as I traversed a vast plain on foot, the clouds that coursed above me in the moonlight began to build into archaic, voiceless pictures. That they could do so in such a manner makes me sure that the reading of such pictures has long preceded what men of today call language. The reading of so endless an alphabet of forms is already beyond the threshold of the animal; man could somehow see a face in a shell or a pointing finger in a cloud. He had both magnified and contracted his person in a way verging on the uncanny. There existed in the growing cortex of man, in its endless ramifications and prolonged growth, a place where, paradoxically, time both flowed and lingered, where mental pictures multiplied and transposed themselves. One is tempted to believe, whether or not it is literally true, that the moment of first speech arrived in a star burst like a supernova. To be sure, the necessary auditory discrimination and memory tracts were a biological preliminary, but the "invention" of language—and I put this carefully, having respect for both the biological and cultural elements involved—may have come, at the last, with rapidity.

Certainly the fossil record of man is an increasingly strange one. Millions of years were apparently spent on the African and Asiatic grasslands, with little or no increase in brain size, even though simple tools were in use. Then quite suddenly in the million years or so of Ice Age time the brain cells multiply fantastically. One prominent linguist would place the emergence of true language at no more than forty thousand years ago. I myself would accord it a much longer history, but all scholars would have to recognize biological preparation for its

emergence. What the fossil record, and perhaps even the studies of living primates, will never reveal is how much can be at-
tributed to slow incremental speech growth associated directly
with the expanding brain, and how much to the final cultural
innovation spreading rapidly to other biologically prepared
groups.

Language, wherever it first appeared, is the cradle of the
human universe, a universe displaced from the natural in the
common environmental sense of the word. In this second
world of culture, forms arise in the brain and can be transmitted in speech as words are found for them. Objects and men
are no longer completely within the world we call natural—
they are subject to the transpositions which the brain can
evoke or project. The past can be remembered and caused to
haunt the present. Gods may murmur in the trees, or ideas of
cosmic proportions can twine a web of sustaining mathematics
around the cosmos.

In the attempt to understand his universe, man has to give
away a part of himself which can never be regained—the certainty of the animal that what it senses is actually there in the
shape the eye beholds. By contrast, man finds himself in Plato's
cave of illusion. He has acquired an interest in the whole of the
natural world at the expense of being ejected from it and returning, all too frequently, as an angry despoiler.

A distinction, however, should be made here. In his first
symbol making, primitive man—and indeed even the last simple hunting cultures of today—projected a friendly image
upon animals: animals talked among themselves and thought
rationally like men; they had souls. Men might even have been
fathered by totemic animals. Man was still existing in close in-
terdependence with his first world, though already he had developed a philosophy, a kind of oracular "reading" of its nature.
Nevertheless he was still inside that world; he had not turned
it into an instrument or a mere source of materials.

Christian man in the West strove to escape this lingering illusion that the primitives had projected upon nature. Intent
upon the destiny of his own soul, and increasingly urban, man
drew back from too great intimacy with the natural, its fertility
and its orgiastic attractions. If the new religion was to survive,
Pan had to be driven from his hillside or rendered powerless

by incorporating him into Christianity—to be baptized, in other words, and allowed to fade slowly from the memory of the folk. As always in such intellectual upheavals, something was gained and something lost.

What was gained intellectually was a monotheistic reign of law by a single deity so that man no longer saw distinct and powerful spirits in every tree or running brook. His animal confreres slunk like pariahs soulless from his presence. They no longer spoke, their influence upon man was broken; the way was unconsciously being prepared for the rise of modern science. That science, by reason of its detachment, would first of all view nature as might a curious stranger. Finally it would, while giving powers to man, turn upon him also the same gaze that had driven the animal forever into the forest. Man, too, would be subject to what he had evoked; he, too, in a new fashion, would be relegated soulless to the wood with all his lurking irrationalities exposed. He would know in a new and more relentless fashion his relationship to the rest of life. Yet as the growing crust of his exploitive technology thickened, the more man thought that he could withdraw from or recast nature, that by drastic retreat he could dispel his deepening sickness.

Like that of one unfortunate scientist I know—a remorseless experimenter—man's whole face had grown distorted. One eye, one bulging eye, the technological, scientific eye, was willing to count man as well as nature's creatures in terms of megadeaths. Its objectivity had become so great as to endanger its master, who was mining his own brains as ruthlessly as a seam of coal. At last Ortega y Gasset was to remark despairingly, "There is no human nature, there is only history." That history, drawn from man's own brain and subject to his power to transpose reality, now looms before us as future on all the confines of the world.

Linguists have a word for the power of language: displacement. It is the way by which man came to survive in nature. It is also the method by which he created and entered his second world, the realm that now encloses him. In addition, it is the primary instrument by which he developed the means to leave the planet earth. It is a very mysterious achievement whose source is none other than the ghostly symbols moving along the ramifying pathways of the human cortex, the gray enfolded matter of the brain. Displacement, in simple terms, is the ability to talk about what is absent, to make use of the imaginary in order to control reality. Man alone is able to manipulate time into past and future, transpose objects or abstract ideas in a similar fashion, and make a kind of reality which is not present, or which exists only as potential in the real world.

From this gift comes his social structure and traditions and even the tools with which he modifies his surroundings. They exist in the dark confines of the cranium before the instructed hand creates the reality. In addition, and as a corollary of displacement, language is characterized by the ability to receive constant increments and modifications. Words drop into or out of use, or change their meanings. The constant easy ingestion of the new, in spite of the stability of grammatical structure, is one of the prime characteristics of language. It is a structured instrument which at the same time reveals an amazing flexibility. This flexibility allows us a distant glimpse of the endlessly streaming shadows that make up the living brain.

III

There is another aspect of man's mental life which demands the utmost attention, even though it is manifest in different degrees in different times and places and among different individuals; this is the desire for transcendence—a peculiarly human trait. Philosophers and students of comparative religion have sometimes remarked that we need to seek for the origins of the human interest in the cosmos, "a cosmic sense" unique to man. However this sense may have evolved, it has made men of the greatest imaginative power conscious of human inadequacy and weakness. There may thus emerge the desire for "rebirth" expressed in many religions. Stimulated by his own uncompleted nature, man seeks a greater role, restructured beyond nature like so much in his aspiring mind. Thus we find the Zen Buddhist, in the words of the scholar Suzuki, intent upon creating "a realm of Emptiness or Void where no conceptualism prevails" and where "rootless trees grow." The Buddhist, in a true paradox, would empty the mind in order that the mind may adequately receive or experience the world. No other creature than man would question his way of thought or feel the need of sweeping the mind's cloudy mirror in order to unveil its insight.

Man's life, in other words, is felt to be unreal and sterile. Perhaps a creature of so much ingenuity and deep memory is almost bound to grow alienated from his world, his fellows, and the objects around him. He suffers from a nostalgia for which there is no remedy upon earth except as it is to be found in the enlightenment of the spirit—some ability to have a perceptive rather than an exploitive relationship with his fellow creatures.

After man had exercised his talents in the building of the first neolithic cities and empires, a period mostly marked by architectural and military triumphs, an intellectual transformation descended upon the known world, a time of questioning. This era is fundamental to an understanding of man, and has engaged the attention of such modern scholars as Karl Jaspers and Lewis Mumford. The period culminates in the first millennium before Christ. Here in the great centers of civilization, whether Chinese, Indian, Judaic, or Greek, man had begun to abandon inherited gods and purely tribal loyalties in favor of an inner world in which the pursuit of earthly power was ignored. The destiny of the human soul became of more significance than the looting of a province. Though these dreams are expressed in different ways by such divergent men as Christ, Buddha, Lao-tse, and Confucius, they share many things in common, not the least of which is respect for the dignity of the common man.

The period of the creators of transcendent values—the axial thinkers, as they are called—created the world of universal thought that is our most precious human heritage. One can see it emerging in the mind of Christ as chronicled by Saint John. Here the personalized tribal deity of earlier Judaic thought becomes transformed into a world deity. Christ, the Good Shepherd, says: "Other sheep I have, which are not of this fold: them also I must bring, and they shall hear my voice; and there shall be one fold and one shepherd. . . . My sheep hear my voice . . . and they follow me."

These words spoken by the carpenter from Nazareth are

those of a world changer. They passed boundaries, whispered in the ears of galley slaves: "One fold, one shepherd. Follow me." These are no longer the wrathful words of a jealous city ravager, a local potentate god. They mark instead, in the high cultures, the rise of a new human image, a rejection of purely material goals, a turning toward some inner light. As these ideas diffused, they were, of course, subject to the wear of time and superstition, but the human ethic of the individual prophets and thinkers has outlasted empires.

Such men speak to us across the ages. In their various approaches to life they encouraged the common man toward charity and humility. They did not come with weapons; instead they bespoke man's purpose to subdue his animal nature and in so doing to create a radiantly new and noble being. These were the dreams of the first millennium B.C. Tormented man, arising, falling, still pursues those dreams today.

Earlier I mentioned Plato's path into the light that blinds the man who has lived in darkness. Out of just such darkness arose the first humanizing influence. It was genuinely the time of the good shepherds. No one can clearly determine why these prophets had such profound effects within the time at their disposal. Nor can we solve the mystery of how they came into existence across the Euro-Asiatic land mass in diverse cultures at roughly the same time. As Jaspers observes, he who can solve this mystery will know something common to all mankind.

In this difficult era we are still living in the inspirational light of a tremendous historical event, one that opened up the human soul. But if the neophytes were blinded by the light, so, perhaps, the prophets were in turn confused by the human darkness they encountered. The scientific age replaced them. The common man, after brief days of enlightenment, turned once again to escape, propelled outward first by the world voyagers, and then by the atom breakers. We have called up vast powers which loom menacingly over us. They await our bidding, and we turn to outer space as though the solitary answer to the unspoken query must be flight, such flight as ancient man engaged in across ice ages and vanished game trails—the flight from nowhere.

The good shepherds meantime have all faded into the

darkness of history. One of them, however, left a cryptic message: "My doctrine is not mine but his that sent me." Even in the time of unbelieving this carries a warning. For He that sent may still be couched in the body of man awaiting the end of the story.

IV

When I was a small boy I once lived near a brackish stream that wandered over the interminable salt flats south of our town. Between occasional floods the area became a giant sunflower forest, taller than the head of a man. Child gangs roved this wilderness, and guerrilla combats with sunflower spears sometimes took place when boys from the other side of the marsh ambushed the hidden trails. Now and then, when a raiding party sought a new path, one could see from high ground the sunflower heads shaking and closing over the passage of the life below. In some such manner nature's green barriers must have trembled and subsided in silence behind the footsteps of the first man-apes who stumbled out of the vine-strewn morass of centuries into the full sunlight of human consciousness.

The sunflower forest of personal and racial childhood is relived in every human generation. One reaches the high ground, and all is quiet in the shaken reeds. The nodding golden flowers spring up indifferently behind us, and the way backward is lost when finally we turn to look. There is something unutterably secretive involved in man's intrusion into his second world, into the mutable domain of thought. Perhaps he questions still his right to be there.

Some act unknown, some propitiation of unseen forces, is demanded of him. For this purpose he has raised pyramids and temples, but all in vain. A greater sacrifice is demanded, the act of a truly great magician, the man capable of transforming himself. For what, increasingly, is required of man is that he pursue the paradox of return. So desperate has been the human emergence from fen and thicket, so great has seemed the virtue of a single magical act carried beyond nature, that man hesitates, as long ago I had similarly shuddered to confront a phantom on a stair.

Written deep in the human subconscious is a simple terror of what has come with us from the forest and sometimes haunts our dreams. Man does not wish to retrace his steps down to the margin of the reeds and peer within, lest by some magic he be permanently recaptured. Instead, men prefer to hide in cities of their own devising. I know a New Yorker who, when she visits the country, complains that the crickets keep her awake. I knew another who had to be awakened screaming from a nightmare of whose nature he would never speak. As for me, a long-time student of the past, I, too, have had my visitants.

The dreams are true. By no slight effort have we made our way through the marshes. Something unseen has come along with each of us. The reeds sway shut, but not as definitively as we would wish. It is the price one pays for bringing almost the same body through two worlds. The animal's needs are very old; it must sometimes be coaxed into staying in its new discordant realm. As a consequence all religions have realized that the soul must not be allowed to linger yearning at the edge of the sunflower forest.

The curious sorcery of sound symbols and written hieroglyphs in man's new brain had to be made to lure him farther and farther from the swaying reeds. Temples would better contain his thought and fix his dreams upon the stars in the night sky. A creature who has once passed from visible nature into the ghostly insubstantial world evolved and projected from his own mind will never cease to pursue thereafter the worlds beyond this world. Nevertheless the paradox remains: man's crossing into the realm of space has forced him equally to turn and contemplate with renewed intensity the world of the sunflower forest—the ancient world of the body that he is doomed to inhabit, the body that completes his cosmic prison.

Not long ago I chanced to fly over a forested section of country which, in my youth, was still an unfrequented wilderness. Across it now suburbia was spreading. Below, like the fungus upon a fruit, I could see the radiating lines of transport gouged through the naked earth. From far up in the wandering air one could see the lines stretching over the horizon. They led to cities clothed in an unmoving haze of smog. From my remote, abstract position in the clouds I could gaze upon all below and watch the incipient illness as it spread with all its slimy tendrils through the watershed.

Farther out, I knew, on the astronauts' track, the earth would hang in silver light and the seas hold their ancient blue. Man would be invisible; the creeping white rootlets of his urban growth would be equally unseen. The blue, cloud-covered planet would appear still as when the first men stole warily along a trail in the forest. Upon one thing, however, the scientists of the space age have informed us. Earth is an inexpressibly unique possession. In the entire solar system it alone possesses water and oxygen sufficient to nourish higher life. It alone contains the seeds of mind. Mercury bakes in an inferno of heat beside the sun; something strange has twisted the destiny of Venus; Mars is a chill desert; Pluto is a cold wisp of reflected light over three billion miles away on the edge of the black void. Only on earth does life's green engine fuel the oxygen-devouring brain.

For centuries we have dreamed of intelligent beings throughout this solar system. We have been wrong; the earth we have taken for granted and treated so casually—the sunflower-shaded forest of man's infancy—is an incredibly precious planetary jewel. We are all of us—man, beast, and growing plant—aboard a space ship of limited dimensions whose journey began so long ago that we have abandoned one set of gods and are now in the process of substituting another in the shape of science.

The axial religions had sought to persuade man to transcend his own nature; they had pictured to him limitless perspectives of self-mastery. By contrast, science in our time has opened to man the prospect of limitless power over exterior nature. Its technicians sometimes seem, in fact, to have proffered us the power of the void as though flight were the most important value on earth.

"We have got to spend everything we have, if necessary, to get off this planet," one such representative of the aerospace industry remarked to me recently.

"Why?" I asked, not averse to flight, but a little bewildered by his seeming desperation.

"Because," he insisted, his face turning red as though from some deep inner personal struggle, "because"—then he flung at me what I suspect he thought my kind of science would take seriously—"because of the ice—the ice is coming back, that's why."

Finally, as though to make everything official, one of the space agency administrators was quoted in *Newsweek* shortly after the astronauts had returned from the moon: "Should man," this official said, "fall back from his destiny . . . the confines of this planet will destroy him."

It was a strange way to consider our planet, I thought, closing the magazine and brooding over this sudden distaste for life at home. Why was there this hidden anger, this inner flight syndrome, these threats for those who remained on earth? Some powerful, not totally scientific impulse seemed tugging at the heart of man. Was it fear of his own mounting numbers, the creeping of the fungus threads? But where, then, did these men intend to flee? The solar system stretched bleak and cold and crater-strewn before my mind. The nearest, probably planetless star was four light-years and many human generations away. I held up the magazine once more. Here and here alone, photographed so beautifully from space, was the blue jewel compounded of water and of living green. Yet upon the page the words repeated themselves: "This planet will destroy him."

No, I thought, this planet nourished man. It took four million years to find our way through the sunflower forest, and after that only a few millennia to reach the moon. It is not fair to say this planet will destroy us. Space flight is a brave venture, but upon the soaring rockets are projected all the fears and evasions of man. He has fled across two worlds, from the windy corridors of wild savannahs to the sunlit world of the mind, and still he flees. Earth will not destroy him. It is he who threatens to destroy the earth. In sober terms we are forced to reflect that by enormous expenditure and effort we have ventured a small way out into the planetary system of a minor star, but an even smaller way into the distances, no less real, that separate man from man.

Creatures who evolve as man has done sometimes bear the scar tissue of their evolutionary travels in their bodies. The human cortex, the center of high thought, has come to dominate, but not completely to suppress, the more ancient portions of the animal brain. Perhaps it was from this last wound that my engineer friend was unconsciously fleeing. We know that within our heads there still exists an irrational restive ghost that can whisper disastrous messages into the ear of reason.

During the axial period, as we have noted, several great religions arose in Asia. For the first time in human history man's philosophical thinking seems to have concerned itself with universal values, with man's relation to man across the barriers of empire or tribal society. A new ethic, not even now perfected, struggled to emerge from the human mind. To these religions of self-sacrifice and disdain of worldly power men were drawn in enormous numbers. Though undergoing confused erosion in our time, they still constitute the primary allegiance of many millions of the world's population.

Today man's mounting numbers and his technological power to pollute his environment reveal a single demanding necessity: the necessity for him consciously to reenter and preserve, for his own safety, the old first world from which he originally emerged. His second world, drawn from his own brain, has brought him far, but it cannot take him out of nature, nor can he live by escaping into his second world alone. He must now incorporate from the wisdom of the axial thinkers an ethic not alone directed toward his fellows, but extended to the living world around him. He must make, by way of his cultural world, an actual conscious reentry into the sunflower forest he had thought merely to exploit or abandon. He must do this in order to survive. If he succeeds he will, perhaps, have created a third world which combines elements of the original two and which should bring closer the responsibilities and nobleness of character envisioned by the axial thinkers who may be acclaimed as the creators, if not of man, then of his soul. They expressed, in a prescientific era, man's hunger to transcend his own image, a hunger not entirely submerged even beneath the formidable weaponry and technological triumphs of the present.

The story of the great saviors, whether Chinese, Indian, Greek, or Judaic, is the story of man in the process of enlightening himself, not simply by tools, but through the slow inward growth of the mind that made and may yet master them through knowledge of itself. "The poet, like the lightning rod," Emerson once stated, "must reach from a point nearer the sky than all surrounding objects down to the earth, and into the dark wet soil, or neither is of use." Today that effort is demanded not only of the poet. In the age of space it is demanded of all of us. Without it there can be no survival of mankind, for man himself must be his last magician. He must seek his own way home.

The task is admittedly gigantic, but even Halley's flaming star has rounded on its track, a pinpoint of light in the uttermost void. Man, like the comet, is both bound and free. Throughout the human generations the star has always turned homeward. Nor do man's inner journeys differ from those of that far-flung elliptic. Now, as in earlier necromantic centuries, the meteors that afflicted ignorant travelers rush overhead. In the ancient years, when humankind wandered through briars and along windy precipices, it was thought well, when encountering comets or firedrakes, "to pronounce the name of God with a clear voice."

This act was performed once more by many millions when the wounded Apollo 13 swerved homeward, her desperate crew intent, if nothing else availed, upon leaving their ashes on the winds of earth. A love for earth, almost forgotten in man's roving mind, had momentarily reasserted its mastery, a love for the green meadows we have so long taken for granted and desecrated to our cost. Man was born and took shape among earth's leafy shadows. The most poignant thing the astronauts had revealed in their extremity was the nostalgic call still faintly ringing on the winds from the sunflower forest.

Bibliography

Alfvén, Hannes. *Atom, Man and the Universe: The Long Chain of Complications.* San Francisco: W. H. Freeman, 1969.
———. *Worlds—Antiworlds: Antimatter in Cosmology.* San Francisco: W. H. Freeman, 1966.
Auden, W.H. *The Dyer's Hand, and Other Essays.* New York: Random House, 1962.
Bates, Marston. *The Jungle in the House.* New York: Walker, 1970.
Benedict, Ruth, *Patterns of Culture.* Boston: Houghton Mifflin, 1934.
Bengelsdorf, Irving S. *Spaceship Earth: People and Pollution.* Los Angeles: Fox-Mathis Publications, 1969.
Berndt, R. M. and C. H. *The World of the First Australians.* Chicago: University of Chicago Press, 1964.
Bilanink, O., and Sudarshan, E.C.G. "Beyond the Light Barrier," *Physics Today*, 22 (1969), 43–51.
Bird, David, "Pollution Fight Gains in Colleges Here," *The New York Times*, February 23, 1970.
Bonner, John Tyler. *The Cellular Slime Mold.* 2nd rev. ed. Princeton: Princeton University Press, 1967.
———. "How Slime Molds Communicate," *Scientific American*, 209 (1963), 84–93.
Bridgman, P. W. "On the Nature and the Limitations of Cosmical Inquiries," *Scientific Monthly*, 37 (November 1933), 385–397.
Burton, Richard. *The Anatomy of Melancholy* [1612]. New York: Tudor, 1951.
Campbell, Joseph. *The Masks of God: I. Primitive Mythology.* New York: Viking Press, 1959.
Chamberlain, George F. *The Story of the Comets.* London: Oxford University Press, 1909.
Childe, V. Gordon. *Man Makes Himself.* New York: New American Library, 1952.
Christensen, Clyde M. *The Molds and Man: An Introduction to the Fungi.* Minneapolis: University of Minnesota Press, 1951.

Conklin, H. C. "The Relation of Hanunóo Culture to the Plant World." Doctoral dissertation. Yale University, 1954.

Cottrell, Fred. *Energy and Society*. New York: McGraw-Hill, 1955.

Cousins, Norman. "Needed: A New Dream," *Saturday Review*, June 20, 1970.

Crowther, J. G. *Francis Bacon, the First Statesman of Science*. Chester, Pa.: Dufour Editions, 1960.

Eiseley, Loren. "Francis Bacon," *Horizon*, 6 (Winter 1964), 33–47.

———. "The Paleo Indians: Their Survival and Diffusion," *New Interpretations of Aboriginal American Culture History*. Washington, D.C.: Anthropological Society of Washington, 1955.

Elder, Frederick. *Crisis in Eden: A Religious Study of Man and Environment*, New York: Abingdon Press, 1970.

Eliade, Mircea. *Patterns in Comparative Religion*. New York: Sheed and Ward, 1958.

Forbes, R. J. *The Conquest of Nature: Technology and Its Consequences*. New York: Frederick A. Praeger, 1968.

Garrett, Garet. *Ouroboros or the Mechanical Extension of Mankind*. New York: E. P. Dutton, 1925.

Gartmann, Heinz. *Science as History*. London: Hodder and Stoughton, 1961.

Glanvill, Joseph. *Scepsis Scientifica* [1665]. London: Kegan Paul, Trench, 1885.

Gold, Thomas. "Observations of a Remarkable Glazing Phenomenon on the Lunar Surface," *Science*, 165 (1969), 1345–1349.

Gooddy, William. "Outside Time and Inside Time," *Perspectives in Biology and Medicine*, 12 (1969), 239–253.

Gregory, William K. *Evolution Emerging. A Survey of Changing Patterns from Primeval Life to Man*. 2 vols. New York: Macmillan, 1951.

Grobstein, Clifford. *The Strategy of Life*. San Francisco: W.H. Freeman, 1964.

Halacy, D. S. *Cyborg: Evolution of the Superman*. New York: Harper and Row, 1965.

Hartland, E. S. *The Science of Fairy Tales: An Inquiry into Fairy Mythology* [1891]. Detroit: Singing Tree Press, 1968.

Hassan, Selim. "The Solar Boats of Khafra: Their Origin and Development Together with the Mythology of the Universe Which They Are Supposed to Traverse," *Excavations at Giza*, vol. 6, part I. Cairo, Egypt: Government Press, 1960.

Herschel, Sir John. *A Preliminary Discourse on the Study of Natural Philosophy* [1830]. New York: Johnson Reprints, 1967.

Humphreys, Christmas. *Buddhism*. New York: Penguin Books, 1951.

Jaspers, Karl. *The Origin and Goal of History*. New Haven: Yale University Press, 1953.

Jepsen, Glenn L. "Time, Strata, and Fossils: Comments and Recommendations," in *Time and Stratigraphy in the Evolution of Man*. Washington, D.C.: National Academy of Sciences, National Research Council, 1967, 88–97.

Juenger, Friedrich George. *The Failure of Technology*. Chicago: Henry Regnery, 1949.

Kazantzakis, Nikos. *Report to Greco*. Translated by P. O. Bien. New York: Simon and Schuster, 1965.

Kroeber, Alfred. *Anthropology*. 2nd rev. ed. New York: Harcourt, Brace, 1948.

Lethaby, W. R. *Architecture: An Introduction*. New York: Oxford University Press, 1955.

Lévi-Strauss, Claude. *The Savage Mind*. Chicago: University of Chicago Press, 1966.

Lovell, A. C. B. *The Individual and the Universe*. New York: New American Library, 1959.

Maritain, Jacques. *Existence and the Existent*. New York: Pantheon Books, 1948.

Martz, Louis L. *The Paradise Within: Studies in Vaughn, Traherne, and Milton*. New Haven: Yale University Press, 1964.

Melville, Herman. *Moby Dick* [1851]. New York: Oxford University Press, 1947.

Mumford, Lewis. *The City in History*. New York: Harcourt, Brace & World, 1961.

———. *The Myth of the Machine*. New York: Harcourt, Brace & World, 1966.

———. *The Transformations of Man*. New York: Harcourt, Brace, 1956.

National Research Council, Committee on Resources, National

Academy of Sciences. *Resources and Man: A Study and Recommendations*. San Francisco: W. H. Freeman, 1969.

Neihardt, John G. *The Stranger at the Gate*. New York: Mitchell Kennerley, 1912.

Ortega y Gasset, José. *Concord and Liberty*. New York: W. W. Norton, 1946.

Osgood, Ernest Staples (ed.). *The Field Notes of Captain William Clark 1803–1805*. New Haven: Yale University Press, 1964.

Pallatino, Massimo. *The Meaning of Archaeology*. New York: Harry N. Abrams, 1968.

Peacock, Thomas Love. "The Four Ages of Poetry," in *The Works of Thomas Love Peacock*, H. Brett-Smith and C. E. Jones (eds.), vol. 8. London: Constable, 1934.

Phillips, Henry. "On the Nature of Progress," *American Scientist*, 33 (October 1945), 253–259.

Plato. *Republic*. In *Five Great Dialogues*. Classics Club ed. Translated by Benjamin Jowett. New York: Walter J. Black, 1942.

Portman, Adolf. *New Paths in Biology*. New York: Harper and Row, 1964.

Pucetti, Roland. *Persons: A Study of the Possible Moral Agents in the Universe*. New York: Herder and Herder, 1969.

Radin, Paul, *The World of Primitive Man*. New York: Abelard Schuman, 1953.

Ritner, Peter. *The Society of Space*. New York: Macmillan, 1961.

Robertson, John M. (ed.). *The Philosophical Works of Francis Bacon*. New York: E. P. Dutton, 1905.

Ronan, Colin A. *Edmond Halley: Genius in Eclipse*. Garden City, N.Y.: Doubleday, 1969.

Santayana, George. *The Birth of Reason and Other Essays*. Daniel Cory (ed.). New York: Columbia University Press, 1968.
———. *Realms of Being*. 1 vol. ed. New York: Charles Scribner's Sons, 1942.

Shapley, Harlow. *Of Stars and Men*. Boston: Beacon Press, 1958.

Shepard, Odell. *Heart of Thoreau's Journals*. Boston: Houghton Mifflin, 1927.

Shepard, Paul, and McKinley, David (eds.). *The Subversive*

Science: Essays Toward an Ecology of Man. Boston: Houghton Mifflin, 1969.

Speck, Frank G. *Naskapi*. Norman: University of Oklahoma Press, 1935.

Spengler, Oswald. *The Decline of the West*. 1 vol. ed. New York: Alfred A. Knopf, 1932.

Spiller, Robert E. *The Cycle of American Literature: An Essay in Historical Criticism*. New York: New American Library, 1957.

Still, Henry. *The Dirty Animal*. New York: Hawthorn Books, 1967.

Strehlow, T. G. H. *Aranda Traditions*. Landmarks in Anthropology Series. New York: Johnson Reprints, 1968.

Stuart, Don. "Twilight," in *The Pocket Book of Science Fiction*. Donald Wollheim (ed.). New York: Pocket Books, 1943.

Sullivan, Walter. "Moon Deposits Linked to Solar Flare," *The New York Times*, September 26, 1969.

Suzuki, Daisetz T. *The Essentials of Zen Buddhism: An Autobiography of the Writings of D. T. Suzuki*. Bernard Phillips (ed.). New York: E. P. Dutton, 1961.

Swinburne, Richard. *Space and Time*. New York: St. Martin's Press, 1968.

Thomas, Elizabeth Marshall. *The Harmless People*. New York: Random House, 1958.

Tocqueville, Alexis de. *Journey to America*. J. P. Mayer (ed.). London: Faber, 1959.

Walsh, William. *Coleridge: The Work and the Relevance*. New York: Barnes and Noble, 1967.

Whewell, William. *On the Philosophy of Discovery*. London: John W. Parker and Son, 1860.

Wilson, John A. *The Burden of Egypt*. Chicago: University of Chicago Press, 1951.

Zawodny, Janos. *Death in the Forest*. South Bend, Indiana: University of Notre Dame Press, 1962.

THE NIGHT COUNTRY

Foreword

LAST WEEK, scuffing the turf while waiting for a plane flight to begin, I turned up a broken wheel from a child's toy. It had once been painted a golden yellow. On impulse I pushed it into the pocket of my topcoat. "For luck," I said to myself and shivered. My mind ran instantly back along a dimension hazy even to myself.

This volume, as all my readers will recognize, has been drawn from many times and places in the wilderness of a single life. Though I sit in a warm room beneath a lamp as I arrange these pieces, my thoughts are all of night, of outer cold and inner darkness. These chapters, then, are the annals of a long and uncompleted running. I leave them here lest the end come on me unawares as it does upon all fugitives.

There is a shadow on the wall before me. It is my own; the hour is late. I write in a hotel room at midnight. Tomorrow the shadow on the wall will be that of another.

LOREN EISELEY

ONE

The Gold Wheel

I N THE WASTE FIELDS strung with barbed wire where the thistles grow over hidden mine fields there exists a curious freedom. Between the guns of the deployed powers, between the march of patrols and policing dogs there is an uncultivated strip of land from which law and man himself have retreated. Along this uneasy border the old life of the wild has come back into its own. Weeds grow and animals slip about in the night where no man dares to hunt them. A thin uncertain line fringes the edge of oppression. The freedom it contains is fit only for birds and floating thistledown or a wandering fox. Nevertheless there must be men who look upon it with envy.

The imagination can grasp this faint underscoring of freedom but there are few who realize that precisely similar lines run in a delicate tracery along every civilized road in the West, or that these hedges of thorn apple and osage orange are the last refuge of wild life between the cultivated fields of civilization. It takes a refugee at heart, a wistful glancer over fences, to sense this one dimensional world, but it is there. I can attest to it for I myself am such a fugitive.

This confession need alarm no one. I am relatively harmless. I have not broken or entered, or passed illegally over boundaries. I am not on the lists of the police. The only time that I have gazed into the wrong end of a gun I have been the injured party. Even this episode, however, took place many years ago and was in another country at the hands of foreigners. In spite of this, I repeat that I am a fugitive. I was born one.

The world will say that this is impossible, that fugitives are made by laws and acts of violence, that without these preliminaries no man can be called a fugitive, that without pursuit no man can be hunted. It may be so. Nevertheless I know that there are men born to hunt and some few born to flee, whether physically or mentally makes no difference. That is purely a legal quibble. The fact that I wear the protective coloration of sedate citizenship is a ruse of the fox—I learned it long ago. The facts of my inner life are quite otherwise. Early, very early, the consciousness of this difference emerges. This is how it began for me.

It begins in the echoing loneliness of a house with no other children; in the silence of a deafened mother; in the child head growing strangely aware of itself as it prattled over immense and solitary games. The child learned that there were shadows in the closets and a green darkness behind the close-drawn curtains of the parlor; he was aware of a cool twilight in the basement. He was afraid only of noise.

Noise is the Outside—the bully in the next block by whose house you had to pass in order to go to school. Noise is all the things you did not wish to do. It is the games in which you were pummeled by other children's big brothers, it is the sharp, demanding voices of adults who snatch your books. Noise is day. And out of that intolerable sunlight your one purpose has been given—to escape. Few men have such motivations in childhood, few are so constantly seeking for the loophole in the fern where the leaves swing shut behind them. But I anticipate. It is in the mind that the flight commences. It is there that the arc lights lay their shadows. It is there, down those streets past unlit houses that the child runs on alone.

II

We stood in a wide flat field at sunset. For the life of me I can remember no other children before them. I must have run away and been playing by myself until I had wandered to the edge of the town. They were older than I and knew where they came from and how to get back. I joined them.

They were not going home. They were going to a place called Green Gulch. They came from some other part of town, and their clothes were rough, their eyes worldly and sly. I think, looking back, that it must have been a little like a child following goblins home to their hill at nightfall, but nobody threatened me. Besides, I was very small and did not know the way home, so I followed them.

Presently we came to some rocks. The place was well named. It was a huge pool in a sandstone basin, green and dark with the evening over it and the trees leaning secretly inward above the water. When you looked down, you saw the sky. I remember that place as it was when we came there. I remember the quiet and the green ferns touching the green water. I remember we played there, innocently at first.

But someone found the spirit of the place, a huge old turtle, asleep in the ferns. He was the last lord of the green water before the town poured over it. I saw his end. They pounded him to death with stones on the other side of the pool while I looked on in stupified horror. I had never seen death before.

Suddenly, as I stood there small and uncertain and frightened, a grimy, splattered gnome who had been stooping over the turtle stood up with a rock in his hand. He looked at me, and around that little group some curious evil impulse passed like a wave. I felt it and drew back. I was alone there. They were not human.

I do not know who threw the first stone, who splashed water over my suit, who struck me first, or even who finally, among that ring of vicious faces, put me on my feet, dragged me to the roadside, pointed and said, roughly, "There's your road, kid, follow the street lamps. They'll take you home."

They stood in a little group watching me, nervous now, ashamed a little at the ferocious pack impulse toward the outsider that had swept over them.

I never forgot that moment.

I went because I had to, down that road with the wind moving in the fields. I went slowly from one spot of light to another and in between I thought the things a child thinks, so that I did not stop at any house nor ask anyone to help me when I came to the lighted streets.

I had discovered evil. It was a monstrous and corroding knowledge. It could not be told to adults because it was the evil of childhood in which no one believes. I was alone with it in the dark. And in the dark henceforth, in some fashion, I was destined to stay until, two years later, I found the gold wheel. I played alone in those days, particularly after my rejection by the boys who regarded Green Gulch as their territory. I took to creeping up alleys and peering through hedges. I was not miserable. There was a wonderful compensating secrecy about these activities. I had little shelters in hedgerows and I knew and perfected secret entrances and exits into the most amazing worlds.

There was, for example, the Rudd mansion. I never saw the inside of it, but I made the discovery that in a stone incinerator, back of the house and close up to the immense hedge through which I had worked a passage, there were often burned toys. Apparently the Rudd family lived with great prodigality and cast recklessly away what to me were invaluable possessions. I got in the habit of creeping through the hedge at nightfall and scratching in the ashes for bits of Meccano sets and other little treasures which I would bear homeward.

One frosty night in early fall I turned up a gold wheel. It was not gold really, but I pretended it was. To me it represented all those things—perhaps in a dim way life itself—that are denied by poverty. The wheel had been part of a child's construction set of some sort. It was grooved to run on a track and it had a screw on the hub to enable it to be fitted adjustably to an axle. The amalgam of which it was made was hard and golden and it had come untouched through the incinerator fires. In my childish world it was a wonderful object and I haunted the incinerator for many nights thereafter hoping I might secure the remaining wheel. The flow of toys declined, however, and I never found the second gold wheel. The one I had found became a sort of fetish which I carried around with me. I had become very conscious of gold wheels and finally I made up my mind to run away upon a pair of them.

My decision came about through the appearance in our neighborhood of a tea wagon which used to stop once or twice a week at the house next door. This was not an ordinary delivery wagon. It was a neatly enclosed cart and at the rear beneath a latched door was a little step for the convenience of the driver when he wished to come around in back and secure the packages of tea which he sold.

Two things occupied my attention at once. First, the little footboard was of just the right height and size to permit a small boy to sit upon it and ride away unseen once the driver had taken the reins and seated himself at the front of the cart. In addition, the wheels of the cart were large and long-spoked and painted a bright golden yellow. When the horse broke into a spanking trot those wheels spun and glittered in the equally golden air of autumn with an irresistible attraction. Upon that rear step I had made my decision to launch out into the world. It was not the product of a momentary whim. I studied for several days the habits of the tea man until I knew the moment to run forward and perch upon the step. It never crossed my mind to concern myself with where he was going. Such adult matters happily never troubled me. It was enough to be gone between a pair of spinning golden wheels.

On the appointed day, without provision for the future and with a renewed sublime trust in the permanence of sunshine and all good and golden things, I essayed my first great venture into the outer wilderness. My mother was busy with her dishes in the kitchen. As the tea wagon drew up to the house next door I loitered by a bush in the front yard. When the driver leapt once more upon his box I swung hastily upon the little step at the rear. There was no flaw in my escape. The horse trotted with increasing speed over the cobbles, the wheels spun on either side of me in the sunshine, and I was off through the city traffic, followed by the amused or concerned stares of adults along the street. I jounced and bumped but my hold was secure. Horseshoes rang and the whole bright world was one glitter of revolving gold. I had never clearly dealt with the problem of what I would do if the driver continued to make stops, but now it appeared such fears were groundless. There were no more stops. The wheels spun faster and faster. We were headed for the open countryside.

It was, I think, the most marvelous ride I shall ever make in this life. I can still hear the pounding echo of the horse's hoofs over wooden bridges. Shafts of light—it was growing cloudy now—moved over the green meadows by the roadside. I have traversed that road many times since, but the green is faded, the flowers ordinary. On that day, however, we were moving through the kind of eternal light which exists only in the minds of the very young. I remember one other queer thing about that journey: the driver made no impression on my mind at all. I do not recall a cry, a crack of the whip, anything to indicate his genuine presence. We went clopping steadily down a long hill, and up, up against the sky where black clouds were beginning to boil and billow with the threat of an oncoming storm. Far up on that great hill I had a momentary flash of memory. We were headed for the bishop's house.

The bishop's house, which lay thus well out into the country beside an orphanage, was a huge place of massive stone so well set and timeless that it gave the appearance of having been there before the city was built. I had heard my elders speak of it with a touch of awe in their voices. It had high battlements of red granite and around the yard ran a black iron fence through which, according to story, only the baptised might pass. Inside, accepted by my childish mind, was another somewhat supernatural world shut off by hedges.

As we wound higher along the skyline I could see the ruts in the road wriggle, diverge, and merge beneath my dangling feet. Because of my position at the rear of the cart it was impossible to see ahead. The first drops of rain were beginning to make little puffs of dirt in the road and finally as we slowed to a walk on the drive leading up to the gates the storm caught up to us in a great gust of wind and driving rain.

With scarce a pause the iron gates swung open for the tea wagon. I heard the horseshoes ringing on the stones of the drive as I leaped from my perch on the little step and darted into the safety of the hedge. The thunder from the clouds mingled with the hollow rolling of the wheels and the crash of the closing gates before me echoed through my frightened head with a kind of dreadful finality. It was only then, in the intermittent flashes of the lightning, that I realized I was not alone.

In the hedge where I crouched beside the bishop's gate were many hundreds of brown birds, strangers, sitting immovable and still. They paid no attention to me. In fact, they were immersed in a kind of waiting silence so secret and immense that I was much too overawed to disturb them. Instead, I huddled into this thin world beneath the birds while the storm leaped and flickered as though hesitating whether to harry us out of our refuge into the rolling domain of the clouds. Today I know that those birds were migrating and had sought shelter from exhaustion. On that desolate countryside they had come unerringly down upon the thin line of the bishop's hedge.

The tea wagon had unaccountably vanished. The storm after

a time grumbled its way slowly into the distance and with equal slowness I crept unwillingly out into the wet road and began my long walk homeward. I felt in the process some obscure sense of loss. It was as though I had been on the verge of a great adventure into another world that had eluded me; the green light had passed away from the fields. I thought once wistfully of the gold wheel I had failed to find and that seemed vaguely linked to my predicament. I was destined to see it only once in the years that followed—those mature years in which, slower and slower, dimmer and dimmer, the fancies and passions of childhood fade away into the past. Strangely enough, it returned in a moment of violence.

III

The event was simple. There were three of us, jammed into the seat of a stripped car. We were doing fifty miles an hour over a stretch of open grassland, while ahead of us still flashed the white hindquarters of a running antelope. There was no road; no signs, no warnings. There was only the green fenceless unrolling plain and that elusive steel-hearted beast dancing away before us.

The driver pushed the pedal toward the floor.

"You can kill him from here," he said and gestured toward the rifle on my lap.

I did not want to kill him. I looked for a barrier, a fence, an obstacle to wheels, something to stop this game while it was still fun. I wanted to see that unscathed animal go over a hedge and vanish, leaving maybe a little wisp of fur on a thorn to let us know he had passed unharmed out of our reach.

I shrugged and said carefully, indifferently, for I knew the man I rode with, "Why hurry? We'll get him all right in the end."

The driver grunted and started to shift his weight once more upon the pedal. It was just then, in one final brake-screaming instant, that I saw the barrier. It was there and our beast had already cleared it without changing his stride. It was the barrier between life and death.

There was a gulch five feet wide and maybe eight feet deep coming up to meet us, its edge well hidden in the prairie grass.

As we saw it we struck, the front wheels colliding and exploding against the opposite bank. By some freak of pressures we remained there stunned, the bumper holding us above the pit. In that moment, as my head snapped nearly into blackness, I saw a loose golden wheel rolling and rolling on the prairie grass. In my ears there resounded the thunder of the tea wagon pounding over the cobbles and the clang of the bishop's iron gate in the midst of the storm. Then the rumbling receded into the distance and I wiped the blood from my nose.

"He's gone," someone said stupidly.

I put my hands against the bent dashboard and shook my head to clear it.

"The man with the tea wagon?" I asked before I thought.

"The buck," someone answered a long way off, his voice a little thickened. "That buck stepped across the ravine like it wasn't there. We damn near stopped for good. It's like an invisible wall, a line you can't see."

"Yes," I said.

But I didn't say I had wished for it. I didn't say that I remembered how the birds sit on those lines and you never knew which side the birds were on because they sat so quietly and were waiting. You had to be a fugitive to know this and to know the lines were everywhere—a net running through one's brain as well as the outside world. Someday I would pass through the leaves into the open when I should have stayed under the hedge with the birds.

With an effort I lifted the rifle and climbed stiffly out, looking all around the horizon like a hunter. I wear, you see, the protective coloring of men. It is a ruse of the fox—I learned it long ago.

TWO

The Places Below

I

IF YOU cannot bear the silence and the darkness, do not go there; if you dislike black night and yawning chasms, never make them your profession. If you fear the sound of water hurrying through crevices toward unknown and mysterious destinations, do not consider it. Seek out the sunshine. It is a simple prescription. Avoid the darkness.

It is a simple prescription, but you will not follow it. You will turn immediately to the darkness. You will be drawn to it by cords of fear and of longing. You will imagine that you are tired of the sunlight; the waters that unnerve you will tug in the ancient recesses of your mind; the midnight will seem restful—you will end by going down.

I am a case of this sort. Choices, more choices than we like afterward to believe, are made far backward in the innocence of childhood. It has been so with me, as, doubtless, it has been so with you. There was a Washington eccentric in the 1920s whose underground tunnels caused a great stir in the newspapers when someone stumbled upon them. He constructed them himself in his spare time. At first the reporters and police thought they were the work of spies. Afterward it developed that the secret passages were the harmless hobby of an elderly professor. They led nowhere.

I should like to have met that man. He was one of us. To have set out alone with a shovel shows the depth of his need. But there are easier ways into the earth, and passages that run farther. Let me tell you of one of them.

II

The house lies on the edge of a rolling plain. It is an old, warm farmhouse where people rock on the porch in the starry evenings. There is no shadow on it; it is lived in, complete, normal.

But it has a cellar, and that cellar has a monster in it. Or something that gives you the same feeling as a monster. I have been there many times. I know men who would not live in a house with a cellar like that. The owner and I understand each other. He knows why I come. If it has been a long time, I question him with my eyes. He nods or he shakes his head, depending upon the conditions below.

"It is better now," he ventures as we go around to the back of the house. "If it drops any lower we may be able to get into the Blue Room."

We never actually get there, of course, but we talk that way. Only once in the course of many years did I find him awed by what he had seen. He came down the front steps that time and hurried to meet me. "It came up the cellar stairs last night, Doctor. In the night. A hundred feet in three hours. You can't go down. Not this season."

But mostly it is not like that. The cellar seems ordinary at first glance—a little deeper than some. Then you take another turn at the stairs and the air seems to grow damp. There is a faint sulphur smell, and the steps, you begin to perceive, are cut out of living rock.

There is a mark on them now. The mark where what was below came up the cellar stairs one night. You listen again to his story and feel the creep of some uneasy power in the rocks below. Then you go down.

You go down zigzagging and sliding through some accidental tremendous fissure torn in the bowels of the earth. Great stones teeter over you. This is the country of Charon and Cerberus; from this the pleasant fields draw sustenance. A country wit has scrawled a pointing arrow on a rock: "Ten miles to hell." Oddly, you do not laugh. The sulphur smell grows stronger. There, suddenly, your journey ends.

A great pool of cold blue water lies before you. It is so clear that if you were not warned you would march into it, following the splendor of that vast, blue chamber that glistens and invites you from below. You stand on the edge of a country you will never enter in the flesh. Its pale blue galleries seem to speak faintly of faces you will never see. It invites you as arsenical springs invite the thirsty. Nostalgia fills your soul. You reach a hand into the water. The distances are greater than they seem.

"It always stays about here," your guide explains plaintively, pointing at the water's edge. "Once it went down about ten feet and I thought sure we were going to get in. But it came back. It always does. I been watchin' my whole life an' I'm never going to see it all. Nobody's ever going to get in there— not to the Blue Room.

"And when it came up that time afterward it was almost like it was going to show us. My God, that cold blue came right up the tunnel, over the lights and almost out of the house. You could see the lights glowing way back there in a hundred feet of water, before they went out. I thought the damn thing was comin' out the front door, wasn't ever goin' to stop, but it did. It stopped there on that landing like something fumbling for a key. Then by and by it turned and went back. A little at a time, but slow, slow enough to show us. It was a year before we saw the Blue Room."

The Blue Room. It was his single obsession, as it was that of anyone who came there too often. A corner of his brain was eerie with stalactites and that wavering world of distance and promises. He knows me well now when I come up the steps. We are both older. We will never see that chamber as it really is. There is nothing like it in the fields or in the sunlight. It is a part of the places below. And whether the places below lie in the dark of an old cellar or in the crypts and recesses of the mind, or whether they are a glimmering reflection of both together, he does not know any longer. When we have reached the state of mind of that elderly professor with his shovel, sometime, perhaps, we will wade slowly down and see. . . .

There are blind fish that have chosen this world and prefer to live there. There are crickets as white as the fungi under rotting boards. There are bats that turn their little goblin faces uneasily in the glow of your lamp and squeak down at you protestingly. There is the world of light and the world of darkness. And some in the world of light prefer the darkness. That is why the old man shoveled his way downward beneath a staid Washington mansion; that is why men come and stare at the Blue Room. I have said that the choice begins in the innocence of childhood. This is how it began for me.

III

The brain is a strange instrument. The things it chooses to remember are as fantastic as the things it chooses to forget. I have not been in that city for over a score of years. I have not been below, in the dripping labyrinth that underlies its streets, since I was ten. Yet lying on my bed at midnight, waiting for sleep, I find myself retracing in memory each twist and turn of the storm drains underneath the sunlit streets that other men remember. I know how sounds can amplify in that darkness and become terrible. My skin creeps from the water and the mud.

My memory is a rat's memory scurrying with disembodied alacrity through a hollow maze of tubes that exist now only in my head. It turns right and left unerringly. It knows the one way out of a chamber where four black openings yawn simultaneously. Sometimes in that chamber a candle flickers, lighting momentarily another sweat-streaked desperate face.

I never see the face completely, though once I knew it well. Perhaps it is the reluctance of remembered terror. Perhaps it is because the face itself is gone. Nevertheless if I strain consciously to remember, I can remember. It is the face of the Rat. It is true I could run the corridors and follow candle glimmerings. But the Rat was my lord and master. He created a world—the world I live in—and he died and left me in it.

He spoke to me on the day we moved into that neighborhood. He was deceptively slight of build, with the terrible intensity of a coiled spring. His face, even, had the quivering eagerness of some small, quick animal.

"You can join our gang," he said.

I was shy with the shyness of many rebuffs. "Thanks," I said. And then, warily, "I don't play ball very good. D'ya 'spose they'll let me?"

"C'mon," he said.

We went down through a waste of weeds in a back lot. There were big red granite boulders from an old house lying there. A couple of kids were pecking pictures and signs on them with a stone.

"Why don't you use a knife or a chisel?" I asked, trying to be helpful.

"Can't," said the Rat scornfully.

"Why not?" I protested.

"We're cavemen. Those are cave pictures. See? 'Stinct animals. You can't use a chisel. Cavemen didn't have no chisels. We're gonna be just like 'em. No chisels either."

He eyed me challengingly. "You wanta be a caveman?"

"Why sure," I said, "only I don't know how. Where are the caves? What do we do?" I belonged to something at last. I was a caveman.

"C'mon," they said. "And don't ever tell your mother."

"Okay," I said. That was easy.

"Cross your heart?"

"Okay," I said. And that one was for always, though I didn't know it.

"I'm the Rat," said the Rat, dispassionately. "This is my gang and you're in it. But 'fore you get a name you gotta prove your guts. You gotta go down under the ground. You ain't afraid of the dark, are ya?"

The group measured me. I eyed them back uneasily. "No," I said. "What do you mean undertheground?"

"Undertheground's real, you'll see all right," said the Rat.

"C'mon," they clamored, moving off through the weeds. My heart knocked a little. "I'm coming," I said.

That did it, you see. It wasn't too late then. I could have gone home, and I wouldn't have these dreams now or go down to look at the Blue Room. But I ran along after them through the weeds, shouting. It takes just that, in some unwary instant, to telescope fifty thousand years. Afterward you wonder how to get back. I doubt if the Rat ever managed it for himself. Not after seeing the way he went into that pipe.

It was the vent to twenty square miles of sky when the rains blew up. It emptied at the edge of a lake, and it ran two miles back under the town before the labyrinth began. It was dark as a vanished geological era, and in a heavy rain the pipes filled and thundered like Niagara. You didn't stay in the sewers then—not if you wanted to live—and you took good care not to get caught where you couldn't scramble out a few minutes after a storm hit.

That was the world we lived in. We never told Mother, and

we avoided Father. We scrounged our own candles; we dragged food into these abysses. We scratched tribal symbols on the big tiles by candlelight, as the Rat directed. We raided other bands and retreated through the sewer network. We lived as men may sometime live in the ruins of New York.

I learned from the Rat what it was about. It seemed that a long time ago everybody had lived this way. Why they had quit was a mystery to me. The Rat couldn't answer that one. His reading hadn't progressed that far.

IV

It dawned a clear day on the morning the world ended. I suppose it must have happened that way when the Neanderthals left their caves for the last time, with the big ice moving down. I figure they expected to come back, but something happened. It was like that with the Rat and myself.

We were down there after breakfast exploring through a side tunnel a quarter filled with sand and standing water. It was bad going in a place we wouldn't have been in without a glance at the sky before coming down. But the sky had been cloudless. The Rat was a good leader that way, foresighted and sharp. I'd guess we were six hundred feet into this tunnel, laboring along on hands and knees and sometimes writhing our way over a sand bar when the Rat, who was ahead of me with the candle, held it up and said, "Listen." We stopped dead. I could see the straining intensity flicker on the Rat's thin face as he turned the candle toward me.

The tunnel dripped a little and at first I heard nothing. The Rat jabbed at me. "Listen, for Christ's sake," he said. I heard it then all right, and my heart gave a big jump and almost stopped altogether, but it was a very little sound. It was nothing but a little murmuring in the water, a little whisper, a little complaint as though the water were growing restless and wanted to go somewhere.

"It's moving," I said, and the Rat said nothing at all except to cast the light forward as far as he could. The darkness swallowed it up and the murmuring sounded louder, except maybe it was the blood in our ears.

We listened again and there seemed to be a far-off pouring

noise muffled by sand and distance. We were only ten years old, and suddenly this place was very narrow and we were tired of playing like men. I might have cried, but the Rat just poked me and said "C'mon" as he always did. I scrambled at his heels and somehow we kept the candle burning looking for something bigger to get into. This was a low pipe. It wouldn't take much to fill it, and you couldn't move very fast. Just try to crawl a couple of blocks on your stomach sometime and you'll see what I mean.

The water kept on moving with us and talking to us in that slightly, just slightly, sinister way it had. This was just the starter. Where had that instantaneous storm come from, and what was blowing it up? There was no use asking down there. The water would tell us soon enough. It did. It began to rise a little. One could hear a pouring thunder in the outer drain.

We were just a couple of leaping automatons now, going forward in the only way we could go—with the pipe. Our clothes were shredded by the little needles in the tile. Our knees were raw flesh, our hands were bruised. We whimpered to ourselves through the talking water, but there was little breath to whimper with. We were almost spent, and the minute we were spent we would go under. It was, by actual measurement, a hundred and fifty feet to where we reached the chamber. Speeding as we were, it could hardly have taken us five minutes. By heart and lungs and brain it was an hour.

The Rat got there first and reached and dragged me. We staggered upright in the water. The water was low and spread out over the chamber floor. It only lapped our ankles. Time was still with us, but what should we do? The candle sputtered and showed us three openings the size of the one out of which we had lunged. We could go into one of them and still be trapped and drowned.

Overhead in the darkness was a street manhole. Was it locked shut, and if not, could we lift it? If we didn't and this was a big rain, the chamber would fill. I knew it would. I had peered many times into such maelstroms. It would not only fill, the suction would be irresistible. We would be swept back into the underground.

Staggering, I got the Rat on my shoulders. He heaved against the cover lid. It stirred a little, but nothing happened. The Rat slid panting down into my arms.

We leaned against the pipes a moment, too exhausted to speak. The stream continued to pour past our ankles. Finally in one despairing burst of energy we each climbed on the protruding edge of a pipe and pushed against the manhole cover. Reluctantly it gave. We shoved it aside and sprawled gasping into the street.

Spent as we were, we leaped from the pavement in amazement. The hot tar burned. The sky was clear. The gutters in the street were dry. Astounded, we looked farther along the street. A city employee had opened a fire hydrant for testing. The water was spouting from a plug into a nearby drain.

A voice interrupted us. It was the voice of my father. "Son," he said with suppressed fury, "I've told you before to keep out of sewers. Look at you. Look at both of you. You're coming home. By god I'm telling you you're *staying* home."

"Yes, Father," I said. I took his hand as he marched me along. *If you fear the sound of water hurrying toward unknown and mysterious destinations do not go there.* "Undisciplined, completely undisciplined," my father muttered, still unmollified. "Yes, Father," I said, clutching him more tightly. I liked the strength in his hand.

v

It is a simple prescription, but you will not follow it. The urge returns. The same voices speak to you. It was true we had left the sewers. Our parents had seen to that. Besides, we were growing too big to navigate them successfully. Down by an old orphanage, however, I found a new opening. It was a drain all right, and it was old. It was not part of the system we knew. It was brick and big enough to stand upright in. There was green moss around the entrance and on the bricks inside. I went in just a few steps and saw that it was dark and that it ran off under the hill. Then I went and got the Rat.

It was like a green door and the air blew out of it cold and smelling of water on a hot midsummer day. We dropped into the brush that covered the opening from sight and listened, looking at each other. Then we went in on tiptoe, the Rat leading as always. We went over the moss like velvet and that air kept blowing cool and clean.

We lost sight of the entrance. It just faded out in a kind of green twilight. We must have been a hundred yards in when the Rat stopped. I knew what he wanted, and we listened. From somewhere up ahead I could hear it—the vibration of falling water—and the air now with a little chill.

I didn't want to see it, where that water was going, I mean, or why it was in this hill. I had had enough of the green door. But I would follow the Rat anywhere. So I waited, standing on one foot and then the other.

"We can come back later with the gang," I ventured. The Rat turned and looked at me and through me and beyond me. He tapped my arm, and I could see the thin, quizzical line deepen on his sharp forehead. "We'll keep it for our own," he said. "Just us." Then he turned, and we left that sound vibrating in the air and went back to the world.

A few weeks later he was dead—dead of some casual childhood illness. All that consuming energy and passionate intellectual hunger had come to nothing. In later years places of learning would become familiar sights to me. I never met a mind like his again.

Once after his death I went to the green door. It was still secret and that cool draft came up from below. I stood a moment on the moss at the entrance. "Just us," the Rat had said, and it warmed me a little. But it was no use without the Rat. I backed out and turned away.

Something had swung behind me then, but I didn't know it. You find those things out by degrees in the passage to manhood, in the way you continue to pick up stones or linger at dark openings. If there is any truth to the story that at death men return to the period they have loved best in life, I know well where I will awake. It will be somewhere on the cold, bleak uplands of the ice-age world, by the fire in the cave, and the watching eyes without. It was the Rat who left me there.

I knew it finally in the Hall of Shadows, in a cave I had no business to be in. By what road I had crept there I had scant knowledge. By what way I would get out I was not sure. Yet something drew me—it was drawing me more and more. My work did not demand that I take that turning under the ledge or chance that passage downward. I was an archaeologist—not a fool.

Yet there I lay on my back, finally, and the outside world seemed far away and infinitely wearying as a place to which to return. I was in a room meant for a king's burial. I lay on the floor of an enormous chamber, but a chamber across which one could not move except by crawling. For that great hall was hung with vast tapestries and heavy curtains over which my lantern played. And those tapestries were iridescent stone. The powers that had built that chamber in the depths of the mountain were closing it again. I had come as the curtains lingered above the floor. If I stayed they would descend.

Some in the world of light desire the darkness. I saw that then more clearly than before. The whole infinite ladder of life was filled with this backward yearning. There were the mammals who had given up the land and returned to the sea; there were fish that slept in the mud, birds that no longer flew. Probably also there had been hairy men who wept when fashion tore them from their caverns. I loved the darkness. I feared it, yet returned to it. It was the mother out of which I came. From somewhere under those moveless curtains, in that utter darkness beneath the stone, a small breeze blew. It was cool, and there was a faint sound of water in it—far off, menacing, and sweet. It invited me forward. It urged me to crawl on. For the first time in years I remembered the green door and the Rat as he stood there listening. The Rat who had eyed me shrewdly, saying "just us." What had he heard there in the falling water? And what had he meant?

The same door was before me now—that door through which the Rat had gone. I waited as I had waited long ago at that other entrance, but this time, as before, no one came over the green moss like velvet; no one touched my arm. I waited, but there was no one to help me. One entered by oneself, or not at all.

The air blew cool on my sweating forehead, and that far-off murmur that might have come through remote distances of stone still urged me onward, but I think it whispered in my brain and not the Hall. Slowly, with agonizing reluctance, I drew backward. Slowly I began to crawl toward daylight, along the way I had come.

THREE

Big Eyes and Small Eyes

"THIS IS your house," says the poet Conrad Aiken, and you know he is talking about the human skull. "On one side there is darkness," he warns you; "on one side there is light." He wrote better than he knew, that sad-voiced man, for nature had pondered the problem before him. On the table as I write lies the skull of a relative of ours, a spectral creature which flits from tree to tree in the night-time forests of Borneo. It is about the size of a kitten's skull, but it possesses a most remarkable feature. If the human cranium were built to similar proportions, every aspect of the human face would be squeezed to provide for two great bony saucers with projecting rims. These saucers would occupy and extend far beyond the area now represented by our eye sockets. We would then possess the enormous owl eyes of a creature who is totally nocturnal but who must leap and spring about in the midnight darkness of a tropical rain forest.

This is your house, said nature, in essence, to the spectral tarsier, and the light, what there is of it, must be made to come in. This is your house, she also said to man, but your eyes will be day eyes. You will not need to cherish every beam of moonlight, or the spark of a star through a leaf. You see what you must, but leave the dark alone. So this far-off relative of ours, with the thin and delicate fingers of a man, lives the life of a ghost. And man, who bumps his head and fumbles in the dark because of his small day-born eyes, fears the ghosts of the dark above all things. As a consequence, my confession is that of a man with night fear, and it is also the confession of a very large proportion of the human race.

In a way it is the fear of the tide, the night tide, I call it, because that is the way you come to feel it—invisible, imperceptible almost, unless it is looked for—and yet, as you grow older you realize that it is always there, swirling like vapor just beyond the edge of the lamp at evening and similarly out to the ends of the universe. Or at least it gives you that kind of sensation—a need to huddle in somewhere with a light.

Maybe that is the real reason why men string lamps far out into country lanes and try to run down everything with red eyes that happens to waddle across the road in front of their headlights. It is cruel but revelatory: we are insecure, and this is our warfare with the dark. It began when man first lit a fire at a cave mouth and the eyes he feared—very big eyes they were then—began to blink and draw back. So he lights and lights in a passion for illumination that is insatiable—a poor day-born thing contending against one of the greatest powers in the universe. Even man's own domestic animals, the creatures he has chosen to bring in to the fire beside him, grow suspect in the evening. His cat hunts alone through the weeds, and his dog whines and snuffles at the door. They all have that allegiance to the dark. They are never wholly his.

When you really get the swell of the night tide, however, is at the moment it comes right in upon you and swirls, figuratively at least, around your ankles. Rats play a good part in such episodes, because they are the real agents of the night and there is a sort of malign intelligence about them that is frightening. Also, they have a particularly bold way of paying visits to men after nightfall, as though they wanted to remind us of something waiting and not very pleasant.

Many years ago a friend of mine took a room in an obscure hotel in the heart of a great city. There was a blaze of street lights outside, and a few shadows. He had opened the window and retired, he told me, when something soft and heavy dropped on his feet as he lay stretched out in bed. Though he admittedly was startled, it occurred to him that the creature on his legs might be a friendly tomcat from the fire escape. He tried to estimate the weight of the crouched body from under his blankets and resisted a frightened impulse to spring up. He spoke soothingly into the dark, for he liked cats, and reached for a match at his bedside table.

The match flared, and in that moment a sewer rat as big as a house cat sat up on its haunches and glared into the match flame with pink demoniac eyes. That one match flare, so my friend told me afterward, seemed to last the lifetime of the human race. Then the match went out and he simultaneously hurtled from the bed. From his incoherent account of what happened afterward I suspect that both rat and man left by the window but fortunately, perhaps, not at the same instant. That sort of thing, you know, is like getting a personal message from the dark. You are apt to remember it a lifetime.

Or, speaking of rats, take what happened to me. It is true I was not confronted by this rat eye to eye in a match flame, but on the other hand there was an even more frightening *intellectual* quality about the situation. And again the creature arrived, like a messenger from space, at an appropriate point in a very significant conversation. He addressed himself to me for the very simple reason that I was the only one in a position to see him. I will call him Conlin's rat, in order to protect the good name of a distinguished American novelist who was unaware of the low company he kept.

This man is capable of most eloquent discourse. The evening was perfect. The light was just fading on the faces of the company, and the perfectly clipped lawns and hedges fell away before us on the terrace as only the very rich and the very powerful can afford to have them. My friend held a mint julep in his hand and gestured toward me.

"Man," he said, "will turn the whole earth into a garden for his own enjoyment. It is just a question of time. I admit the obstacles you have mentioned, but I have tremendous faith in man. He will win through. I drink to him."

He poised his glass and said other happy and felicitous things to which the company present raised their glasses. Even my own glass—and I am a weak and doubting character—was somewhat dubiously being lifted, when I saw an incredible and revolting sight. There, under my friend's white canvas chair, and outlined against the stuccoed wall at his back, a thin, greasy, wet-backed rat upreared himself and twitched his whiskers with a cynical contempt for all that white-gowned, well-clothed company.

I say he addressed himself to me, for I have never seen anything so peculiarly appropriate. He had obviously emerged from a drain a bit farther on in the wall, perhaps a little prematurely, along with the rising tide of evening. I stared in unbelief and waited for the ladies to scream. I wanted to lift my glass. I wanted to set it down. I waited for my novelist friend to come to his senses and spring away from that bewhiskered mocking animal that crouched beneath him. The novelist made no move. He spoke on as eloquently as always, while the rat sniffed his shoe and listened, stretching up to his full height. I felt for an uneasy moment that the creature might ironically applaud. He looked across at me, and it seemed best not to warn the company. Anyway, it was unlikely that I could warn them sufficiently.

For that this was a message I felt certain. I alone saw him and I never spoke. He listened a while with great attentiveness to our voices, and then he went back into the rising darkness and the drain pipe swallowed him up. But you can see what I mean by a tide and how with the dark it comes in around your ankles. Light the lights, I always say, but I have found that even this is no real security—not in the night. Because in the end you may find that the remaining light has only allowed you to see something it would have been better not to see at all.

Take, for instance, the time that I saw the black beetle. I repeat that "light the lights" is my motto, so when this fit of insomnia came on me I retired to the living room, my wife not sharing my view of the universe at three in the morning, and settled down to read T. K. Oesterreich's *Demoniacal Possession*.

This was assuredly not bravado on my part. It was just that as a professor I had to give a lecture on primitive religion in the morning, and this seemed as good a time as any to "get on with it," as my dean is always saying. I snapped my small reading lamp to the lowest power and crept into what I thought in my innocence was its charmed and healthy circle of light.

I had been reading for about an hour and was pretty well into the business of familiars, recognizable signs of demonic intrusion, shape-shifting and other supernatural phenomena, when I saw something moving along the edge of my vision. Now to understand this episode you have to realize where it took place. We then lived high up in an apartment house that is noted for its cleanliness; I had never seen a mouse or a roach in the kitchen, let alone in the living room.

I did not believe, therefore, in what began to march across my circle of green light as though conjuring itself into existence. In fact, I pulled my feet up in the chair and leaned down until I was practically standing on my head. It was no illusion. A huge black bug—not a roach, but a fat-bodied and particularly odious beetle of dubious affinities—was marching right across the carpet under my nose. Before I could adjust my bifocals and marshal the necessary militancy which my wife is generally on hand to supply, the creature waddled with the most absolute surety under the heavy green chest in the living room and disappeared.

I have never seen the beetle since, and my wife, who forced me to move the chest, absolutely denies his existence. I know better, but I am willing to admit now that people may have a point who refuse to turn on lights in the dark. At three in the morning, my wife says, what do you expect to see if you turn on a light? My only retort to this is the rather obvious one that she will have to think what may be there if she does not turn on the light.

This does not affect my wife in the least, but it does me. I get to thinking about it and feel impelled to snap on a switch and look, just in case . . . My wife, however, is a genuinely daylight person. She has never had the slightest interest in this dark world over the border except to pursue all its manifestations, like the black beetle, with an exterminating broom. My own viewpoint, I like to think, is rather more traveled. I have been over into that nocturnal country, as I will presently recount, and though I stand in awe of it, it has also stimulated my curiosity. In fact, though I hesitate to speak openly of the matter, I have a faint, though not too secure, feeling of kinship with certain creatures I have encountered there.

Sometimes in a country lane at midnight you can sense their eyes upon you—the eyes that by daylight may be the vacuous protuberant orbs of grazing cattle or the good brown eyes of farm dogs. But there, in the midnight lane, they draw off from you or silently watch you pass from their hidden coverts in the hedgerows. They are back in a secret world from which man has been shut out, and they want no truck with him after nightfall. Perhaps it is because of this that more and more we employ machines with lights and great noise to rush by these watchful shadows. My experience, therefore, may be among the last to be reported from that night world, which, with our machines to face, is slowly ebbing back into little patches of wilderness behind lighted signboards. It concerns a journey. I will not say where the journey began, but it took place in the years that have come to be called the Great Depression and was made alone and on foot. Finally I had come to a place where, far off over an endless blue plain, I could see the snow on the crests of the mountains. The city to which I was journeying lay, I knew, at the foot of the highest peak. I would keep the mountains in sight, I thought to myself, and find my own path to the city.

I climbed over a barbed-wire fence and marched directly toward the city through the blue air under the great white peak. I think sometimes now, long afterward, that it was the happiest, most independent day of my entire life. No one waited for me anywhere. I was complete in myself like a young migrating animal whose world exists totally in the present moment. The range with its drifting cattle and an occasional passing bird began to unroll beneath my stride. I meant to be across that range and over an escarpment of stone to the city at the mountain's foot before the dawn of another morning. During the entire walk I was never to meet another human being. The lights and showers of that high landscape, the moving shadows of clouds, shone upon or darkened my face alternately, but I was destined to share the experience with no one.

In the later afternoon, after descending into innumerable arroyos and scrambling with difficulty up the vertical bank of the opposite side, I began to grow tired. Coming out finally into a country that was less trenched and eroded, I was trudging steadily onward when I came upon a pond. At least for all purposes it could answer to the name, though it was only a few inches deep—mere standing rainwater caught in a depression of impermeable soil and interspersed with tufts of brown buffalo grass. I hesitated by it for a moment, somewhat disturbed by a few leeches which I could see moving among the grass stems. Then, losing my scruples, I crouched and drank the bitter water. The hollow was sheltered from the wind that had been sweeping endlessly against me as I traveled. I found a dry spot by the water's edge and stretched out to rest a moment. In my exhaustion the minute must have stretched into an hour. Something, some inner alarm, brought me to my senses.

Long shadows were stealing across the pond water and the light was turning red. One of those shadows, I thought dimly as I tried to move a sleep-stiffened elbow from under my head, seemed to be standing right over me. Drowsily I focused my sight and squinted against the declining sun. In the midst of the shadow I made out a very cold yellow eye and then saw that the thing looming over me was a great blue heron.

He was standing quietly on one foot and looking, like an expert rifleman, down the end of a bill as deadly as an assassin's dagger. I had seen, not long before, a man with his brow split open by a half-grown heron which he had been rash enough to try and capture. The man had been fortunate, for the inexperienced young bird had driven for his eye and missed, gashing his forehead instead.

The bird I faced was perfectly mature and had come softly down on a frog hunt while I slept. Why was he now standing over me? It was certain that momentarily he did not recognize me for a man. Perhaps he was merely curious. Perhaps it was only my little brown eye in the mud that he wanted. As this thought penetrated my sleeping brain I rolled, quick as a frog shrieking underfoot, into the water. The great bird, probably as startled as I, rose and beat steadily off into the wind, his long legs folded gracefully behind him.

A little shaken, I stood up and looked after him. There was nothing anywhere for miles and he had come to me like a ghost. How long he had been standing there I did not know. The light was dim now and the cold of the high plains was rising. I shivered and mopped my wet face. The snow on the peak was still visible. I got my bearings and hurried on, determined to make up for lost time. Again the long plain seemed to pass endlessly under my hastening footsteps. For hours I moved under the moon, not too disastrously, though once I fell. The sharp-edged arroyos had appeared again and were a menace in the dark. They were very hard to see, and some were deep.

It was some time after the moon rose that I began to realize that I was being followed. I stopped abruptly and listened. Something, several things, stopped with me. I heard their feet put down an instant after mine. Dead silence. "Who's there?" I said, trying to make the words adjustable and appeasing to any kind of unwanted companion. There was no answer, though I had the feel of several shapes just beyond my range of vision. There was nothing to do. I started on again and the footsteps began once more, but always they stopped and started with mine. Finally I began to suspect that the number of my stealthy followers was growing and that they were closing up the distance between us by degrees.

I had a choice then: I had been realizing it for some time. I could lose my nerve, run, and invite pursuit, possibly breaking my leg in a ditch, or, like a sensible human being a little out of his element perhaps, I could go back and see what threatened me. On the instant, I stopped and turned.

There was a little clipclop of sound and dead silence once more, but this time I heard a low uneasy snuffling that could only come from many noses. I groped in the dark for a stick or a stone but could find none. I ran three steps back in a threatening manner and raised a dreadful screech that caused some shifting of feet and a little rumble of menacing sound. The screech had nearly shattered my own nerves. My heart thumped as I tried to recover my poise.

Cattle.

What was it that gave them this eerie behavior in the dark?

I affected to ignore them. I started on again, whistling, but my mouth was dry. Range cattle, something spelled out in my mind—wild, used to horsemen—what are they like to a man on foot in the dark? They were getting ominously close—that was certain; even if they were just curious, that steady trampling bearing down on my heels was nerve-wracking. Ahead of me at that instant I saw a section of barbed wire against the moon, and behind it a wide boulder-strewn stream bed.

The stream was dry and the starlight shone on the white stones. I swung about and yelled, making a little rush back. Then, without waiting to observe the effects, I turned and openly ran for it. There was a growing thunder behind me. I heard it as I vaulted the fence and landed eight feet down in the sand of the stream bed. Above me I heard a sound like a cavalry troop wheeling off into the night. A braver man might have stood by the fence and waited to see what would happen, but in the night there is this difference that comes over things. I sat on a stone in the stream bed and breathed hard for a long time. Then the chill forced me up again. The arroyo twisted in the direction I was headed. I wandered down it, feeling safer among the stones that reflected light and half-illuminated my path.

Somewhere along a section of damp sand I encountered several large toads who were also making a night journey and who hopped clumsily for a little way with me. There was something so attractive about their little bursts of energy that, tired as I was, I began to skip with them. I was delighted now to have even lowly company. First one would hop and then another, and I began to take my turn automatically with the rest. I do not know where they might have eventually led me, though I had a feeling that if I stayed and hopped with them long enough I might acquire this knowledge in some primordial manner.

With this thought I parted from them at a turn in the stream bed and made my way again over open rolling foothills in the dark. The land was rising. I was approaching the escarpment which I knew overhung the deeper valley in which lay the city at the mountain's foot.

I met nothing living now except small twisted pines. Boulders swelled up from the turf like huge white puff balls, and there was a flash of lightning off to the south that lit for one blue, glistening instant a hundred miles of churning, shifting landscape. I have thought since that each stone, each tree, each ravine and crevice echoing and re-echoing with thunder tells us more at such an instant than any daytime vision of the road we travel. The flash hangs like an immortal magnification in the brain, and suddenly you know the kind of country you pass over, and the powers abroad in it. It was at that moment that I reached the edge of the escarpment and looked down.

The night lights of the city glistened in hundreds far out on the plain, but I had chosen a bad pathway. I was high up on a clifflike eminence, and a straight descent in the dark was dangerous. As far as I could see there was no break to right or left. I was tired and hungry—too weary to go on circling in the dark and too cold to sit and wait for dawn. I decided to climb down, though with the utmost caution. Those faraway street lights beneath me were an irresistible attraction. They were the world I knew. The mind inside us is vaster than the world outside and I had been wrestling with its terrors for a long time now.

I began with discretion, working my way by inches down a precipitous gully. After a while the gully ended and I seemed to be looking out through a tree root at a solitary light on a mine tipple still far beneath me. I must have stayed there an hour groping about in the dark. Then I found a ledge along which I edged farther until I knocked over a stone that went rolling and grinding downward. Gaining momentum, it began to leap and volley against unseen stumps and boulders, making a hideous din.

As soon as the echoes died I knew I was in for trouble. I was well down from the summit now and there was no way back up that mountain wall in the dark. I heard them coming before I saw them—two huge watch dogs from the mine property. They barked with great night-foggy voices and leaped and slavered at me up the cliff. The sound was enough to wake the dead, and I expected at first that someone, a watchman, might appear. I hoped to be able to explain myself and have the man get the dogs under control. The dogs, however, happened to be alone. Nobody came and there I hung, a few feet up the cliff, while that formidable chorus played up and down my spine. The grip of my hands was growing tired and I thought with sudden careful prevision: If you wait till you're tired and fall, you won't be able to fight them off.

I climbed on then, slowly working downward along the ledge. I didn't want to have to drop suddenly in their midst. It would startle them, and if I were unlucky enough to fall they might spring on me. In fact, it looked as though they were going to spring in any case. But there it was. I tried to choose a moment when they seemed tired from their own great bellowing exertions. In a pause I vaulted down onto their own level from the wall.

I said something in a voice I tried to keep confident and friendly. I held one arm over my throat and stood stock-still. They came up to me warily, but one made a small woofing sound in his throat and I could see the motion of his tail in the dark. Seeing this, I dropped one hand on his head and the other on the other beast whose jaws had closed with surprising gentleness about my ankle. I stood there for long minutes talking and side-thumping and trying all the dog language I knew.

At the end of that time my foot was reluctantly released and the great hounds, with the total irrationality that prevails over the sheer cliff of Chaos, leaped and bounded about me, as though I were their returning master. Did they take me finally, because of my successful descent, as a demon like themselves— for, if I had fallen, they had given every indication of devouring me—or are the dogs of Cerberus, the hoarse-voiced, much feared guardian of Darkness, actually abysmally lonely and friendly creatures?

Since that long agonizing descent before I reached the city on the plain, I have never been quite sure. When I come to the Final Pit in which they howl, I shall, without too great a show of confidence, put out my hand and speak once more. Perhaps the great hounds of fear may wait with wagging tails for a voice which knows them. And what dog is there that knows how to tell one demon from another in the dark?

By the eyes, some will say, but I think not, really, for to the spectral tarsier in the bush, or to the owl in the churchyard tower, man and his lights must truly hold a demonic menace. Having journeyed once along the dark side of the planet, I am willing to testify that it is a shifting and unmapped domain of terrors. But as one demon to another, in memory of that hour on a cliff wall, I have helped a bat to escape from a university classroom, and I have never told on a frightened owl I once saw perched on the curtain rod above a Pullman berth. Somewhere in the blasts over the roaring cliff of Chaos I may meet their like again. It will be all one in that place, light and dark, big and small eyes, and the true demon will not fear his brother from another element. No. I think now the great dogs will know me. At least I shall put out my hand and speak.

FOUR

Instruments of Darkness

THE Nature in which Shakespeare's Macbeth dabbles so unsuccessfully with the aid of witchcraft, in the famous scene on the heath, is unforgettable in literature. We watch in horrified fascination the malevolent change in the character of Macbeth as he gains a dubious insight into the unfolding future—a future which we know to be self-created. This scene, fearsome enough at all times, is today almost unbearable to the discerning observer. Its power lies in its symbolic delineation of the relationship of Macbeth's midnight world to the realm of modern science—a relationship grasped by few.

The good general, Banquo, who, unlike Macbeth, is wary of such glimpses into the future as the witches have allowed the two companions, seeks to restrain his impetuous comrade. "'Tis strange," Banquo says,

> "And oftentimes, to winne us to our harme
>   The Instruments of Darknesse tell us Truths
>   Winne us with honest trifles, to betray's
>   In deepest consequence."

Macbeth, who has immediately seized upon the self-imposed reality induced by the witches' prophecies, stumbles out of their toils at the last, only to protest in his dying hour:

> "And Be these Jugling Fiends no more believ'd . . .
>   That keep the word of promise to our eare,
>   And breake it to our hope."

Who, we may now inquire, are these strange beings who waylaid Macbeth, and why do I, who have spent a lifetime in the domain of science, make the audacious claim that this old murderous tale of the scientific twilight extends its shadow across the doorway of our modern laboratories? These bearded, sexless creatures who possess the faculty of vanishing into air or who reappear in some ultimate flame-wreathed landscape only to mock our folly, are an exteriorized portion of ourselves. They are projections from our own psyche, smoking wisps of mental vapor that proclaim our subconscious intentions and bolster them with Delphic utterances—half-truths which we consciously accept, and which then take power over us. Under the spell of such oracles we create, not a necessary or real future, but a counterfeit drawn from within ourselves, which we then superimpose, through purely human power, upon reality. Indeed, one could say that these phantoms create a world that is at the same time spurious and genuine, so complex is our human destiny.

Every age has its style in these necromantic projections. The corpse-lifting divinations of the Elizabethan sorcerers have given way, in our time, to other and, at first sight, more scientific interpretations of the future. Today we know more about where man has come from and what we may expect of him—or so we think. But there is one thing which identifies Macbeth's "Jugling Fiends" in any age, whether these uncanny phantoms appear as witches, star readers, or today's technologists. This quality is their claim to omniscience—an omniscience only half stated on the basis of the past or specious present and always lacking in genuine knowledge of the future. The leading characteristic of the future they present is its fixed, static, inflexible quality.

Such a future is fated beyond human will to change, just as Macbeth's demons, by prophecy, worked in him a transformation of character which then created inevitable tragedy. Until the appearance of the witches on the heath gave it shape, that tragedy existed only as a latent possibility in Macbeth's subconscious. Similarly, in this age, one could quote those who seek control of man's destiny by the evocation of his past. Their wizardry is deceptive because their spells are woven out of a genuine portion of reality, which, however, has taken on this always identifiable quality of fixity in an unfixed universe. The ape is always in our hearts, we are made to say, although each time a child is born something totally and genetically unique enters the universe, just as it did long ago when the great ethical leaders—Christ, the Buddha, Confucius—spoke to their followers.

Man escapes definition even as the modern phantoms in militarist garb proclaim—as I have heard them do—that man will fight from one side of the solar system to the other, and beyond. The danger, of course, is truly there, but it is a danger which, while it lies partially in what man is, lies much more close to what he chooses to believe about himself. Man's whole history is one of transcendence and self-examination, which has led him to angelic heights of sacrifice as well as into the bleakest regions of despair. The future is not truly fixed but the world arena is smoking with the caldrons of those who would create tomorrow by evoking, rather than exorcising, the stalking ghosts of the past.

Even this past, however, has been far deeper and more pregnant with novelty than the short-time realist can envisage. As an evolutionist I never cease to be astounded by the past. It is replete with more features than one world can realize. Perhaps it was this that led the philosopher George Santayana to speak of men's true natures as not adequately manifested in their condition at any given moment, or even in their usual habits. "Their real nature," he contended, "is what they would discover themselves to be if they possessed self-knowledge, or as the Indian scripture has it, if they became what they are." I should like to approach this mystery of the self, which so intrigued the great philosopher, from a mundane path strewn with the sticks and stones through which the archaeologist must pick his way.

Let me use as illustration a very heavy and peculiar stone which I keep upon my desk. It has been split across and, carbon black, imprinted in the gray shale, is the outline of a fish. The chemicals that composed the fish—most of them at least—are still there in the stone. They are, in a sense, imperishable. They may come and go, pass in and out of living things, trickle away in the long erosion of time. They are inanimate, yet at one time they constituted a living creature.

Often at my desk, now, I sit contemplating the fish. Nor does it have to be a fish. It could be the long-horned Alaskan bison on my wall. For the point is, you see, that the fish is extinct and gone, just as those great heavy-headed beasts are gone, just as our massive-faced and shambling forebears of the Ice have vanished. The chemicals still about me here took a shape that will never be seen again so long as grass grows or the sun shines. Just once out of all time there was a pattern that we call *Bison regius*, a fish called *Diplomystus humlis*, and, at this present moment, a primate who knows, or thinks he knows, the entire score.

In the past there has been armor, there have been bellowings out of throats like iron furnaces, there have been phantom lights in the dark forest and toothed reptiles winging through the air. It has all been carbon and its compounds, the black stain running perpetually across the stone.

But though the elements are known, nothing in all those shapes is now returnable. No living chemist can shape a dinosaur; no living hand can start the dreaming tentacular extensions that characterize the life of the simplest ameboid cell. Finally, as the greatest mystery of all, I who write these words on paper, cannot establish my own reality. I am, by any reasonable and considered logic, dead. This may be a matter of concern, or even a secret, but if it is any consolation, I can assure you that all men are as dead as I. For on my office desk, to prove my words, is the fossil out of the stone, and there is the carbon of life stained black on the ancient rock.

There is no life in the fossil. There is no life in the carbon in my body. As the idea strikes me, and it comes as a profound shock, I run down the list of elements. There is no life in the iron, there is no life in the phosphorus, the nitrogen does not contain me, the water that soaks my tissues is not I. What am I then? I pinch my body in a kind of sudden desperation. My heart knocks, my fingers close around the pen. There is, it seems, a semblance of life here.

But the minute I start breaking this strange body down into its constituents, it is dead. It does not know me. Carbon does not speak, calcium does not remember, iron does not weep. Even if I hastily reconstitute their combinations in my mind, rebuild my arteries, and let oxygen in the grip of hemoglobin go hurrying through a thousand conduits, I have a kind of machine, but where in all this array of pipes and hurried flotsam is the dweller?

From whence, out of what steaming pools or boiling cloudbursts, did he first arise? What forces can we find which brought him up the shore, scaled his body into an antique, reptilian shape and then cracked it like an egg to let a soft-furred animal with a warmer heart emerge? And we? Would it not be a good thing if man were tapped gently like a fertile egg to see what might creep out? I sometimes think of this as I handle the thick-walled skulls of the animal men who preceded us or ponder over those remote splay-footed creatures whose bones lie deep in the world's wastelands at the very bottom of time.

With the glooms and night terrors of those vast cemeteries I have been long familiar. A precisely similar gloom enwraps the individual life of each of us. There are moments, in my bed at midnight, or watching the play of moonlight on the ceiling, when this ghostliness of myself comes home to me with appalling force, when I lie tense, listening as if removed, far off, to the footfalls of my own heart, or seeing my own head on the pillow turning restlessly with the round staring eyes of a gigantic owl. I whisper "Who?" to no one but myself in the silent, sleeping house—the living house gone back to sleep with the sleeping stones, the eternally sleeping chair, the picture that sleeps forever on the bureau, the dead, also sleeping, though they walk in my dreams. In the midst of all this dark, this void, this emptiness, I, more ghostly than a ghost, cry "Who? Who?" to no answer, aware only of other smaller ghosts like the bat sweeping by the window or the dog who, in repeating a bit of his own lost history, turns restlessly among nonexistent grasses before he subsides again upon the floor.

"Trust the divine animal who carries us through the world," writes Ralph Waldo Emerson. Like the horse who finds the way by instinct when the traveler is lost in the forest, so the divine within us, he contends, may find new passages opening into nature; human metamorphosis may be possible. Emerson wrote at a time when man still lived intimately with animals and pursued wild, dangerous ways through primeval forests and prairies. Emerson and Thoreau lived close enough to nature to know something still of animal intuition and wisdom. They had not reached that point of utter cynicism, that distrust of self and of the human past which leads finally to total entrapment in that past, "man crystallized," as Emerson again was shrewd enough to observe.

This entrapment is all too evident in the writings of many concerned with the evolutionary story of man. Their gaze is fixed solely upon a past into which, one begins to suspect, has been poured a certain amount of today's frustration, venom,

and despair. Like the witches in *Macbeth*, these men are tempt-
ing us with seeming realities about ourselves until these realities take shape in our minds and become the future. It was not
necessary to break the code of DNA in order to control human
destiny. The tragedy lies in the fact that men are already controlling it even while they juggle retorts and shake vials in
search of a physical means to enrich their personalities. We
would like to contain the uncontainable future in a glass, have
it crystallized out before us as a powder which we might swal-
low. All then, we imagine, would be well.

As our knowledge of the genetic mechanism increases, our
ears are bombarded with ingenious accounts of how we are to
control, henceforth, our own evolution. We who have recourse
only to a past which we misread and which has made us cynics
would now venture to produce our own future. Again I judge
this self-esteem as a symptom of our time, our powerful mis-
used technology, our desire not to seek the good life but to
produce a painless mechanical version of it—our willingness to
be good if goodness can, in short, be swallowed in a pill.

Once more we are on the heath of the witches, or, to come
closer to our own day, we are in the London laboratory where
the good Doctor Jekyll produced a potion and reft out of his
own body the monster Hyde.

Nature, as I have tried to intimate, is never quite where we
see it. It is a becoming as well as a passing, but the becoming is
both within and without our power. This lesson, with all our
hard-gained knowledge, is difficult to grasp. All along the
evolutionary road it could have been said, "This is man," if
there had then been such a magical self-delineating and
mind-freezing word. It could have immobilized us at any step
of our journey. It could have held us hanging to the bough
from which we actually dropped; it could have kept us cower-
ing, small-brained and helpless, whenever the great cats came
through the reeds. It could have stricken us with terror before
the fire that was later to be our warmth and weapon against
ice-age cold. At any step of the way, the word "man," in retrospect, could be said to have encompassed just such final
limits.

Each time the barrier has been surmounted. Man is not
man. He is elsewhere. There is within us only that dark, divine

animal engaged in a strange journey—that creature who, at midnight, knows its own ghostliness and senses its far road. "Man's unhappiness," brooded Thomas Carlyle, "comes of his Greatness; it is because there is an Infinite in him, which with all his cunning he cannot quite bring under the Finite." This is why hydrogen, which has become the demon element of our time, should be seen as the intangible dagger which hung before Macbeth's vision, but which had no power except what was lent to it by his own mind.

The terror that confronts our age is our own conception of ourselves. Above all else this is the potion which the modern Dr. Jekylls have concocted. As Shakespeare foresaw:

> "It hath been taught us from the primal state
> That he which is was wished until he were."

This is not the voice of the witches. It is the clear voice of a great poet almost four centuries gone, who saw at the dawn of the scientific age what was to be the darkest problem of man: his conception of himself. The words are quiet, almost cryptic; they do not foretell. They imply a problem in free will. Shakespeare, in this passage, says nothing of starry influences, machinery, beakers, or potions. He says, in essence, one thing only: that what we wish will come.

I submit that this is the deadliest message man will ever encounter in all literature. It thrusts upon him inescapable choices. Shakespeare's is the eternal, the true voice of the divine animal, piercing, as it has always pierced, the complacency of little centuries in which, encamped as in hidden thickets, men have sought to evade self-knowledge by describing themselves as men.

FIVE

The Chresmologue

I

"FORMER MEN," observed Emerson in the dramatic days of the new geological science, "believed in magic, by which temples, cities and men were swallowed up, and all trace of them gone. We are coming on the secret of a magic which sweeps out of men's minds all vestige of theism and beliefs which they and their fathers held. . . . Nature," he contended clairvoyantly, "is a mutable cloud." Within that cloud is man. He constitutes in truth one of Emerson's most profound questions. Examined closely, he is more than a single puzzle. He is an indecipherable palimpsest, a walking document initialed and obscured by the scrawled testimony of a hundred ages. Across his features and written into the very texture of his bones are the half-effaced signatures of what he has been, of what he is, or of what he may become.

Modern man lives increasingly in the future and neglects the present. A people who seek to do this have an insatiable demand for soothsayers and oracles to assure and comfort them about the insubstantial road they tread. By contrast, I am a person known very largely, if at all, as one committed to the human past—to the broken columns of lost civilizations, to what can be discovered in the depths of tombs, or dredged from ice-age gravels, or drawn from the features of equally ancient crania. Yet as I go to and fro upon my scientific errands I find that the American public is rarely troubled about these antiquarian matters. Instead, people invariably ask, What will man be like a million years from now?—frequently leaning back with complacent confidence as though they already knew the answer but felt that the rituals of our society demanded an equally ritualistic response from a specialist. Or they inquire, as a corollary, what the scientists' views may be upon the colonization of outer space. In short, the cry goes up, Prophesy! Before attempting this dubious enterprise, however, I should like to recount the anecdote of a European philosopher who, over a hundred years ago, sensed the beginnings of the modern predicament.

It seems that along a particularly wild and forbidding section of the English coast—a place of moors, diverging and reconverging trackways, hedges, and all manner of unexpected cliffs and obstacles—two English gentlemen were out riding in the cool of the morning. As they rounded a turn in the road they saw a coach bearing down upon them at breakneck speed. The foaming, rearing horses were obviously running wild; the driver on the seat had lost the reins. As the coach thundered by, the terrified screams of the occupants could be heard.

The gentlemen halted their thoroughbred mounts and briefly exchanged glances. The same thought seemed to strike each at once. In an instant they set off at a mad gallop which quickly overtook and passed the lurching vehicle before them. On they galloped. They distanced it.

"Quick, the gate!" cried one as they raced up before a hedge. The nearest horseman leaped to the ground and flung wide the gate just as the coach pounded around the curve. As the swaying desperate driver and his equipage plunged through the opening, the man who had lifted the bar shouted to his companion, "Thirty guineas they go over the cliff!"

"Done!" cried his fellow, groping for his wallet.

The gate swung idly behind the vanished coach and the two sporting gentlemen listened minute by minute, clutching their purses. A bee droned idly in the heather and the smell of the sea came across the moor. No sound came up from below.

There is an odd resemblance in that hundred-year-old story to what we listen for today. We have just opened the gate and the purse is in our hands. The roads on that fierce coast diverge and reconverge. In some strange manner, in a single instant we are both the sporting gentlemen intent on their wager and the terrified occupants of the coach. There is no sound on all this wild upland. Something has happened or is about to happen, but what? The suspense is intolerable. We are literally enduring a future that has not yet culminated, that has perhaps been hovering in the air since man arose. The lunging, rocking juggernaut of our civilization has charged by. We wait by minutes, by decades, by centuries, for the crash we have engendered.

The strain is in our minds and ears. The betting money never changes hands because there is no report of either safety or disaster. Perhaps the horses are still poised and falling on the great arc of the air.

We shift our feet uneasily and call to the first stranger for a word, a sanctified guess, an act of divination. As among the ancient Greeks, chresmologues, dealers in crumbling parchment and uncertain prophecy, pass among us. I am such a one. But the chresmologue's profession demands that he be alert to signs and portents in both the natural and human worlds—events or sayings that others might regard as trivial but to which the gods may have entrusted momentary meaning, pertinence, or power. Such words may be uttered by those unconscious of their significance, casually, as in a bit of over-heard conversation between two men idling on a street, or in a bar at midnight. They may also be spoken upon journeys, for it is then that man in the role of the stranger must constantly confront reality and decide his pathway.

It was on such an occasion not long ago that I overheard a statement from a ragged derelict which would have been out of place in any age except, perhaps, that of the Roman twilight or our own time. A remark of this kind is one that a knowledgeable Greek would have examined for a god's hidden meaning and because of which a military commander, upon overhearing the words, might have postponed a crucial battle or recast his auguries.

I had come into the smoking compartment of a train at midnight, out of the tumult of a New York weekend. As I settled into a corner I noticed a man with a paper sack a few seats beyond me. He was meager of flesh and his cheeks had already taken on the molding of the skull beneath them. His thread-bare clothing suggested that his remaining possessions were contained in the sack poised on his knees. His eyes were closed, his head flung back. He drowsed either from exhaustion or liquor, or both. In that city at midnight there were many like him.

By degrees the train filled and took its way into the dark. After a time the door opened and the conductor shouldered his way in, demanding tickets. I had one sleepy eye fastened on the dead-faced derelict. It is thus one hears from the gods.

"Tickets!" bawled the conductor.

I suppose everyone in the car was watching for the usual thing to occur. What happened was much more terrible.

Slowly the man opened his eyes, a dead man's eyes. Slowly a sticklike arm reached down and fumbled in his pocket, producing a roll of bills. "Give me," he said then, and his voice held the croak of a raven in a churchyard, "give me a ticket to wherever it is."

The conductor groped, stunned, over the bills. The dead eyes closed. The trainman's hastily produced list of stations had no effect. Obviously disliking this role of Charon he selected the price to Philadelphia, thrust the remaining bills into the derelict's indifferent hand, and departed. I looked around. People had returned to their papers, or were they only feigning?

In a single sentence that cadaverous individual had epitomized modern time as opposed to Christian time and in the same breath had pronounced the destination of the modern world. One of the most articulate philosophers of the twentieth century, Henri Bergson, has dwelt upon life's indeterminacy, the fact that it seizes upon the immobile, animates, organizes, and hurls it forward into time. In a single poignant expression this shabby creature on a midnight express train had personalized the terror of an open-ended universe. I know that all the way to Philadelphia I fumbled over my seat check and restudied it doubtfully. It no longer seemed to mean what it indicated. As I left the train I passed the bearer of the message. He slept on, the small brown sack held tightly in his lap. Somewhere down the line the scene would be endlessly repeated. Was he waiting for some final conductor to say, "This is the place," at a dark station? Or was there money in the paper sack and had he been traveling for a hundred years in these shabby coaches as a stellar object might similarly wander for ages on the high roads of the night?

All I can assert with confidence is that I was there. I heard the destination asked for, I saw the money taken. I was professionally qualified to recognize an oracle when I heard one. It does not matter that the remark was cryptic. Good prophecy is always given in riddles, for the gods do not reveal their every secret to men. They only open a way and wait for mortal nobility or depravity to take its natural course. "A ticket to wherever it is" carries in the phrase itself the weight of a moral judgment. No civilization professes openly to be unable to declare its destination. In an age like our own, however, there comes a time when individuals in increasing numbers unconsciously seek direction and taste despair. It is then that dead men give back answers and the sense of confusion grows. Soothsayers, like flies, multiply in periods of social chaos. Moreover, let us not confuse ourselves with archaic words. In an age of science the scientist may emerge as a soothsayer.

II

There is one profound difference which separates psychologically the mind of the classical world from that of the present: the conception of time. The Ancient World was, to use Frank Manuel's phrase, bound to the wheel of Ixion, to the maxim: what has been is, passes, and will be. By contrast, the Christian thinkers of Western Europe have, until recently, assumed a short time scale of a few millennia. In addition, Christianity replaced the cyclical recurrences of Greek and Roman history by the concept of an unreturning past. History became the drama of the Fall and the Redemption, and therefore, as drama, was forewritten and unrepeatable. Novelty was its essence, just as duration and repetition lay at the heart of classical thinking.

Between the earlier conception of time and its reordering in the phrase, what is will *not* be, lies an irredeemable break with the past even though, in the course of two thousand years, much has changed and conceptions derived originally from both realms of thought have interpenetrated. Western philosophy has been altered under the impact of science and become secularized, but history as the eternally new, as "progress," repeats the millenarianism of Christianity. As for the time scale, which modern science has enormously extended, the intuitions of the ancients have proved correct, but biology has contributed an unreturning novelty to the shapes of life. Thus the great play has lengthened and become subject to the mysterious contingencies which are the proper matter of genetics. The play itself remains, however, just as the anthropologist has similarly demonstrated in the social realm, a performance increasingly strange, diverse, and unreturning. In the light of this distinction, the role of the oracle in the ancient world can be seen to differ from that of his modern analogue, whether the latter be disguised as a science-fiction writer, a speculative scientist engaged in rational extrapolation, or a flying-saucer enthusiast replacing the outmoded concept of the guardian angel by the guarding intelligence of extraterrestrial beings.

Since emergent creativity went largely unnoticed in the living world, the kind of future in which Western man now participates was also neglected. Men lived amidst the ruins of past civilizations or epochs, indifferently wedging great sculpture or invaluable inscriptions into the wall of a peasant's hut or a sheep corral. Few indeed were the attempts to probe the far future or the remote past. Men requested of the oracles what men have always desired: the cure for illness, the outcome of battle, the wisdom or unwisdom of a sea journey, the way to a girl's heart. They asked, in effect, next day's or next year's future because, save for the misfortunes that beset the individual's pathway, all lives and all generations were essentially the same.

There was, in the words of the Old Testament, "nothing new under the sun." The wind went about its circuits; the wave subsided on the beach only to rise again. The generations of men were like the wave—endless but the same. It was a wave of microcosmic futures, the difference between the emperor in purple and the slave under the lash. Each man was mortal; roles could be reversed and sometimes were. This was important to the buffeted individual but not to the wave. Men's individual fates resembled the little dance of particles under the microscope that we call the Brownian movement.

Perhaps, over vaster ranges of time than man has yet endured, the dance of civilizations may seem as insignificant. Indeed it must have seemed so intuitively to the ancients, for, in the endless rising and falling of the wave, lost palace and lost throne would all come round again. It was of little use, therefore, to trouble one's heart over the indecipherable inscription on a fallen monument. Let the immortal gods on the mountain keep their own accounts. In the sharp cold of midwinter one asked only if next year's pastures would be green.

But with the agony in the garden at Gethsemane came the

concern for last things, for the end of the story of man. A solitary individual, one who prayed sleepless that his fate might pass, had spoken before the Pharisees, "I know whence I have come and whither I am going." No man had said such a thing before and none would do so after him. For our purpose here it does not matter what we believe or disbelieve; whether we are pagan, Jew, Christian, or Marxist. The voice and the words were those of a world-changer.

At the place Golgotha they say the earth shook. It is true in retrospect, for the mind of Western man was there shaken to its foundations. It had gained the courage to ask the final terrible question: For what end was it made? Not the insignificant queries long addressed to wandering magicians, but such a question as a man could ask only in a desert: What is the end? Not of me, not of my neighbor or my generation, but the end of man. For what was the lime engendered in our bones, our bodies made to rise in the bright sun and again in dust to be laid down? It may well be that rocks were torn when that cry escaped on human breath. With it man had entered unknowingly upon history, upon limitless time, and equally limitless change. Nothing would be what it had been. The wave would fall no longer idly on the shore. It would loom vaster, bluer, darker until lightning played along its summit, the deepest, most dangerous wave in the entire universe—the wave of man.

The play upon which man had entered would at first be confined to a tiny immovable stage. Its acts would be centered within the brief time span then humanly conceivable. The very compression and foreshortening thus achieved, however, would heighten the intensity of the drama and whet man's concern with the unique course of events. The ancient cyclical conceptions of the pagan world would seem wearisome and banal, its gods without dignity. By contrast, a historic event, the mallet strokes upon a hill outside Jerusalem, would echo in men's minds across nineteen hundred years.

The Crucifixion was not an act that could be re-endured perpetually. "God forbid," wrote Saint Augustine, "that we should believe this. Christ died once for our sins, and, rising again, dies no more." The magnitude of the universe remained unknown, its time depths undiscovered, its evolutionary transformations unguessed. One thing alone had changed: the drama of man's life. It now had force, direction, and significance beyond the purely episodic. The power of a single divinity sustained the stage, the drama, and the actors. Men had arrived at true historicity. Acts of evil and of good would run long shadows out into eternity. Self-examination and self-knowledge would be intensified.

On this scene of increasing cosmic order would also emerge eventually a heightened interest in nature as a manifestation of that same divinity. In time, nature would be spoken of as the second look of God's revelation. Some would regard it as the most direct communication of all, less trammeled by words, less obscured by human contention. There would begin, by degrees, the attentive, innocent examination that would lead on through doubts and questionings to the chill reality of the ever-wandering stars, to time stretched across millions of light-years, or read in the erosion of mountain systems or by virtue of unexpected apparitions in the stratified rocks. Finally, Jean Baptiste Lamarck, in 1809, the year of Darwin's birth, would venture dryly, "Doubtless nothing exists but the will of the sublime Author of all things, but can we set rules for him in the execution of his will, or fix the routine for him to observe? Could not his infinite power create an order of things which gave existence *successively* to all that we see . . . ?"

The tragedy on a barren hill in Judea, which for so long had held human attention, would seem to shrink to a minuscule event on a sand-grain planet lost in a whirl of fiery galaxies. Reluctantly men would peer into the hollow eye sockets of the beasts from which they had sprung. The Christian dream would linger but the surety of direction would depart. Nature, the second book of the theologians, would prove even more difficult of interpretation than the first. Once launched upon the road into the past, man's insatiable hunger to devour eternity would grow. He would seek to live in past, present, and future as one, one eternity of which he might be the intellectual master.

Over fifty years ago it was possible to catch something of this feeling in the musings of the archaeologist Arthur Weigall, wandering in the upper Egyptian deserts. In an abandoned quarry he came upon many hewn stones addressed, as he says, "to the Caesars, but never dispatched to them; nor is there anything in this time-forsaken valley which so brings the past before one as do these blocks awaiting removal to vanished cities. . . . Presently," he continues, "a door seems to open in the brain. Two thousand years have the value of the merest drop of water."

III

Like Weigall, the desert wanderer, I have done much walking in my younger years. When I climbed I almost always carried seeds with me in my pocket. Often I liked to carry sunflower seeds, acorns, or any queer "sticktight" that had a way of gripping fur or boot tops as if it had an eye on Himalayas and meant to use the intelligence of others to arrive at them. I have carried such seeds up the sheer walls of mesas and I have never had illusions that I was any different to them from a grizzly's back or a puma's paw.

They had no interest in us, bear, panther, or man—but they were endowed with a preternatural knowledge that at some point we would lie down and there they would start to grow. I have, however, aided their machinations in a way they could scarcely have intended. I have dropped sunflower seed on stony mesa tops and planted cactus in alpine meadows amidst the sounds of water and within sight of nodding bluebells. I have sowed northern seeds south and southern seeds north and crammed acorns into the most unlikely places. You can call it a hobby if you like. In a small way I, too, am a world-changer and hopefully tampering with the planetary axis. Most of my experiments with the future will come to nothing but some may not.

Life is never fixed and stable. It is always mercurial, rolling and splitting, disappearing and re-emerging in a most unpredictable fashion. I never make a journey to a wood or a mountain without experiencing the temptation to explode a puffball in a new clearing or stopping to encourage some sleepy monster that is just cracking out of the earth mold. This is, of course, an irresponsible attitude, since I cannot tell what will come of it, but if the world hangs on such matters it may be well to act boldly and realize all immanent possibilities at once. Shake the seeds out of their pods, I say, launch the milkweed down, and set the lizards scuttling. We are in a creative universe. Let us then create. After all, man himself is the unlikely consequence of such forces. In the spring when a breath of wind sets the propellers of the maple seeds to whirring, I always say to myself hopefully, "After us the dragons."

To have dragons one must have change; that is the first principle of dragon lore. Otherwise everything becomes stale, commonplace, and observed. I suspect that it is this unimaginative boredom that leads to the vulgar comment that evolution may be all very well as a theory but you can never really see anything in the process of change. There is also the even more obtuse and less defensible attitude of those who speak of the world's creative energies as being exhausted, the animals small and showing no significant signs of advance. "Everything is specialized in blind channels," some observers contend. "Life is now locked permanently in little roadside pools, or perching dolefully on television aerials."

Such men never pause to think how *they* might have looked gasping fishily through mats of green algae in the Devonian swamps, but that is where the *homunculus* who preceded them had his abode. I have never lost a reverent and profound respect for swamps, even individually induced ones. I remember too well what, on occasion, has come out of them. Only a purblind concern with the present can so limit men's views, and it is my contention that a sympathetic observer, even at this moment, can witness such marvels of transitional behavior, such hoverings between the then and the now, as to lay forever to rest the notion that evolution belongs somewhere in the witch world of the past.

One may learn much in those great cemeteries of which Weigall spoke, those desolate Gobis and wind-etched pinnacles that project like monuments out of the waste of time itself. One must learn, however, to balance their weight of shards and bones against a frog's leap, against a crow's voice, against a squeak in the night or something that rustles the foliage and is gone. It is here that the deception lies. The living are never seen like the dead and the living appear to be so surely what they are. We lack the penetration to see the present and the onrushing future contending for the soft feathers of a flying bird, or a beetle's armor, or shaking painfully the frail confines of the human heart.

We are in the center of the storm and we have lost our sense of direction. It is not out of sadistic malice that I have carried cockleburs out of their orbit or blown puffball smoke into new worlds. I wanted to see to what vicissitudes they might adapt or in what mountain meadows the old thorns might pass away. One out of all those seeds may grope forward into the future and writhe out of its current shape. It is similarly so on the windswept uplands of the human mind.

Evolution is far more a part of the unrolling future than it is of the past, for the past, being past, is determined and done. The present, in the words of Karl Heim, "is still in the molten phase of becoming. It is still undecided. It is still being fought for." The man who cannot perceive that battleground looks vaguely at some animal which he expects to transform itself before his eyes. When it does not, he shrugs and says, "Evolution is all very well but you cannot see it. Besides it does not direct you. It only teaches you that you are an animal and had better act like one."

Yet even now the thing we are trying to see is manifesting itself. Missing links, partial adaptations, transitions from one environmental world to another, animals caught in slow motion half through some natural barrier are all about us. They literally clamor for our attention. We ourselves are changelings. Like Newton, those who possess the inclination and the vision may play on the vast shores of the universe with the living seeds of future worlds. Who knows, through the course of unimaginable eons, how the great living web may vibrate slightly and give out a note from the hand that plucked it long ago? In the waste dumps at city edges bloom plants that have changed and marched with man across the ages since he sat by hill barrows and munched with the dogs. A hand there, brown with sun, threw a seed and the world altered. Perhaps, in some far meadow, a plant of mine will survive the onset of an age of ice. Perhaps my careless act will root life more firmly in the dying planetary days when man is gone and the last seeds shower gallantly against the frost.

What is true biologically is also true along the peripheries of the mind itself. We possess our own alpine meadows, excoriating heat, and freezing cold. There have been, according to philosophers, political man, religious man, economic man. Today there are, variously, psychological man, technological man, scientific man. Dropped seeds, all of them, the mind's response to its environments, its defense against satiety. He who seeks naïvely to embrace his own time will accept its masks and illusions. The men of one period may turn completely to religious self-examination and become dogmatically contentious. Our own age, by contrast, turns outward, as if in the flight from self of which its rockets have become the symbol. It has been well said by Philip Rieff that every personality cure seems to expose man to a new illness. I believe it is because man always chooses to rest on his cure.

We have forgotten the greatest injunction of the wise traveler from Galilee. He did not say before the Pharisees, "I know where I am *staying*." Instead he observed that he knew where he was going. As is true of all great prophets, he left something unspoken hanging in the air. Men have chosen to assume that Jesus had knowledge of his physical fate or that he was bound to some safe haven beyond mortal reach. It seldom occurs to us that he was definitely engaged on a journey. If, in traveling that road, it led incidentally to a high place called Golgotha, it was because his inward journey was higher and more dangerous still.

Five centuries ago an unknown Christian mystic spoke thus of heaven, which his contemporaries assumed to be a definable place: "Heaven ghostly, is as high down as up, and up as down: behind as before, before as behind, on one side as another. Insomuch, that whoso had a true desire for to be at heaven, then that same time he were in heaven ghostly. For the high and the next way thither is run by desires and not by paces of feet."

Today our glimpses of heaven have become time-projected. They are secular; they are translated into paces measured by decades and centuries. Science is the assumed instrument and progress a dynamic flow, as is the heaven we seek to create or abjure. In final analysis we deceive ourselves. Our very thought, through the experimental method, is outwardly projected upon time and space until it threatens to lose itself, unexamined, in vast distances. It does not perform the contemplative task of inward perception.

The mysterious author of *The Cloud of Unknowing* spoke rightly and his words apply equally to that future we seek to conjure up. The future is neither ahead nor behind, on one side or another. Nor is it dark or light. It is contained within ourselves; it is drawn from ourselves; its evil and its good are perpetually within us. The future that we seek from oracles, whether it be war or peace, starvation or plenty, disaster or happiness, is not forward to be come upon. Rather its gestation is now, and from the confrontation of that terrible immediacy we turn away to spatial adventures and to imagination projected into time as though the future were fixed, unmalleable to the human will, and to be come upon only as a seventeenth-century voyager might descry, through his spyglass, smoke rising from an unknown isle.

Not so is the human future. It is made of stuff more immediate and inescapable—ourselves. If our thought runs solely outward and away upon the clever vehicles of science, just so will there be in that future the sure intellectual impoverishment and opportunism which flight and anonymity so readily induce. It will be, and this is the difficult obstacle of our semantics, not a future come upon by accident with all its lights and shadows, guiltless, as in a foreign sea. It will be instead the product of our errors, hesitations, and escapes, returning inexorably as the future which we wished only to come upon like a geographical discoverer, but to have taken no responsibility in shaping.

If, therefore, it is my occasional task to cast auguries, I will add as pertinent some further words of that long-vanished seer: "Be wary that thou conceive not bodily, that which is meant ghostly, although it be spoken bodily in bodily words as be these, up or down, in or out, behind or before. This thought may be better felt than seen; for it is full blind and full dark to them that have but little while looked thereupon." If we banish this act of contemplation and contrition from our midst, then even now we are dead men and the future dead with us. For the endurable future is a product not solely of the experimental method, or of outward knowledge alone. It is born of compassion. It is born of inward seeing. The unknown one called it simply "All," and he added that it was not in a bodily manner to be wrought.

IV

A former colleague of mine, who was much preoccupied with travel and who suffered from absent-mindedness, once turned timidly to his wife as he set forth upon a long journey. "Is the place where I am going," he asked her anxiously, for he depended much upon her notes of instruction, "in my pocket?" It strikes me now that in few centuries has the way seemed darker or the maps we carry in our separate pockets more contradictory, if not indecipherable.

The Russians in their early penetration of space saw fit to observe irreverently that they had not seen heaven or glimpsed the face of God. As for the Americans, in our first effort we could only clamorously exclaim, "Boy, what a ride!" During those words on a newscast I had opened a window on the night air. It was moonrise. In spite of the cynical Russian pronouncement, my small nephew had just told me solemnly that he had seen God out walking. Concerned as adults always are lest children see something best left unseen, I consulted his mother. She thought a moment. Then a smile lighted her face. "I told him God made the sun and the stars," she explained. "Now he thinks the moon is God."

I went and reasoned gravely with him. The gist of my extemporized remarks came from the medieval seer. "Not up, or down," I cautioned, "nor walking in the sun, nor in the night—above all not that."

There was a moment of deep concentration. An uncertain childish voice reached up to me suddenly. "Then where did God get all the dirt?"

I, in my turn, grew quiet and considered.

"Out of a dark hat in a closet called Night," I parried. "We, too, come from there."

"*Conceive it better as not wrought by hands*," the voice repeated in my head.

"Then how do we see Him?" the dubious little voice trailed up to me. "Where is He then?"

"He is better felt than seen," I repeated. "We do not look up or down but in here." I touched the boy's heart lightly. "In here is what a great man called simply 'All.' The rest is out there"—I gestured—"and roundabout. It is not nearly so important."

The world was suddenly full of a vast silence. Then upon my ear came a sound of galloping, infinitely remote, as though a great coach passed, sustained upon the air. I touched the child's head gently. "We are in something called a civilization," I said, "a kind of wagon with horses. It is running over the black bridge of nothing. If it falls, we fall."

"Thirty guineas," a cricket voice chirped in my brain. I shut it out along with the glimpse of a sea cliff in the English fog.

"Conceive it not bodily," the clear voice persisted like a bell, "for it is meant ghostly." From below a hand gave itself up trustingly to mine.

"I saw Him. I did so," said the child.

"We will go and look all about," I comforted, "for that is good to do. But mostly we will look inside, for that is where we ache and where we laugh and where at last we die. I think it is mostly there that He is very close."

We went out side by side a little shyly onto the lawn and watched the stars. After a while, and carefully, being small, we turned and looked for the first time at our two selves. Not bodily, I mean, but ghostly. And being still the wandering chresmologue, I told him about a very ancient manuscript in which is dimly written: "Wherever thou wilt thou dost assemble me, and in assembling me thou dost assemble thyself."

SIX

Paw Marks and Buried Towns

Many years ago, when the first cement sidewalks were being laid in our neighborhood, we children took the paw of our dog Mickey and impressed it into a kind of immortality even as he modestly floundered and objected. Some time ago after the lapse of many decades, I stood and looked at the walk, now crumbling at the edges from the feet of many passers.

No one knows where Mickey the friendly lies; no one knows how many times the dust that clothed that beautiful and loving spirit has moved with the thistledown across the yards where Mickey used to play. Here is his only legacy to the future—that dabbled paw mark whose secret is remembered briefly in the heart of an aging professor.

The mark of Mickey's paw is dearer to me than many more impressive monuments—perhaps because, in a sense, we both wanted to be something other than what we were. Mickey, I know, wanted very much to be a genuine human being. If permitted, he would sit up to the table and put his paws together before his plate, like the rest of the children. If anyone mocked him at such a time by pretending to have paws and resting his chin on the table as Mickey had to do, Mickey would growl and lift his lip. He knew very well he was being mocked for not being human.

The reminder that he was only a poor dog with paws annoyed Mickey. He knew basically a lot more than he ever had the opportunity to express. Though people refused to take Mickey's ambition seriously, the frustration never affected his temperament. Being of a philosophic cast of mind, he knew that children were less severe in their classifications. And if Mickey found the social restrictions too onerous to enable him quite to achieve recognition inside the house, outside he came very close to being a small boy. In fact, he was taken into a secret order we had founded whose club house was an old piano box in the backyard. We children never let the fact that Mickey walked on four legs blind us to his other virtues.

Now the moral of all this is that Mickey tried hard to be a human being. And as I stood after the lapse of years and looked at the faint impression of his paw, it struck me that every ruined civilization is, in a sense, the mark of men trying to be human, trying to transcend themselves. Like Mickey, none of them has quite made it, but they have each left a figurative paw mark—the Shang bronzes, the dreaming stone faces on Easter Island, the Parthenon, the Sphinx, or perhaps only rusted stilettos, chain mail, or a dolmen on some sea-pounded headland. The archaeologist, it is said, is a student of the artifact. That harsh, unlovely word, as sharply angled as a fist ax or a brick, denudes us of human sympathy. In the eye of the public we loom, I suppose, as slightly befuddled graybeards scavenging in grave heaps. We caw like crows over a bit of jade or a broken potsherd: we are eternally associated in the public mind with sharp-edged flints and broken statues. The utter uselessness of the past is somehow magnificently incorporated into our activities.

No one, I suppose, would believe that an archaeologist is a man who knows where last year's lace valentines have gone, or that from the surface of rubbish heaps the thin and ghostly essence of things human keeps rising through the centuries until the plaintive murmur of dead men and women may take precedence at times over the living voice. A man who has once looked with the archaeological eye will never see quite normally. He will be wounded by what other men call trifles. It is possible to refine the sense of time until an old shoe in the bunch grass or a pile of nineteenth-century beer bottles in an abandoned mining town tolls in one's head like a hall clock. This is the price one pays for learning to read time from surfaces other than an illuminated dial. It is the melancholy secret of the artifact, the humanly touched thing.

Although the successful moon rockets have swung everyone's attention to outer space, a surprising number of archaeological books dealing with the lost city civilizations are still being published. The rapidity of their appearance and the avidity with which they are received suggest that while the public's eye has been forced upward it has also, in the same act, been cast downward toward the earth. Perhaps no great civilization ever before has been more self-consciously aware of the possible doom that confronts it or more curious about those brother thinkers and artists who carved the gods that lie now in temples visited by rain, or who ventured through the Pillars of Hercules when all beyond was wild and unknown as outer space is today. They built in their separate ways, then fell, and we, with one winged foot poised toward the stars, hear in a subdued quiet the old voices out of the grass. Whatever the disease that ate the heart of these lost cultures it was not the affliction of ignorance—not, at least, the technical ignorance of the savage who cannot lay one stone successfully upon another. In every one of the fallen cities which our spades have revealed, there existed the clever artisan, the engineer devoted to the service of the particular human dream that flourished there.

Here is Leonard Cottrell's description of the Ceylonese city of Annadhapura, known to be roughly contemporaneous, in the West, with the conquests of Alexander the Great:

> ". . . the palaces . . . would have made Diocletian's palace seem a poor thing by comparison, their great *dagobas*, artificial hills of masonry supporting shrines and reliquaries, were sometimes over three hundred feet high, and can be compared with the pyramids of Egypt. Their hydraulic engineering has no parallel save in the nineteenth and twentieth centuries; for example, the artificial lake of Mineria, created in the third century A.D. by Maha Sen, has a circumference of twenty miles, and the masonry and earthwork dams which were made to divert the waters of the stream which fills it extend for eighty miles; their average height is eighty feet."

Today, in the ruined tanks of those great Sinhalese cities, the bear alone stands upright, and leopards drink from the few puddles that remain. The cunning workers in stone and gold have long since departed. Given time enough, this is a state of affairs more the rule than the exception among the cities of men, as the studies of the modern urban archaeologists clearly demonstrate. The lesson is felt in the search for a single burial, as in *The Lost Pyramid* by the late Egyptian archaeologist Zakaria Goneim.

The story of Goneim's excavation, ending in the presence of a mysterious alabaster coffin of an unknown third dynasty pharaoh, is a peculiarly fascinating one. The coffin he discovered proved to be totally empty, although a wreath had been laid upon it before it was left in the darkness of its burial vault almost five thousand years ago. The evidence did not suggest the usual tomb robbery, but rather some bit of human drama lost forever in the darkness that shrouds the history of the Old Kingdom. Dr. Goneim, moreover, was not ashamed to acknowledge the feelings of which I have earlier spoken—the increasing sensitivity of the archaeologist to the voices from the ground:

"No one who has not crawled along the galleries beneath a pyramid, and experienced the silence and darkness, can fully appreciate the sensation which, at times, overwhelms one. It may sound fantastic, but I felt that the pyramid had a personality and that this personality was that of the king for whom it was built and which still lingered within it. I know that my workmen, some of whom have spent their whole lives in such work, often experience this feeling. You crawl along some dark corridor on hands and knees, past falls of rock; the light of the lamp gleams on minute crystals in the stratified walls; beyond, the corridor disappears into the blackness. You turn corners, feeling your way with your hands; the workmen have been left behind, and suddenly you realize you are alone in a place which has not heard a footfall for nearly fifty centuries."

After this, the individual vicariously devoted to archaeological adventure is only too eager to wander among the lost cities and buried libraries of Babylonia or to follow spirited accounts of ancient Egypt, including the still fascinating though often-told story of the Tutankhamen discovery. Perhaps, in the end, part of the pathos which the episode holds for us rests in the youth of that young king interred in April, left with a wreath of corn flowers on his breast, to lie alone for three thousand years encased in solid gold.

One may now have a choice of cities like Pompeii, destroyed by the elements, or cities stricken down by war, or cities ruined by the vagaries of trade. Mohenjo-daro and Harappa, whose writing no man can read, perhaps passed to their end by conquest. Today the carefully laid out cities lie in a waste of sand.

The North African cities of Sabratha and Leptis Magna, since they have been extensively photographed, touch us more closely. In these photographs we come back to what I spoke of as the artifact, the humanly touched thing. Here, in a series of clear, sunlit pictures, remnants of the Roman civilization, once as powerful as our own, are seen dissolving under sand and wind. From Libya there came the vast quantities of wheat and olive oil that nourished metropolitan Rome. Now the theaters lie empty and open to the sun, and the baths are waterless; but still the bold Roman letters stand across the entrance to the theater at Leptis Magna, naming, amid the surrounding ruin, one Annobal, the donor.

The columns totter; Annobal has left a paw mark and gone thence, like my dog Mickey. It is the immediacy, across the waste of centuries, that catches the heart—the sculptured head that might have just been finished in a modern studio, the empty seats vacated by a crowd that has just left but is not coming back. As I study such pictures I am reminded of a feeling I once experienced when examining a remarkable Victorian photograph of a girl waif asleep on a London park bench. She was obviously poor, her shoes were scuffed, her young body ill at ease in its graceless Victorian garments. There was despair and beauty in her face—so much beauty that it was like looking through a little window in time and wanting to reach out and touch her shoulder in compassion. I realized with difficulty that I was glancing for one unrelieved instant upon a drama ended before I was born, a drama and a human soul upon whom I would never look again. Where she went upon the evening of that day a century ago or what darkness swallowed her up, it would never be mine to know. Now, looking once more upon the ruined Roman theater, it comes to me that not even the shadow of a shadow remains of the good citizen Annobal. Even worse, the roofs of the town he loved, and to whose arts he contributed, lie open to the stars.

I have said that the ruins of every civilization are the marks of men trying to express themselves, to leave an impression upon the earth. We in the modern world have turned more stones, listened to more buried voices, than any culture before us. There should be a kind of pity that comes with time, when one grows truly conscious and looks behind as well as forward, for nothing is more brutally savage than the man who is not aware he is a shadow. Nothing is more real than the real; and that is why it is well for men to hurt themselves with the past—it is one road to tolerance.

The long history of man, besides its ennobling features, contains also a disruptive malice which continues into the present. Since the rise of the first neolithic cultures, man has hanged, tortured, burned, and impaled his fellow men. He has done so while devoutly professing religions whose founders enjoined the very opposite upon their followers. It is as though we carried with us from some dark tree in a vanished forest, an insatiable thirst for cruelty. Of all the wounds man's bodily organization has suffered in his achievement of a thinking brain, this wound is the most grievous of all, this shadow of madness, which has haunted every human advance since the dawn of history and which may well precipitate the final episode in the existence of the race.

Not many months ago I chanced to be lecturing at a university whose grounds adjoin a depressed area of slums. After the conclusion of the class hour I sauntered out into a courtyard filled with sunshine and some fragments of Greek statuary. As I passed by the inner gate I was confronted by a scene as old as time. Approaching me along the path upon which they had intruded by squirming through a hedge, was a ragged band of children led by a sharp-featured boy with a bow. The arrow he held drawn was pointed with tin. Instinctively we both paused —I because I feared for my eyes. There was no more human recognition in the face of the leader than I might have received from a group of hunting man-apes on the African savanna. We measured each other as mutually powerful and unknown forces, best to be avoided. The band drew in unconsciously about its leader and veered aside, with that wide, momentary animal stare haunting me as they passed. Before my eyes there marched a million years of human history, and I was a stranger and afraid, although, in my own lifetime, I had made that formidable passage from the caves and sewers of my childhood to this deceptively quiet campus across which these ghosts of long ago now persisted in passing. There was no humor in them, no real play. They slunk on, cruel as man's past, deadly, with the bow poised and the sharp, observant eyes alert to spy out any helpless thing.

I sighed with relief as they clambered over an embankment

and disappeared. On a nearby bench with my books spread out before me I did not read. "Man will survive," I said to myself, touched with a slight horror. "God should pity the world. Man will survive." All that passionate energy which in my own life, after many stumblings, had lodged me in these great silent halls, suddenly seemed dissipated and lost. I was as empty and filled with light as a milkweed pod whose substance has evaporated into the silvery autumn air. I thought of the beautiful ruined courts of an Aztec city in which I once had stood. I drew my hand over the bust of Hermes that I knew with surety would find a second burial in the earth. More ghostly, more insubstantial, than that hunting pack which roams the world and its dark thickets forever, I felt the dissolving power of the light which falls across lost columns and bleaching mosaics. Beauty man has, but in the very act of possession he is dissolved. I saw in my hand against the statue the projected shriveling of the skin. Each man repeats that history—endlessly and forever.

Nor has any civilization sustained beauty without returning it to the earth. But perhaps man will eventually achieve this victory, I thought doubtfully, standing a little longer in the timeless eternal light that flowed from the great sculpture. Perhaps. Perhaps he will, I thought again, and went on my way toward the darkness.

SEVEN

Barbed Wire and Brown Skulls

I

ARCHAEOLOGISTS, during the course of their lives, see and hear many strange things, but the fact that they are scientific men keeps them for the most part silent. They have good, if not superior, rationalizations for the things they do. No layman would dare impugn their motives. I, for example, have a certain number of skulls in my possession. As I write I can see four on the shelf above me. At least two are hidden in my filing cabinet, and there is a beautiful fragment on my desk which is often fondled by visitors who are unaware of its human significance.

Now as it happens I am fortunate. I practice a trade which enables me to keep these objects about in a perfectly logical and open manner. I have not murdered to possess them, and if one or two were acquired in dark and musty places, my motives, as I have hinted, are beyond reproach. As an archaeologist I can be both a good citizen and a frequenter of graveyards.

It was different in the case of the man who finally led me to question my own motives as a skull collector. He was a lawyer, but that, perhaps has little to do with the tale. I knew him as an austere, high-collared member of the bar—a moral and upright citizen—but that, I am afraid, has little to do with it either. The truth is that the gentleman left a box.

He had died, and after the passage of a certain number of months during which the box either lay undiscovered in his attic or, as is more likely, circulated uneasily through the hands of his heirs, I received a call about it. There was nothing unusual in this. I was simply not a policeman. When you are the heir to a considerable estate and unfortunately also have a box to be disposed of, you never go to a policeman. You go instead to an archaeologist. He is apt to be more understanding of human frailty, less prone to dark suspicions than a police officer, and above all, he will relieve you of the box.

If you have ever wandered the streets of a strange city with a parcel of this nature, you will appreciate the fact that there are very few human beings who can be trusted to relieve you of such a burden without making some hideous public commotion. Naturally you wish to avoid this. There are only two solutions: bury the box (an act which can lead to serious complications, including the suggestion of guilt) or find an archaeologist, smiling trustingly, and deposit it in his arms.

The heirs in this case pursued the inevitable pathway. They came to me. The legal gentleman and I had had mutual friends. My profession was known. Perhaps the property was really mine. Attics, you know, and the things that get into them. A loan perhaps? Some lodge doings?

I preserved a noncommittal air.

"Uncle Tobias was a church man. He would not tolerate—"

Yes, I said, I knew that.

A nephew toyed uneasily with the strings of the box. "It is very unlikely that his profession would have brought him into contact with—?"

"And him a lawyer?" I said. "Nothing likelier."

The niece's hands twisted. "Show him," she prompted.

It was the real thing, of course, and no lodge fake. As fine a skull as I've ever fondled.

"You recognize it?" they cried hopefully. "We are glad to restore it to your collection." Almost they started up.

"Hmmmm," I said. They subsided nervously. "The jaw, you see. It doesn't—"

"Doesn't what?" the nephew challenged. "I'm sure it's just like you loaned it to him."

"It's not mine," I said bluntly, "and besides that I'll tell you something. There are two of them—individually represented, I mean. The jaw doesn't fit the skull. It belonged to someone else. You can see by the color it's out of a different grave."

"Two of them," murmured the niece.

"Out of a different grave," repeated the nephew.

I waited patiently. After a time he came to the point. Some see it more rapidly than others.

"I guess Uncle Tobias was—uh—uh—a collector," he said. "We should now like to present his collection to you—or your institution—anonymously, of course."

"Of course," I said. "Would you like a receipt? Would you like to take the box back with you?"

"Thank you, no," said the niece. "You're too kind. And it will be an anonymous gift?"

"We have many of them," I said. "Many of them."

As they went down the steps I saw them walking more lightly. Their arms swung better without the burden. They ran to the car at the curb. On the desk the skull waited. It was a rich old brown, I saw as my hand went over it—a rich old mahogany brown. They needn't have been so jittery—that skull had been hundreds of years underground when Uncle Tobias was born. But where had he got it—and that jaw from another body?

"There's no accounting," I said, "for tastes. Tobias must have been a collector." I said it disapprovingly to the nearest cabinet. Then I picked the skull up and put it inside. I was not, you see, a genuine collector. My motivations were purely scientific and unemotional.

Or were they? I went back to the desk and sat down. I could see Uncle Tobias's long-hidden relic staring back vacantly at me through the glass door of the cabinet. It would never tell its secret, but it had one. It had a secret and so had Uncle Tobias. And I? Perhaps I was a keeper of secrets. Or of orphans, I thought, as my eyes ran along the shelf overhead. And at last I knew where it had begun. Behind the steady chipping of the pick that began to sound in my ears was another sound—the creaking of weathered timbers and the uneasy movement of stormy air in a closed place. That would be it, I thought suddenly —the heads in Hagerty's barn.

II

When Grandma was alone in the kitchen we used to bake heads together in the kitchen stove. When I first approached her on this matter she naturally demurred, but in the end her cooking enthusiasm got the better of her and she would line them up like biscuits in a pie tin and put them in the oven. It was before the days of Charles Addams and we never conceived of ourselves as monsters. It is probably true, however, that it was at this time I developed a mild antipathy for the normal human skull.

This was not my grandmother's fault. In fact, at times, out of some lingering religious scruple she would protest the nature of some of the heads in the oven—opening the door now and then and peering in, partly to see that they were properly done and partly to grumble over their strangeness.

They were clay, burnt clay, and modeled as well as a boy could model skulls he had never handled. Some of them had matchstick teeth or bits of pearl shell from broken buttons. The eyes were the hollow eyes of skulls and the mandibles were shaped as I thought they should be shaped, from drawings in the red-brick museum that I frequented. As for the cranium itself, practically everything I made was slope-browed and primitive. Even today I am apt to be faintly repelled by skulls with no brow ridges or teeth of too delicate a cast.

"Mind you," Grandma would protest, tapping me with a roasting fork, "this is getting out of hand. Them's no ordinary heads in there and no young'un can tell me so. They've got that *look*, they have. That Darwin look. You be staying out of that building now. There's things there wasn't intended to be seen—not by anybody.

"You've got to stop it, youngster," she would say finally and swing the range door shut with a great clang. "You've got to stop it 'fore the Devil gets you by the foot. That little one there looks no more'n half a man. Where'd you find him, boy? Speak out now. Not from any book in this house, I'll warrant."

"No Grandma, honest not."

"Where then?"

"The room, Grandma, the room in the museum. I climbed up on the railing and looked close. His head was just like that—no forehead—and there was a big card with long words, and there was another head—ordinary—a plain old ordinary head beside him—"

"That's enough, boy, that's enough. They're done now. Get 'em out of the house. Take 'em away. Out of doors now. And don't touch 'em till they cool."

I never did. When they were cool enough, I put them in a little bag I carried and then I went halfway down the block to Hagerty's barn. It was an old sagging weatherbeaten stable, locked up and unused. I knew where a board could be edged aside, however, and there was just room enough to scrape in and let the board drop in place behind me. I always waited then until my eyes were adjusted to the light that came in through cracks and knotholes. In the spring when the light came in through the leaves outside it made a kind of green-lit secrecy.

Then I would take the bag of heads in my teeth and climb by way of some nailed crosspieces way up into the shadows under the roof. There was a half-loft up there—pretty rickety, but it would still bear a boy's weight. I could see after a while, even in that light, and then I would open the bag and take out the heads.

No one but Grandma and I ever saw them. Though I strove in my modelings for painstaking accuracy, it was only because without it the things seemed less real, less alive somehow. They were smaller than life, the size of big marbles, perhaps. Nevertheless they had a peculiar significance to me, a kind of being—the *anima* that exists in all properly shaped miniatures.

Up there under the barn roof I laid them out in little rows along the cross-beams. It was my museum, like the red-brick museum that my grandmother distrusted. Only in my museum nothing was dead. It was filled with a kind of patient, unwinking persistence—the persistence of a half-bewitched league of jack-o'-lantern faces waiting for me to come and sit with them in the green light high in the loft.

In the end I deserted them. There was no help for it. We moved away in what, to my mother, was a small triumph. I had no luggage of my own and no place to conceal the heads.

I can still remember that white, frosty morning and the cold clatter of hoofs as the cab rolled on its way toward the station. Away over the edge of the trees I could see the broken wind vane on Hagerty's stable, pointing steadily, as it always did, in one direction, no wind ever turning it. The heads were there. They would be there till the building fell.

"We will never come back here, son. Never." My mother's voice rang harshly over the cobblestones. But all the time I could feel the magnetic pull of those heads in Hagerty's stable. They would be there in the gray light and the green light; they would be there till the building fell.

III

Fainter than spider silk to my nearsighted gaze, the map lines run under the magnifying glass across a tumbled expanse of southwestern desert and lava beds. Names like Big Hatchet and Buckhorn still bring that vast and ominous landscape into my mind. Though the white man has taken it, it will never be rid of the ghosts of its last owners—the Apaches. It is their bones that lie in the cold on nameless peaks and in the red clay of the washes. Cochise, Victorio, Nana, and Geronimo will haunt it always. In the seventies of the last century many men died here. Dozens of others, the historians say, were never accounted for—the desert swallowed them up. Old Mr. Harney knew; he had been one of the missing. But it was from his family that I first got a hint of his story.

"He keeps her in the china closet," one of them told me, "right with the dishes."

"Kinfolk," sniffed another, with a gesture of distaste.

"The skull of Aunt Lucinda," explained a grandson with less heat. "He never buried her."

"Oh?" I said, puzzled and tactful, while the relatives all chattered together. They would have to make it clear. I had come at their invitation.

"He liked meeting you," they finally got out in chorus. "We think maybe you could influence him."

"Influence?" I said.

"The skull," they countered. "He won't bury it. But he's curious about your work. Maybe you could persuade him to give it to you. He's restless about it. Old, you know, quite old. We don't like having her there. It isn't right. Nor proper. People say—" They tapped their heads in unison like little marionettes.

"It was barbed wire," Mr. Harney said, "it was barbed wire finished our world." He was eighty years old, and the skull lay on the table before us. We sat silent, gazing out into the clear white desert sunlight. Eighty years, I thought, and reached out and turned the skull gently over. Years of smoking pistols and Apaches riding fast through the narrow canyons.

"You have lived a long life," I said. He sighed then, and

began talking—the merest wisp of a sound. I leaned forward to catch it.

"Six years in that valley after the haul from Texas, and me a youngster of ten. Mother dead on the trail. Her younger sister, Aunt Lucinda, raised me—the old man meanin' well but ridin'—ridin' most of the time. It took plenty ridin' to hold things together without the wire.

"Sure, we knew there was Apaches in the hills, always was. But people had a way of stickin'. A way—" he paused and reached out as if to touch the nearest blue hill—"as though they liked somethin' here—the air, maybe, so clear, or all this land at sunset, or maybe the feel of it, no fence from Texas to the Big Horns. Or maybe, like me, you had just followed along 'cause your people was moving and they was your people and you didn't go askin' 'em why their names changed along those little roads from the East.

"Lucinda was young and pretty with hair like the sheen on a blackbird's feather, and as good to me as my own mother. Young enough to play and imagine things the way a kid will. When my father was gone she used to play in the yard with me. Aaahh"—the old man got out something between a sigh and a groan—"it didn't last long.

"One night Pa didn't come home. Nobody knows what that means any more. They can't. The miles of darkness creeping in, and a woman and a kid sittin' in a shack waitin' for a man that ain't comin' back no more. You sit there and you dassent light the light for fear of drawin' 'em. And all the time you know they know about you, and it's no good, they'll take their time.

"They got us in the morning, in the first light, with Lucinda standin' out there lookin' for Pa. One of 'em just picked her off out of the mesquite. I'm old, but I've never got it out of my head, so that sometimes I see it like now, with people and things of years later all shadows, and just me with my hand at my mouth, and that shot. She stood there a minute all young and pretty with her hands stretched out to me. And all that love flowed up in her a minute and held her as if she wouldn't fall, and I ran toward her not thinkin' of anything except, as a kid will, that in the circle of such love I must be safe.

"And then she gave a little sigh and that light went out of

her and she pitched face down into a clump of prickly pear. They took me then, squalling and kicking, and put me on a horse. After that I was an Apache till I was fifteen."

The faded old eyes turned slowly over the whole compass of the horizon as though they remembered every peak and gully. He didn't offer to go on.

"Mr. Harney," I chided.

"Mexico," he said. "We rode into Old Mexico. They was Victorio's men. And I learned to be an Apache. Kids learn quick. That's why I lived. Ride, shoot, steal. Live on nothing. Trust nobody, and keep ridin'—keep ridin'. South of the border, north of the border, it was all the same.

"Apaches! Y'know, son, that's a joker. We wasn't Apaches. We was a way of life. We lived so hard that half the kids in camp was stolen. Most of 'em Mexicans, stolen south of the border. Raised Apaches. It was the only way to keep our strength up.

"Maybe I was a little old. Maybe I remembered too much. Anyhow I used to see Victorio watching me." Again he paused, searching his memories. "You know, in the end I didn't hate them. I was beginning to look at it the way they did, and to nurse the same feelings. I'd been shot at a lot and seen Indian families and kids I knew disappear. In the end I would have stayed with them, I guess. I spoke the language by then. I could get along." He stopped and whispered to himself a moment in syllables that were not English. Then he went on.

"Victorio must have thought different. Either that or he'd taken a shine to me—I never knew. He was a great warrior and Geronimo was nothing compared to him. He was hard, but there was a kind of bigness in him. When I was fifteen we were sitting on our horses one day looking down into a little town from the hills. I could see people in the streets, and smoke in chimneys. We watched it like animals must watch people—curious and sharp and wild. I watched like everyone else, ready to vanish at the least sign of danger.

"The next thing I knew, Victorio had edged his horse up beside me. 'Those are your people,' he said soft and low and searching my face with his eyes. 'Do you remember?'

"And I looked at him and was afraid, and suddenly the face

of Lucinda came to me and I looked back at him, speaking Apache, and I said, 'Yes, I remember.'

"And he nodded, a little sad, and said, 'They are your people. Go down to them.' Then he spoke a word behind me and the thirty people of his band were gone.

"'I don't know how—' I said. 'My people,' I said, and stopped. It came to me that all the people I had were Apache, and that I was Apache, too.

"Not a muscle of Victorio's face moved. 'Those are your people,' he said, pointing. 'We killed your father and the black-haired one. The white men will take care of you. You are not one of us.' With that he whirled his horse. I never saw him again.

"After a little while I picked my way down and spoke some words of English. It was slow work, like an old hinge squeaking in the wind. People came up to me and stared at my rags and at the pony."

Harney paused, considering, then he said flatly, "It wasn't so uncommon then—changing sides like that. There was room for two lives, and sometimes you had no choice. I got to be a white man even if I was a little late catchin' up. It was really about the same life: ride, shoot, kill. No difference, really, none to amount to anything. Not then, anyhow."

His eyes came almost shut against the midday heat shimmer that was beginning to roil the air out on the flats. I was afraid he was beginning to lose interest and go to sleep. I pushed the skull toward him. "The skull, Mr. Harney," I prodded. "You promised to tell me about the skull. It's a nice thing. Well cared for, too. A woman, I take it. Young. You can tell by the basilar suture. See?"

His eyes opened a little way, defensively, I thought.

"Aahh," he said again in that voice I was beginning to learn meant something hurt him. "It was afterward, sometime, that the thought came to me. I rode back to the old place. Nobody had been there all those years. And I found her—a few little bits of white bone, that is, and the skull in a drift of sand with the prickly pear grown over it. The hair," and with this he put up a careful, stroking finger, "was all gone. You wouldn't think it would go away so fast. For a while I looked around.

"Then it came on me I should bury her—and she out in the heat and dust and among bone-cracking coyotes so long. But what was there to bury, really? And besides this is a big wide land where you see miles as long as you can see at all. Every day of your life you see that way. And it's hard to be underground afterward. I had lived on the land enough to know.

"In the end I knew I couldn't bury her there. She was the only kin I had, so I took her up carefully and rode back with her. I figured at first maybe I'd have it done in a proper ceremony with a churchyard and a preacher to ease it a little.

"But then I couldn't. I couldn't face up to it. I kept putting it off and getting that feeling that if I did bury her she would go away; that she wouldn't be real any longer. I settled on this place finally and I kept Lucinda safe in the china closet. She never had to be afraid any more, and she could look out through the glass. Sometimes I talked to her.

"I'm a grown man, but that I did not get over, do you see—though I know all's dark in the grave and this is cold bone on the table top. I have a wife and sons, but this I will not bear—that they should put her under the ground with me."

He reached out and clutched my wrist and I cursed my easy juggling with anatomy a moment before. One of the family made a sign to me from the doorway.

I stood up then and took his hand and said quickly, by way of comfort, "She will not want to look through the glass at strange faces. Let her go with you. One can stay too long in the sun."

"Aahh," he said blindly, and took her back into his hands, fumbling. "It's plain you are not one of the open people, or you would not say that. It's the wire," he said, his voice subsiding once more to a thin whisper that seemed to come out of the grass beside us. "It's the wire that's made a difference. No wire from Texas to the Big Horns. It was all space and bright sun."

A granddaughter led him away.

IV

I wouldn't have taken old Mr. Harney's skull, even if he had offered it to me, for anything in the world. He had assumed a personal responsibility there that was not transferable. I knew too much of the story, and yet I was not part of it. Young Aunt Lucinda would have haunted me. Not physically, perhaps, but with that kind of intangible loneliness that comes of knowing about events behind you in time that you can never alter or intrude within, and yet there is somebody there you know or love, or wish greatly to have comforted, but it is back behind you and of all things the loneliest. So I left Harney with that burden as all men are left with it. It was his time, and he would have to deal with it as best he could.

Now, years later, I have some intimation of the emotions that had shaken him. I get out all the skulls. A massive unknown cranium which bears the look of the Cro-Magnon past about it is one I rescued from a medical dissecting room. I touch with fondness a mineralized skull vault whose age I can never prove but that rolled, I well know, for ages in the glacial gravels of the Platte. I look at them all, these silent masks whose teeth I have mended and whose mortal rags I have patched together with preservatives. Where will they go after the years of comfort—these fading, anonymous individuals who have somehow come to have a claim upon me? Scientifically they are worthless, for museums scrutinize with ever greater care the credentials of the bones that are donated to their skull rooms.

What chance has a dissecting room specimen without a pedigree? Should I hide him as Tobias did, in the attic, and hope for a kinder time? Should I seek to protect him by surreptitiously introducing him into a cemetery vault? Well, you see the problem.

And it is a burden, too. I realize it more as I get older, and I know, now, why Tobias the lawyer left that unrecorded legacy in his attic. What else could he do? Most people don't look at these things in the same way, and it's just as well they don't. Otherwise we would be like certain Indian tribes who had to move the cemetery with them when they migrated. The attitude is easier to catch than you think. I know two men who have moved dead wives.

Generally I can't refuse skulls that are offered to me. It is not that I am morbid, or a true collector, or that I need many of them in my work. It is just that in most cases, people being what they are, I know the skulls are safer with me. Call it a kind of respect for the bones, ingrained through long habit. That, I guess, is the reason I keep those two locked in the filing cabinet —they are delicate, and not in a position to defend themselves. So I look out for them. I'd do as much for you.

EIGHT

The Relic Men

I

S O THE REPORTERS follow you into that place without roads and they say: "Give us the story, Doctor. Give us the dope." Overhead a turkey vulture spirals slowly out of sight on an updraft of air.

You look everywhere—at your shoes and the poised, eager pencils—and you say, clearing your throat a little, "A remarkable discovery, gentlemen. These remains constitute a new species from the terminal Pleistocene fauna. The bones are associated with human artifacts. Such remarkable preservation is rare. It is our belief—"

"No, no, Doctor," they protest. "Give us some human interest. Never mind that Pleistocene business. Tell us how many years. And they're broken. How did they get broken? Maybe there was a fight, huh?"

"Human interest, why, uh, yes, human interest," you counter, thinking in the back of your mind, "Look, I'm human, too, and this is my interest." But that doesn't count, you know that. Haven't there been these articles, "We Can't Hear You, Professor"? Brace up, now, this is the press.

"What's the angle, Professor? Who found it?"

You try once more. "The site has been excavated by the State University, which I represent. The site was called to our attention by—" A warm feeling suffuses you. You point. A shabby little man with a tobacco-stained mustache stands at the edge of the trench, peering in. He looks pleased and wondering. "Mr. Johnson, there, he found it. He's been looking for years—"

"Okay, Professor. And you came all the way out here for this thing? Must have been important, all right. How are you going to put it all back together? You fellows always do that from a single bone, don't you?"

"No, we don't. It depends," you protest wearily, but the

man isn't listening. His pencil is busy. Then he turns and taps you with it.

"We might take your picture with that bone, Doctor. It's about the only way to get any human interest into a science like this. We've got to show people what it's all about. If they don't see it, they're not interested. I'll send Ed over for the camera."

The car makes a little splash of sound in the wide prairie silence. The silence flows back minute by minute, the High Plains silence that has swallowed a quarter of your lifetime. You sigh, and your knees feel unaccountably weak. You sit on the edge of the trench and press your hands into the warm soil. The thing is out of your hands now. That reporter—a nice young fellow—knows what he's after. Odd what people are interested in and what they make of it. I remember—

II

I remember the sound of the wind in that country never stopped. I think everyone there was a little mad because of it. In the end I suppose I was like all the rest. It was a country of topsyturvy, where great dunes of sand blew slowly over ranch houses and swallowed them, and where, after the sand had all blown away from under your feet, the beautiful arrowheads of ice-age hunters lay mingled with old whisky bottles that the sun had worked upon. I suppose, now that I stop to think about it, that if there is any place in the world where a man might fall in love with a petrified woman, that may be the place.

In the proper books, you understand, there is no such thing as a petrified woman, and I insist that when I first came to that place I would have said the same. It all happened because bone hunters are listeners. They have to be.

We had had terrible luck that season. We had made queries in a score of towns and tramped as many canyons. The institution for which we worked had received a total of one Oligocene turtle and a bag of rhinoceros bones. A rag picker could have done better. The luck had to change. Somewhere there had to be fossils.

I was cogitating on the problem under a coating of lather in

a barbershop with an 1890 chair when I became aware of a
voice. You can hear a lot of odd conversation in barbershops,
particularly in the back country, and particularly if your trade
makes you a listener, as mine does. But what caught my ear at
first was something about stone. Stone and bone are pretty
close in my language and I wasn't missing any bets. There was
always a chance that there might be a bone in it somewhere
for me.

The voice went off into a grumbling rural complaint in the
back corner of the shop, and then it rose higher.

"It's petrified! It's petrified!" the voice contended excitedly.

I managed to push an ear up through the lather.

"I'm a-tellin' ya," the man boomed, "a petrified woman,
right out in that canyon. But he won't show it, not to nobody.
'Tain't fair, I tell ya."

"Mister," I said, speaking warily between the barber's razor
and his thumb, "I'm reckoned a kind of specialist in these
matters. Where is this woman, and how do you know she's
petrified?"

I knew perfectly well she wasn't, of course. Flesh doesn't
petrify like wood or bone, but there are plenty of people who
think it does. In the course of my life I've been offered objects
that ranged from petrified butterflies to a gentleman's top hat.

Just the same, I was still interested in this woman. You can
never tell what will turn up in the back country. Once I had a
mammoth vertebra handed to me with the explanation that it
was a petrified griddle cake. Mentally, now, I was trying to
shape that woman's figure into the likeness of a mastodon's
femur. This is a hard thing to do when you are young and far
from the cities. Nevertheless, I managed it. I held that shining
bony vision in my head and asked directions of my friend in
the barbershop.

Yes, he told me, the woman was petrified all right. Old man
Buzby wasn't a feller to say it if it 'tweren't so. And it weren't
no part of a woman. It was a *whole* woman. Buzby had said
that, too. But Buzby was a queer one. An old bachelor, you
know. And when the boys had wanted to see it, 'count of it
bein' a sort of marvel around these parts, the old man had
clammed up on where it was. A-keepin' it all to hisself, he was.
But seein' as I was interested in these things and a stranger, he

might talk to me and no harm done. It was the trail to the right and out and up to the overhang of the hills. A little tar-papered shack there.

I asked Mack to go up there with me. He was silent company but one of the best bone hunters we had. Whether it was a rodent the size of a bee or an elephant the size of a house, he'd find it and he'd get it out, even if it meant that we carried a five-hundred-pound plaster cast on foot over a mountain range.

In a day we reached the place. When I got out of the car I knew the wind had been blowing there since time began. There was a rusty pump in the yard and rusty wire and rusty machines nestled in the lee of a wind-carved butte. Everything was leaching and blowing away by degrees, even the tarpaper on the roof.

Out of the door came Buzby. He was not blowing away, I thought at first. His farm might be, but he wasn't. There was an air of faded dignity about him.

Now in that country there is a sort of etiquette. You don't drive out to a man's place, a bachelor's, and you a stranger, and come up to his door and say: "I heard in town you got a petri-fied woman here, and brother, I sure would like to see it." You've got to use tact, same as anywhere else.

You get out slowly while the starved hounds look you over and get their barking done. You fumble for your pipe and explain casually you're doin' a little lookin' around in the hills. About that time they get a glimpse of the equipment you're carrying and most of them jump to the conclusion that you're scouting for oil. You can see the hope flame up in their eyes and sink down again as you explain you're just hunting bones. Some of them don't believe you after that. It's a hard thing to murder a poor man's dream.

But Buzby wasn't the type. I don't think he even thought of the oil. He was small and neat and wore—I swear it—pince-nez glasses. I could see at a glance that he was a city man dropped, like a seed, by the wind. He had been there a long time, certainly. He knew the corn talk and the heat talk, but he would never learn how to come forward in that secure, heavy-shouldered country way, to lean on a car door and talk to strangers while the horizon stayed in his eyes.

He invited us, instead, to see his collection of arrowheads. It looked like a good start. We dusted ourselves and followed him in. It was a two-room shack, and about as comfortable as a monk's cell. It was neat, though, so neat you knew that the man lived, rather than slept there. It lacked the hound-asleep-in-the-bunk confusion of the usual back-country bachelor's quarters.

He was precise about his Indian relics as he was precise about everything, but I sensed after a while a touch of pathos —the pathos of a man clinging to order in a world where the wind changed the landscape before morning, and not even a dog could help you contain the loneliness of your days.

"Someone told me in town you might have a wonderful fossil up here," I finally ventured, poking in his box of arrowheads, and watching the shy, tense face behind the glasses.

"That would be Ned Burner," he said. "He talks too much."

"I'd like to see it," I said, carefully avoiding the word *woman*. "It might be something of great value to science."

He flushed angrily. In the pause I could hear the wind beating at the tarpaper.

"I don't want any of 'em hereabouts to see it," he cried passionately. "They'll laugh and they'll break it and it'll be gone like—like everything." He stopped, uncertainly aware of his own violence, his dark eyes widening with pain.

"We are scientists, Mr. Buzby," I urged gently. "We're not here to break anything. We don't have to tell Ned Burner what we see."

He seemed a little mollified at this, then a doubt struck him. "But you'd want to take her away, put her in a museum."

I noticed the pronoun but ignored it. "Mr. Buzby," I said, "we would very much like to see your discovery. It may be we can tell you more about it that you'd like to know. It might be that a museum would help you save it from vandals. I'll leave it to you. If you say no, we won't touch it, and we won't talk about it in the town, either. That's fair enough, isn't it?"

I could see him hesitating. It was plain that he wanted to show us, but the prospect was half-frightening. Oddly enough, I had the feeling his fright revolved around his discovery, more than fear of the townspeople. As he talked on, I began to see what he wanted. He intended to show it to us in the hope we would confirm his belief that it was a petrified woman. The whole thing seemed to have taken on a tremendous importance in his mind. At that point, I couldn't fathom his reasons.

Anyhow, he had something. At the back of the house we found the skull of a big, long-horned, extinct bison hung up under the eaves. It was a nice find, and we coveted it.

"It needs a dose of alvar for preservation," I said. "The museum would be the place for a fine specimen like this. It will just go slowly to pieces here."

Buzby was not unattentive. "Maybe, Doctor, maybe. But I have to think. Why don't you camp here tonight? In the morning—"

"Yes?" I said, trying to keep the eagerness out of my voice. "You think we might—?"

"No! Well, yes, all right. But the conditions? They're like you said?"

"Certainly," I answered. "It's very kind of you."

He hardly heard me. That glaze of pain passed over his face once more. He turned and went into the house without speaking. We did not see him again until morning.

The wind goes down into those canyons also. It starts on the flats and rises through them with weird noises, flaking and blasting at every loose stone or leaning pinnacle. It scrapes the sand away from pipy concretions till they stand out like strange distorted sculptures. It leaves great stones teetering on wineglass stems.

I began to suspect what we would find, the moment I came there. Buzby hurried on ahead now, eager and panting. Once he had given his consent and started, he seemed in almost a frenzy of haste.

Well, it was the usual thing. Up. Down. Up. Over boulders and splintered deadfalls of timber. Higher and higher into the back country. Toward the last he outran us, and I couldn't hear what he was saying. The wind whipped it away.

But there he stood, finally, at a niche under the canyon wall. He had his hat off and, for a moment, was oblivious to us. He might almost have been praying. Anyhow I stood back and waited for Mack to catch up. "This must be it," I said to him. "Watch yourself." Then we stepped forward.

It was a concretion, of course—an oddly shaped lump of

mineral matter—just as I had figured after seeing the wind at work in those miles of canyon. It wasn't a bad job, at that. There were some bumps in the right places, and a few marks that might be the face, if your imagination was strong. Mine wasn't just then. I had spent a day building a petrified woman into a mastodon femur, and now that was no good either, so I just stood and looked.

But after the first glance it was Buzby I watched. The un-skilled eye can build marvels of form where the educated see nothing. I thought of that bison skull under his eaves, and how badly we needed it.

He didn't wait for me to speak. He blurted with a terrible intensity that embarrassed me, "She—she's beautiful, isn't she?"

"It's remarkable," I said. "Quite remarkable." And then I just stood there not knowing what to do.

He seized on my words with such painful hope that Mack backed off and started looking for fossils in places where he knew perfectly well there weren't any.

I didn't catch it all; I couldn't possibly. The words came out in a long, aching torrent, the torrent dammed up for years in the heart of a man not meant for this place, nor for the wind at night by the windows, nor the empty bed, nor the neighbors twenty miles away. You're tough at first. He must have been to stick there. And then suddenly you're old. You're old and you're beaten, and there must be something to talk to and to love. And if you haven't got it you'll make it in your head, or out of a stone in a canyon wall.

He had found her, and he had a myth of how she came there, and now he came up and talked to her in the long after-noon heat while the dust devils danced in his failing corn. It was progressive. I saw the symptoms. In another year, she would be talking to him.

"It's true, isn't it, Doctor?" he asked me, looking up with that rapt face, after kneeling beside the niche. "You can see it's her. You can see it plain as day." For the life of me I couldn't see anything except a red scar writhing on the brain of a living man who must have loved somebody once, beyond words and reason.

"Now Mr. Buzby," I started to say then, and Mack came up

and looked at me. This, in general, is when you launch into a careful explanation of how concretions are made so that the layman will not make the same mistake again. Mack just stood there looking at me in that stolid way of his. I couldn't go on with it. I couldn't even say it.

But I saw where this was going to end. I saw it suddenly and too late. I opened my mouth while Mr. Buzby held his hands and tried to regain his composure. I opened my mouth and I lied in a way to damn me forever in the halls of science.

I lied, looking across at Mack, and I could feel myself getting redder every moment. It was a stupendous, a colossal lie. "Mr. Buzby," I said, "that—um—er—figure is astonishing. It is a remarkable case of preservation. We must have it for the museum."

The light in his face was beautiful. He believed me now. He believed himself. He came up to the niche again, and touched her lovingly.

"It's okay," I whispered to Mack. "We won't have to pack the thing out. He'll never give her up."

That's where I was a fool. He came up to me, his eyes troubled and unsure, but very patient.

"I think you're right, Doctor," he said. "It's selfish of me. She'll be safer with you. If she stays here somebody will smash her. I'm not well." He sat down on a rock and wiped his forehead. "I'm sure I'm not well. I'm sure she'll be safer with you. Only I don't want her in a glass case where people can stare at her. If you can promise that, I—"

"I can promise that," I said, meeting Mack's eyes across Buzby's shoulder.

"And if I come there I can see her!"

I knew I would never meet him again in this life.

"Yes," I said, "you can see her there." I waited, and then I said, "We'll get the picks and plaster ready. Now that bison skull at your house . . ."

It was two days later, in the truck, that Mack spoke to me. "Doc."

"Yeah."

"You know what the Old Man is going to say about shipping that concretion. It's heavy. Must be three hundred pounds with the plaster."

"Yes, I know."

Mack was pulling up slow along the abutment of a bridge. It was the canyon of the big Piney, a hundred miles away. He got out and went to the rear of the truck. I didn't say anything, but I followed him back.

"Doc, give me a hand with this, will you?"

I took one end, and we heaved together. It's a long drop in the big Piney. I didn't look, but I heard it break on the stones.

"I wish I hadn't done that," I said.

"It was only a concretion," Mack answered. "The old geezer won't know."

"I don't like it," I said. "Another week in that wind and I'd have believed in her myself. Get me the hell out of here—maybe I do, anyhow. I tell you I don't like it. I don't like it at all."

"It's a hundred more to Valentine," Mack said.

He put the map in the car pocket and slid over and gave me the wheel.

III

The devil, in the eyes of many devout Fundamentalists, occupies the bone lands, the waste places at the margins of everyday existence. I never knew but one who thought differently and who considered that God, too, might have something to do with the edges of the known.

Most men proceed under a burden of fear, and the majority of them in my youth could not stand the sight of a bone hunter. He was associated in their minds with the search for the missing link. At the very least he was apt to appear triumphantly waving the thigh bone of some creature not mentioned in Holy Writ. His ways were queer ways, and he suffered for them. At best he was regarded with tolerant amusement; at the worst, he would have screen doors latched against him.

It was that way in the Valley of the Pumpkin Seed. To compound our misery, we learned at the local newspaper office that the only impressive bone in the whole county—a veritable giant of a bone—was owned by a devout member of a sect which lent no ear to modern geological heresies. This word was carried on lagging and pessimistic feet to the Director.

"Well," he said, kicking dubiously at a loose rock in the roadway, "there's nothing to do but try. He must have found it somewhere around here. What do you suppose he saved it for—a chap like that? Most of them would have broken it up for lime fertilizer. Something curious there. Let's go and ask him. I'll go myself."

There was a road up through the outliers of the Wildcat hills, and by and by, a gate. We descended self-consciously while a horde of screeching children broke and ran for the house. It was a soddy. Forty years ago you could still see them sometimes on the Pumpkin Seed.

First there was the woman with the children peering from behind her skirts. She didn't latch the door, but she stood there with about the same stance that I remember my grandmother used to assume when she described the time she chased some Pawnee out of the hayfield.

We stopped at a respectful distance. "Good morning, ma'am," we said. "Would the Mister be about now? We would like to see that fossil bone he's got. The one people think is an elephant's bone. We heard about it in town. We're from the University."

The woman's expression did not change. There was silence for a moment, while we fingered our hats uneasily. Then she lifted her voice. "Pa," she called uncertainly. "There's some men in the yard. Relic men, I reckon."

Mr. Mullens's moustache was straw yellow, and it contrasted oddly with his face, which was blue-tinged because of an aggravated heart condition. The moment I saw him I felt a great sense of relief. He had the merry, wondering, wandering eye of the born naturalist. Poor, uneducated, reared in a sect which frowns upon natural science, some wind out of the Pliocene had touched him. He threw wide the door and made a great, sweeping gesture that propelled us almost bodily into the room.

"Boys," he chuckled, "I been a-waitin' for ye. I knew ye'd come to see the bone. Didn't I tell ye, Carrie? Didn't I tell ye they'd be here?" Carrie, her fortress taken, retired with a weary air into the kitchen.

"Now boys," said Mr. Mullens, "we'll git to business. That bone ain't all, not by a jugful. There's more where that come from." He breathed heavily with the excitement of the moment. His lips turned a little more blue.

"There it is, boys." He swept the parlor curtain aside. A great brown bone, seamed with the weather onslaughts of more than a million years, stood upright on a pedestal behind the curtain.

"My god," said the Director, "there's no mistaking that—it's elephant." He used the term generically, as the bone men always do, to cover a score of species of vanished mammoth and mastodon.

"That's what I told 'em," said Mr. Mullens proudly. "I told 'em it were too big for a cow, or a horse or anythin' in these parts. It's an elyphant."

His breathing began to labor like something whispering apart from him. "I found it in the hills out there, just twenty years ago. I fixed it, too, brought it home and poured glue in her to harden. Been keepin' her ever since."

"Twenty years," murmured the Director dejectedly. I knew what was in his mind. "Look, Mr. Mullens, do you think you can remember how to get there?"

"Can I remember how to get there?" said Mr. Mullens with proper contempt for our lack of faith. "Listen to the man. I been over these hills for fifty years! I'll take ye there myself."

The breathing began to whisper behind his words again. "I tell you this ain't all. I saw bones all over that valley, scattered up and down a-shinin' in the sun. I tell ye," he paused to gather strength for his intellectual effort, "I got a notion about it.

"I figure it's a place where all the garbage from the Flood— them big animals and pore human sinners—all got carried along and dropped when the water went back where it came from. The sinners and the elyphants all a-tossin' and a-rollin' and a-grindin' together till the good Lord saw fit to lay 'em down.

"It's the garbage from the biggest Flood of all, lads—come right down to our valley. And that there thigh bone I picked up off the hillside. The marvel of it struck me. I was scared green, but I wanted it. I wanted to bring it home. I ain't never been back and I ain't never told nobody. But it's all there, boys—I know it's there, an' I'll show ye. I'm not what you'd call spry, but I'll show ye. I got a hankerin'—a hankerin' to see it once more—that hill and that valley and them big bones all a-shinin' in the sun."

I don't know whether anybody said "Amen," but I knew what we were all thinking: Could the old man deliver? Ninety-nine out of every hundred of these stories are spun out of thin air. It is the hundredth chance that the bone hunter plays for. That means, really, that he cannot afford to neglect anybody's story; the worst of them may contain a germ of truth or, at the very least, a bone. But a quarry is the thing, a real fossil quarry. They don't come easy. There are only a few in the world.

It was in this frame of mind that we started with Dad Mullens. We were warmed by his unexpected friendliness, but as for his tall tale, that was another matter entirely. I didn't believe a word of it, and I knew that the Director, as befitted a seasoned veteran, believed even less. All we hoped was that the old boy might just possibly remember and be able to guide us to the spot where he had found that huge thigh bone. The big, brown femur, the color of oakwood, was real enough, whatever else may have ballooned out of Mr. Mullens's obviously powerful gift for words.

We drove across sand banks and up dry arroyos; we rose and fell over short grass prairie. We lowered fence wire and intruded on the privacy of range cattle so remote from civilization that they never went to market. We came at last to a barrier of stony hills.

"What do we do now, Dad?" some one asked. The old man stood upright beside me. He peered at those hills like a Spaniard pursuing the seven golden cities. "It's somewhere hereabouts," he called to us. "But I jist can't seem to recollect quite where. I think we'd better press on a piece back into the hills. Just leave the cars here, boys. They won't go no further."

There were five hills, there were ten hills, and at the last it may have been twenty-five. We scaled them all, you see. Our Stout Cortez had to look for an appropriate point, a commanding view. I tell you he scared the daylights out of me. He wheezed and choked and he turned bluer and blacker, but he kept picking bigger hills to climb.

"Dad," I croaked, "for the love of God, you're killing yourself. We can't go on like this. You've just forgotten in twenty years, that's all. There's nothing to be ashamed of. Let's just go back now and forget it. We believe you, all right. We know you saw it. But maybe the good Lord didn't aim for you to see it again."

"Hallelujah!" the old man bellowed, ignoring me. "This is the place."

There was a late afternoon sun playing on that hillside, and I can remember still the way my eyes traveled down it from boulder to gray boulder between the spines of Spanish bayonet. And then I saw it. Maybe this won't mean anything to you. Maybe you don't understand this game, or why men follow it. But I saw it. I tell you I saw five million years of the planet's history lying there on that hillside with the yucca growing over it and the roots working through it, just the way the old man had remembered it from a day long ago in the sun.

I saw the ivory from the tusks of elephants scattered like broken china that the rain has washed. I saw the splintered, mineralized enamel of huge unknown teeth. I paused over the bones of ferocious bear-dog carnivores. I saw, protruding from an eroding gully, the jaw of a shovel-tusked amebelodont that has been gone twice a million years into the night of geologic time. I tell you I saw it with my own eyes and I knew, even as I looked at it, that I would never see anything like it again.

The old man stood there muttering about the Flood and seeing, no doubt, his own fearful visions. And while he breathed and whispered, and we waited for the others to explore, I wandered aimlessly as a daisy picker along the hillsides. I touched and picked up and dropped again. I stuffed my pockets and cleaned them out in indecision. We were rich at last—as the bone hunter reckons riches.

The fragments on that weathered hillside were only a hint of the undisturbed treasures that lay in those slowly eroding strata. The place must have represented the shallows of some great Pliocene lake. Into it, over millennia, had washed the bones of hunter and hunted, themselves drifting in the vaster eddies of time. The sabretooth cats had paced there. In the shallows, the shovel-tuskers had scooped and crunched strange water lilies. Now the desert hills lay over their watery kingdom. The bark of the bone hunters' dynamite would disturb their sleep.

A couple of years later we again drove up the path to the old man's shanty. The kids were still there by the gate—a little bigger, but not much. They recognized us and turned and broke for the house.

"Ma," I heard them yell, "Ma, Ma, the relic men are here again." This time they screeched in expectant enthusiasm.

The old man was not at the door to greet us. The great brown bone still stood on its pedestal in the parlor, but its discoverer was gone. I looked at Carrie mutely. She shook her head. I touched the massive relic. Its owner was safe with Noah now, beyond the waters of the greatest flood of all.

As I went down the hill in the gathering darkness, I shivered. Old Mullens had lived in a small, tight world of marvels, and they had lasted him till the end. Never by word or deed had we intruded upon his beliefs.

A great river of stars spilled southward over the low hills, and a cold wind began to race me onward. Bone hunters were lonely people, I thought briefly, as I turned on the car heat for comfort. It had something to do with time. Perhaps, in the end, we did not know where we belonged.

Perhaps . . . Take that young reporter, for instance; he knew where he belonged, and what to do with human interest. He was tapping me again with his pencil. "All right, Doctor, we're all set. Ed's come back with the camera."

"Never mind the big words, never mind the names of little people. Stand right there and point. It's the best sort of human interest, you and that bone. Just keep pointing. That's it—keep pointing, and we'll have something. Good-by now. Good-by."

NINE

Strangeness in the Proportion

"I MAY TRULY SAY," wrote Sir Francis Bacon, in the time of his tragic fall in 1621, "my soul hath been a stranger in the course of my pilgrimage. I seem to have my conversation among the ancients more than among those with whom I live." I suppose, in essence, this is the story of every man who thinks, though there are centuries when such thought grows painfully intense, as in our own. Bacon's contemporary Shakespeare also speaks of it from the shadows when he says:

> "Sir, in my heart there was a kinde of fighting,
>   That would not let me sleepe."

In one of those strange, elusive stories upon which Walter de la Mare exerted all the powers of his marvelous poetic gift, a traveler musing over the quaint epitaphs in a country cemetery suddenly grows aware of the cold on a bleak hillside, of the onset of a winter evening, of the miles he has yet to travel, of the solitude he faces. He turns to go and is suddenly confronted by a man who has appeared from no place our traveler can discover, and who has about him, though he is clothed in human garb and form, an unearthly air of difference. The stranger, who appears to be holding a forked twig like that which diviners use, asks of our traveler, the road. "Which," he queries, "is the way?"

The mundane, though sensitive, traveler indicates the high road to town. The stranger, with a look of revulsion upon his face, almost as though it flowed from some secret information transmitted by the forked twig he clutches, recoils in horror. The way—the human way—that the traveler indicates to him is obviously not his way. The stranger has wandered, perhaps like Bacon, out of some more celestial pathway.

When the traveler turns from giving directions, the stranger has gone, not necessarily supernaturally, for de la Mare is careful to move within the realm of the possible, but in a manner that leaves us suddenly tormented with the notion that our

road, the road to town, the road of everyday life, has been rejected by a person of divinatory powers who sees in it some
disaster not anticipated by ourselves. Suddenly in this magical
and evocative winter landscape, the reader asks himself with an
equal start of terror, "What *is* the way?" The road we have
taken for granted is now filled with the shadowy menace and
the anguished revulsion of that supernatural being who exists
in all of us. A weird country tale—a ghost story if you will—has
made us tremble before our human destiny.

Unlike the creatures who move within visible nature and are
indeed shaped by that nature, man resembles the changeling of
medieval fairy tales. He has suffered an exchange in the safe
cradle of nature, for his earlier instinctive self. He is now susceptible, in the words of theologians, to unnatural desires. Equally, in
the view of the evolutionist, he is subject to indefinite departure,
but his destination is written in no decipherable tongue.

For in man, by contrast with the animal, two streams of evolution have met and merged: the biological and the cultural.
The two streams are not always mutually compatible. Sometimes
they break tumultuously against each other so that, to a degree
not experienced by any other creature, man is dragged hither
and thither, at one moment by the blind instincts of the forest, at
the next by the strange intuitions of a higher self whose rationale
he doubts and does not understand. He is capable of murder
without conscience. He has denied himself thrice over, and is as
familiar as Judas with the thirty pieces of silver.

He has come part way into an intangible realm determined
by his own dreams. Even the dreams he doubts because they
are not fanged and clawed like the life he sees about him. He is
tormented, and torments. He loves, and sees his love cruelly
rejected by his fellows. Far more than the double evolutionary
creatures seen floundering on makeshift flippers from one medium to another, man is marred, transitory, and imperfect.

Man's isolation is even more terrifying if he looks about at
his fellow creatures and searches for signs of intelligence behind the universe. As Francis Bacon saw, "all things . . . are
full of panic terrors; human things most of all; so infinitely
tossed and troubled as they are with superstition (which is in
truth nothing but a panic terror) especially in seasons of hard-
ship, anxiety, and adversity."

Unaided, science has little power over human destiny save in a purely exterior and mechanical way. The beacon light of truth, as Hawthorne somewhere remarks, is often surrounded by the flapping wings of ungainly night birds drawn as unerringly as moths toward candlelight. Man's predicament is augmented by the fact that he is alone in the universe. He is locked in a single peculiar body; he can compare observations with no other form of life.

He knows that every step he takes can lead him into some unexplored region from which he may never return. Each individual among us, haunted by memory, reveals this sense of fear. We cling to old photographs and letters because they comfort our intangible need for location in time. For this need of our nature science offers cold comfort. To recognize this, however, is not to belittle the role of science in our world. In his enthusiasm for a new magic, modern man has gone far in assigning to science—his own intellectual invention—a role of omnipotence not inherent in the invention itself. Bacon envisioned science as a powerful and enlightened servant—but never the master—of man.

One of the things which must ever be remembered about Francis Bacon and the depth of his prophetic insight is that it remains, by the nature of his time, in a sense paradoxical. Bacon was one of the first time-conscious moderns. He felt on his brow as did no other man—even men more skilled in the devising of experiment—the wind of the oncoming future, those far-off airs blowing, as he put it in the language of the voyagers, "from the new continent." Ironically, neither king, lawyer, nor scientist could tolerate Bacon's vision of the oncoming future. Because William Harvey was a scientist whose reputation has grown with the years, he is sometimes quoted by scholars even today as demonstrating that Bacon was a literary man who need not be taken seriously by historians of science.

That Bacon was a writer of great powers no one who has read his work would deny. He exercised, in fact, a profound stylistic influence both upon English writers who followed him and upon the scientists of the Royal Society. To say, for this reason, that he is of no scientific significance is to miss his importance as a statesman and philosopher of science as well as to deny to the scientist himself any greater role in discovery than the casual assemblage of facts. Harvey's attitude serves only to illustrate that great experimental scientists are not necessarily equally great philosophers, and that there may be realms denied to them. Similar able but particulate scientists, it could easily be pointed out, wrote disparagingly of Darwin in his time.

The great synthesizer who alters the outlook of a generation, who suddenly produces a kaleidoscopic change in our vision of the world, is apt to be the most envied, feared, and hated man among his contemporaries. Almost by instinct they feel in him the seed of a new order; they sense, even as they anathematize him, the passing away of the sane, substantial world they have long inhabited. Such a man is a kind of lens or gathering point through which past thought gathers, is reorganized, and radiates outward again into new forms.

"There are . . . minds," Emerson once remarked, "that deposit their dangerous unripe thoughts here and there to lie still for a time and be brooded in other minds, and the shell not to be broken until the next age, for them to begin, as new individuals, their career." Francis Bacon was such a man, and it is perhaps for this very reason that there has been visited upon him, by both moralist and scientist alike, so much misplaced vituperation and rejection.

He has been criticized, almost in the same breath, as being falsely termed a scientist and on the other hand as being responsible for all the technological evils from which we suffer in the modern age. His vision was, to a degree, paradoxical. The reason lies in the fact that even the great visionary thinker never completely escapes his own age or the limitations it imposes upon him. Thus Bacon, the weather-tester who held up a finger to the winds of time, was trapped in an age still essentially almost static in its ideas of human duration and in the age and size of man's universe.

A man of the Renaissance, Bacon, for all his cynicism and knowledge of human frailty, still believed in man. He argues well and lucidly that to begin with doubt is, scientifically, to end in certainty, while to begin in certainty is to end in doubt. He failed to see that science, the doubter, might end in metaphysical doubt itself—doubt of the rationality of the universe, doubt as to the improvability of man. Today the "great machine" that

Bacon so well visualized, rolls on, uncontrolled and infinitely devastating, shaking the lives of people in the remote jungles of the Congo as it torments equally the hearts of civilized men.

It is evident from his *New Atlantis* (1624), the Utopian fragment begun toward the end of his life and left unfinished, that this attempt to picture for humanity the state it might attain under science and just rule retains a certain static quality. Bacon is sure about the scientific achievements of his ideal state, but, after all, his pictured paradise is an island without population problems, though medicine there is apparently a high art. Moreover, like most of the Utopias of this period, it is hidden away from the corrupting influence of the world. It is an ideal and moving presentation of men going about their affairs under noble and uplifting circumstances. It is, as someone has remarked, "ourselves made perfect."

But as to how this perfection is to come into being, Bacon is obscure. It is obvious that the wise men of *The New Atlantis* must keep their people from the debasing examples of human behavior in the world outside. Bacon, in other words, has found it easier to picture the growth of what he has termed "experiments of fruit" than to establish the reality of a breed of men worthy to enjoy them. Even the New Atlantis has had to remain armed and hidden, like Elizabethan England behind its sea fogs.

The New Atlantis cannot be read for solutions to the endless permutations and combinations of cultural change, the opened doorway through which Bacon and his followers have thrust us, and through which there is no return. To Bacon all possible forms of knowledge of the world might be accumulated in a few scant generations. With education the clouded mirror of the mind might be cleansed. "It is true," he admitted in an earlier work, *Valerius Terminus*, "that there is a limitation rather potential than actual which is when the effect is possible, but the time or place yieldeth not matter or basis whereupon man should work."

In this statement we see the modern side of Bacon's mind estimating the play of chance and time. We see it again a few pages later when, in dealing with the logical aspects of contingency, he writes, "our purpose is not to stir up men's hopes but to guide their travels." "Liberty," he continues, speaking in a scientific sense, "is when the direction is not restrained to some definite means, but comprehendeth all the means and ways possible." For want of a variety of scientific choice, he is attempting to say, you may be prevented from achieving a scientific good, some desirable direction down which humanity might travel. The bewildering multiplicity of such roads, the recalcitrance of even educated choice, are not solely to be blamed on Bacon's four-hundred-year gap in experience. In fact, it could readily be contended that science, as he intended to practice it, has not been practiced at all.

Although he has been hailed with some justice as the prophet of industrial science, it is often forgotten that he wished from the beginning to press forward on all scientific fronts at once, instead of pursuing the piecemeal emergence of the various disciplines in the fashion in which investigation was actually carried out. Three centuries have been consumed in establishing certain anthropological facts that he asserted from the beginning. He distinguished cultural and environmental influences completely from the racial factors with which they have been confused down to this day. He advocated the careful study and emulation of the heights of human achievement. Today scientific studies of "creativity" and the conditions governing the release of such energies in the human psyche are just beginning to be made. He believed and emphasized that it was within man's latent power to draw out of nature, as he puts it, "a second world."

It is here, however, that we come back upon that place of numerous crossroads where man has lifted the lantern of his intellect hopefully to many ambiguous if not treacherous sign posts. There is, we know now to our sorrow, more than one world to be drawn out of nature. When once drawn, like some irreplaceable card in a great game, that world leads on to others. Bacon's "second world" becomes a multiplying forest of worlds in which man's ability to choose is subdued to frightened day-to-day decisions.

One thing, however, becomes ever more apparent: the worlds drawn out of nature are human worlds, and their imperfections stem essentially from human inability to choose intelligently among those contingent and intertwined roads which Bacon hoped would enhance our chances of making a proper and intelligent choice. Instead of regarding man as a corresponding problem, as Bacon's insight suggested, we chose, instead, to concentrate upon that natural world which he truthfully held to be protean, malleable, and capable of human guidance. Although worlds can be drawn out of that maelstrom, they do not always serve the individual imprisoned within the substance of things.

I have often had occasion to comment on the insights of D'Arcy Thompson, the late renowned British naturalist. He saw, long after, in 1897, that with the coming of industrial man, contingency itself is subjected to a kind of increasing tempo of evolution. The simplicity of the rural village of Shakespeare's day, or even the complex but stabilized and harmonious life of a very ancient civilization, is destroyed in the dissonance of excessive and rapid change. "Strike a new note," said Thompson, "import a foreign element to work and a new orbit, and the one accident gives birth to a myriad. Change, in short, breeds change, and chance—chance. We see indeed a sort of *evolution* of chance, an ever-increasing complexity of accident and possibilities. One wave started at the beginning of eternity breaks into component waves, and at once the theory of interference begins to operate." This evolution of chance is not contained within the human domain. Arising within the human orbit it is reflected back into the natural world where man's industrial wastes and destructive experiments increasingly disrupt and unbalance the world of living nature.

Bacon shared in some part with his age a belief in the biform nature of the worldly universe. "There is no nature," he says, "which can be regarded as simple; every one seeming to participate and be compounded of two." Man has something of the brute; the brute has something of the vegetable, the vegetable something of the inanimate body; and so, Bacon emphasizes, "all things are in truth biformed and made up of a higher species and a lower." Strange though it may seem, in this respect Bacon, though existing on the brief Elizabethan stage of a short-term universe, was perhaps better prepared for the protean writhings of external nature and the variability manifest in the interior world of thought than many a specialist in the physical and biological sciences who would follow him.

Patrick Cruttwell, in his study of Shakespeare, comments on

how frequently war within the individual, a sense of divided personality, is widespread in the spirit of that age, as it also is in ours:

> "Within my soul there doth conduce a fight
> Of this strange nature, that a thing inseparate,
> Divides more widely than the skie and earth."

How much more we would see, I sometimes think, if the world were lit solely by lightning flashes from the Elizabethan stage. What miraculous insights and perceptions might our senses be trained to receive amidst the alternate crash of thunder and the hurtling force that give a peculiar and momentary shine to an old tree on a wet night. Our world might be transformed interiorly from its staid arrangement of laws and uniformity of expression into one where the unexpected and blinding illumination constituted our faith in reality.

Nor is such a world as incredible as it seems. Physicists, it now appears, are convinced that a principle of uncertainty exists in the submicroscopic realm of particles and that out of this queer domain of accident and impact has emerged, by some kind of mathematical magic, the sustaining world of natural law by which we make our way to the bank, the theater, to our homes, and finally to our graves. Perhaps, after all, a world so created has something still wild and unpredictable lurking behind its more sober manifestations. It is my contention that this is true, and that the rare freedom of the particle to do what most particles never do is duplicated in the solitary universe of the human mind.

The lightning flashes, the smashed circuits through which, on occasion, leaps the light of universes beyond our ken, exist only in rare individuals. But the flashes from such minds can fascinate and light up through the arts of communication the intellects of those not necessarily endowed with genius. In a conformist age science must, for this reason, be wary of its own authority. The individual must be re-created in the light of a revivified humanism which sets the value of man the unique against that vast and ominous shadow of man the composite, the predictable, which is the delight of the machine. The polity we desire is that ever-creative polity which Robert Louis Stevenson had in mind when he spoke of each person as containing a group of incongruous and ofttimes conflicting citizenry. Bacon himself was seeking the road by which the human mind might be opened to the full image of the world, not reduced to the little compass of a state-manipulated machine.

It is through the individual brain alone that there passes the momentary illumination in which a whole human countryside may be transmuted in an instant. "A steep and unaccountable transition," Thoreau has described it, "from what is called a common sense view of things, to an infinitely expanded and liberating one, from seeing things as men describe them, to seeing them as men cannot describe them." Man's mind, like the expanding universe itself, is engaged in pouring over limitless horizons. At its heights of genius it betrays all the miraculous unexpectedness which we try vainly to eliminate from the universe. The great artist, whether he be musician, painter, or poet, is known for this absolute unexpectedness. One does not see, one does not hear, until he speaks to us out of that limitless creativity which is his gift.

The flash of lightning in a single brain also flickers along the horizon of our more ordinary heads. Without that single lightning stroke in a solitary mind, however, the rest of us would never have known the fairyland of *The Tempest*, the midnight world of Dostoevsky, or the blackbirds on the yellow harvest fields of Van Gogh. We would have seen blackbirds and endured the depravity of our own hearts, but it would not be the same landscape that the act of genius transformed. The world without Shakespeare's insights is a lesser world, our griefs shut more inarticulately in upon themselves. We grow mute at the thought—just as an element seems to disappear from sunlight without Van Gogh. Yet these creations we might call particle episodes in the human universe—acts without precedent, a kind of disobedience of normality, unprophesiable by science, unduplicable by other individuals on demand. They are part of that unpredictable newness which keeps the universe from being fully explored by man.

Since this elusive "personality" of the particle may play a role in biological change and diversity, there is a way in which the mysterious world of particles may influence events within the realm of the living. It is just here, within the human domain of infinite variability and the individual act, that the role of the artist lies. Here the creative may be contrasted to the purely scientific approach to nature, although we must bear in mind that a man may be both a scientist and artist—an individual whose esthetic and humanistic interests are as much a part of his greatness in the eyes of the world as the technical skills which have brought him renown.

Ordinarily, however, there is between the two realms a basic division which has been widened in the modern world. Granted that the great scientific discoverer may experience the esthetic joy of the true artist, a substantial difference still remains. For science seeks essentially to naturalize man in the structure of predictable law and conformity, whereas the artist is interested in man the individual.

"This is your star," says science. "Accept the world we describe to you." But the escaping human mind cries out, in the words of G. K. Chesterton, "We have come to the wrong star. . . . That is what makes life at once so splendid and so strange. The true happiness is that we don't fit. We come from somewhere else. We have lost our way."

A few years ago I chanced to write a book in which I had expressed some personal views and feelings upon birds, bones, spiders, and time, all subjects with which I had some degree of acquaintance. Scarcely had the work been published when I was sought out in my office by a serious young colleague. With utter and devastating confidence he had paid me a call in order to correct my deviations and to lead me back to the proper road of scholarship. He pointed out to me the time I had wasted—time which could have been more properly expended upon my own field of scientific investigation. The young man's view of science was a narrow one, but it illustrates a conviction all too common today: namely, that the authority of science is absolute.

To those who have substituted authoritarian science for authoritarian religion, individual thought is worthless unless it is the symbol for a reality which can be seen, tasted, felt, or thought about by everyone else. Such men adhere to a dogma as rigidly as men of fanatical religiosity. They reject the world of the personal, the happy world of open, playful, or aspiring thought.

Here, indeed, we come upon a serious aspect of our discussion. For there is a widespread but totally erroneous impression that science is an unalterable and absolute system. It is supposed that other institutions change, but that science, after the discovery of the scientific method, remains adamant and inflexible in the purity of its basic outlook. This is an iron creed which is at least partly illusory. A very ill-defined thing known as the scientific method persists, but the motivations behind it have altered from century to century.

The science of the seventeenth century, as many historians have pointed out, was essentially theoretical and other-worldly. Its observations revolved largely about a world regarded as under divine control and balance. As we come into the nineteenth century, cosmic and organic evolution begin to effect a change in religious outlook. The rise of technology gave hope for a Baconian Utopia of the New Atlantis model. Problem solving became the rage of science. Today problem solving with mechanical models, even of living societies, continues to be popular. The emphasis, however, has shifted to power. From a theoretical desire to *understand* the universe, we have come to a point where it is felt we *must* understand it to survive. Governments expend billions upon particle research, cosmic-ray research, not because they have been imbued suddenly with a great hunger for truth, but for the very simple, if barbarous, reason that they know the power which lies in the particle. If the physicist learns the nature of the universe in his cyclotron, well and good, but the search is for power.

One period, for reasons of its own, may be interested in stability, another in change. One may prefer morphology, another function. There are styles in science just as in other institutions. The Christianity of today is not totally the Christianity of five centuries ago; neither is science impervious to change. We have lived to see the technological progress that was hailed in one age as the savior of man become the horror of the next. We have observed that the same able and energetic minds which built lights, steamships, and telephones turn with equal facility to the creation of what is euphemistically termed the "ultimate weapon."

It is in this reversal that the modern age comes off so badly. It does so because the forces which have been released have tended to produce an exaggerated conformity and, at the same time, an equally exaggerated assumption that science, a tool for manipulating the outside, the material universe, can be used to create happiness and ethical living. Science can be—and is—used by good men, but in its present sense it can scarcely be said to create them. Science, of course, in discovery represents the individual, but in the moment of triumph, science creates uniformity through which the mind of the individual once more flees away.

It is the part of the artist—the humanist—to defend that eternal flight, just as it is the part of science to seek to impose laws, regularities and certainties. Man desires the certainties but he also transcends them. Thus, as in so many other aspects of life, man inhabits a realm half in and half out of nature, his mind reaching forever beyond the tool, the uniformity, the law, into some realm which is that of mind alone. The pen and the brush represent that eternal search, that conscious recognition of the individual as the unique creature beyond the statistic.

Modern science itself tacitly admits the individual, as in this statement from P. B. Medawar: "We can be sure that, identical twins apart, each human being alive today differs genetically from any other human being; moreover, he is probably different from any other human being who has ever lived or is likely to live in thousands of years to come. The potential variation of human beings is enormously greater than their actual variation; to put it in another way, the ratio of possible men to actual men is overwhelmingly large."

So far does modern science spell out for us that genetic indeterminacy which parallels, in a sense, the indeterminacy of the subatomic particle. Yet all the vast apparatus of modern scientific communication seems fanatically bent upon reducing that indeterminacy as quickly as possible into the mold of rigid order. Programs which do not satisfy in terms of millions vanish from the air. Gone from most of America is the kind of entertainment still to be found in certain of the world's pioneer backlands where a whole village may gather around a little company of visitors. The local musician hurries to the scene, an artist draws pictures to amuse the children, stories are told with gestures across the barrier of tongues, and an enormous release of creative talent goes on into the small hours of the night.

The technology which, in our culture, has released urban and even rural man from the quiet before his hearth log has debauched his taste. Man no longer dreams over a book in which a soft voice, a constant companion, observes, exhorts, or sighs with him through the pangs of youth and age. Today he is more likely to sit before a screen and dream the mass dream which comes from outside.

No one need object to the elucidation of scientific principles in clear, unornamental prose. What concerns us is the fact that there exists a new class of highly skilled barbarians—not representing the very great in science—who would confine men entirely to this diet. Once more there is revealed the curious and unappetizing puritanism which attaches itself all too readily to those who, without grace or humor, have found their salvation in "facts."

There has always been violence in the world. A hundred years ago the struggle for existence among living things was much written upon and it was popular for even such scholars as Darwin and Wallace to dwell upon the fact that the vanquished died quickly and that the sum of good outweighed the pain. Along with the rising breed of scientific naturalists, however, there arose a different type of men. Stemming from the line of parson naturalists represented by Gilbert White, author of *The Natural History of Selborne*, these literary explorers of nature have left a powerful influence upon English thought. The grim portrait of a starving lark cracking an empty snail shell before Richard Jefferies' window on a bleak winter day is from a world entirely different from that of the scientist. Jefferies' observation is sharp, his facts accurate, yet there is, in his description, a sense of his own poignant hunger—the hunger of a dying man—for the beauty of an earth insensible to human needs. Here again we are in the presence of an artist whose vision is unique.

Even though they were not discoverers in the objective sense, one feels at times that the great nature essayists had more individual perception than their scientific contemporaries. Theirs was a different contribution. They opened the minds of men by the sheer power of their thought. The world of nature, once seen through the eye of genius, is never seen in quite the same manner afterward. A dimension has been added, something that lies beyond the careful analyses of professional biology. Something uncapturable by man passes over

W. H. Hudson's vast landscapes. They may be touched with the silvery light from summer thistledown, or bleaker weathers, but always a strange nostalgia haunts his pages—the light of some lost star within his individual mind.

This is a different thing from that which some scientists desire, or that many in the scientific tradition appreciate, but without this rare and exquisite sensitivity to guide us the truth is we are half blind. We will lack pity and tolerance, not through intent, but from blindness. It is within the power of great art to shed on nature a light which can be had from no other source than the mind itself. It was from this doorway, perhaps, that de la Mare's celestial visitant had intruded. Nature, Emerson knew, is "the immense shadow of man." We have cast it in our image. To change nature, mystical though it sounds, we have to change ourselves. We have to draw out of nature that ideal second world which Bacon sought. The modern world is only slowly beginning to realize the profound implications of that idea.

Perhaps we can amplify to some degree certain of our observations concerning man as he is related to the natural world. In Western Europe, for example, there used to be a strange old fear, a fear of mountains, of precipices, of wild untrodden spaces which, to the superstitious heart, seemed to contain a hint of lurking violence or indifference to man. It is as though man has always felt in the presence of great stones and rarified air something that dwarfed his confidence and set his thoughts to circling—an ice age, perhaps, still not outlived in the human mind.

There is a way through this barrier of the past that can be taken by science. It can analyze soil and stones. It can identify bones, listen to the radioactive tick of atoms in the lattices of matter. Science can spin the globe and follow the age-long marchings of man across the wastes of time and space.

Yet if we turn to the pages of the great nature essayists we may perceive once more the role which the gifted writer and thinker plays in the life of man. Science explores the natural world and thereby enhances our insight, but if we turn to the pages of *The Maine Woods*, regarded by critics as one of Henry David Thoreau's minor works, we come upon a mountain ascent quite unparalleled in the annals of literature.

The effect does not lie in the height of the mountain. It does not lie in the scientific or descriptive efforts made on the way up. Instead the cumulative effect is compounded of two things: a style so appropriate to the occasion that it evokes the shape of earth before man's hand had fallen upon it and, second, a terrible and original question posed on the mountain's summit. Somewhere along the road of that spiritual ascent— for it *was* a spiritual as well as a physical ascent—the pure observation gives way to awe, the obscure sense of the holy.

From the estimate of heights, of geological observation, Thoreau enters what he calls a "cloud factory" where mist was generated out of the pure air as fast as it flowed away. Stumbling onward over what he calls "the raw materials of a planet" he comments: "It was vast, titanic, and such as man never inhabits. Some part of the beholder, even some vital part, seems to escape through the loose grating of his ribs as he ascends. His reason is dispersed and shadowy, more thin and subtile, like the air. Vast, inhuman nature has got him at disadvantage, caught him alone, and pilfers him of some of his divine faculty." Thoreau felt himself in the presence of a force "not bound to be kind to man." "What is it," he whispers with awe, "to be admitted to a Museum, compared with being shown some star's surface, some hard matter in its home."

At this moment there enters into his apprehension a new view of substance, the heavy material body he had dragged up the mountain the while something insubstantial seemed to float out of his ribs. Pausing in astonishment, he remarks: "I stand in awe of my body, this matter to which I am bound has become so strange to me. I fear not spirits, ghosts, of which I am one—*that* my body might—but I fear bodies, I tremble to meet them. What is this Titan that has possession of me? Talk of mysteries!—think of our life in nature—daily to be shown matter, to come in contact with it—rocks, trees, wind on our cheeks! the solid earth, the actual world." Over and over he muses, his hands on the huge stones, "*Who* are we? Where are we?"

The essayist has been struck by an enormous paradox. In that cloud factory of the brain where ideas form as tenuously as mist streaming from mountain rocks, he has glimpsed the truth that mind is locked in matter like the spirit Ariel in a cloven pine. Like Ariel, men struggle to escape the drag of the matter they inhabit, yet it is spirit that they fear. "A Titan grasps us," argues Thoreau, confronting the rocks of the great mountain, a mass solid enough not to be dragged about by the forces of life. "Think of our life in nature," he reiterates. "Who are we?"

From the streaming cloud-wrack of a mountain summit, the voice floats out to us before the fog closes in once more. In that arena of rock and wind we have moved for a moment in a titanic world and hurled at stone titanic questions. We have done so because a slight, gray-eyed man walked up a small mountain which, by some indefinable magic, he transformed into a platform for something, as he put it, "not kind to man."

I do not know in the whole of literature a more penetrating expression of the spirit's horror of the substance it lies trapped within. It is the cry of an individual genius who has passed beyond science into a high domain of cloud. Let it not be forgotten, however, that Thoreau revered and loved true science, and that science and the human spirit together may find a way across that vast mountain whose shadow still looms menacingly above us.

"If you would learn the secrets of nature," Thoreau insisted, "you must practice more humanity than others." It is the voice of a man who loved both knowledge and the humane tradition. His faith has been ill kept within our time.

Mystical truths, however, have a way of knowing neither time nor total death. Many years ago, as an impressionable youth, I found myself lost at evening in a rural and obscure corner of the United States. I was there because of certain curious and rare insects that the place afforded—beetles with armored excrescences, stick insects which changed their coloration like autumn grass. It was a country which, for equally odd and inbred reasons, was the domain of people of similar exuberance of character, as though nature, either physically or mentally, had prepared them for odd niches in a misfit world.

As I passed down a sandy backwoods track where I hoped to obtain directions from a solitary house in the distance, I was overtaken by one of the frequent storms that blow up in that region. The sky turned dark and a splatter of rain struck the ruts of the road. Standing uncertainly at the roadside I heard a sudden rumble over a low plank bridge beyond me. A man high on a great load of hay was bearing down on me through the lowering dark. I could hear through the storm his harsh cries to the horses. I stepped forward to hail him and ask directions. Perhaps he would give me a ride.

There happened then, in a single instant, one of those flame-lit revelations which destroy the natural world forever and replace it with some searing inner vision which accompanies us to the end of our lives. The horses, in the sound and fury of the elements, appeared, even with the loaded rick, to be approaching at a gallop. The dark figure of the farmer with the reins swayed high above them in some limbo of lightning and storm. At that moment I lifted my hand and stepped forward. The horses seemed to pause—even the rain.

Then, in a bolt of light that lit the man on the hayrick, the waste of sodden countryside, and what must have been my own horror-filled countenance, the rain plunged down once more. In that brief, momentary glimpse within the heart of the lightning, haloed, in fact, by its wet shine, I had seen a human face of so incredible a nature as still to amaze and mystify me as to its origin. It was—by some fantastic biological exaggeration —two faces welded vertically together along the midline, like the riveted iron toys of my childhood. One side was lumpish with swollen and malign excrescences; the other shone in the blue light, pale, ethereal, and remote—a face marked by suffering, yet serene and alien to that visage with which it shared this dreadful mortal frame.

As I instinctively shrank back, the great wagon leaped and rumbled on its way to vanish at what spot I knew not. As for me, I offer no explanation for my conduct. Perhaps my eyes deceived me in that flickering and grotesque darkness. Perhaps my mind had spent too long a day on the weird excesses of growth in horned beetles. Nevertheless I am sure that the figure on the hayrick had raised a shielding hand to his own face.

One does not, in youth, arrive at the total meaning of such incidents or the deep symbolism involved in them. Only if the event has been frightening enough, a revelation out of the heavens themselves, does it come to dominate the meaning of our lives. But that I saw the double face of mankind in that instant of vision I can no longer doubt. I saw man—all of us —galloping through a torrential landscape, diseased and fungoid, with that pale half-visage of nobility and despair dwarfed but serene upon a twofold countenance. I saw the great horses with their swaying load plunge down the storm-filled track. I saw, and touched a hand to my own face.

Recently it has been said by a great scientific historian that the day of the literary naturalist is done, that the precision of the laboratory is more and more encroaching upon that individual domain. I am convinced that this is a mistaken judgment. We forget—as Bacon did not forget—that there is a natural history of souls, nay, even of man himself, which can be learned only from the symbolism inherent in the world about him.

It is the natural history that led Hudson to glimpse eternity in some old men's faces at Land's End, that led Thoreau to see human civilizations as toadstools sprung up in the night by solitary roads, or that provoked Melville to experience in the sight of a sperm whale some colossal alien existence without which man himself would be incomplete.

"There is no Excellent Beauty that hath not some strangeness in the Proportion," wrote Bacon in his days of insight. Anyone who has picked up shells on a strange beach can confirm his observation. But man, modern man, who has not contemplated his otherness, the multiplicity of other possible men who dwell or might have dwelt in him, has not realized the full terror and responsibility of existence.

It is through our minds alone that man passes like that swaying furious rider on the hayrick, farther and more desperately into the night. He is galloping—this twofold creature whom even Bacon glimpsed—across the storm-filled heath of time, from the dark world of the natural toward some dawn he seeks beyond the horizon.

Across that midnight landscape he rides with his toppling burden of despair and hope, bearing with him the beast's face and the dream, but unable to cast off either or to believe in either. For he is man, the changeling, in whom the sense of goodness has not perished, nor an eye for some supernatural guidepost in the night.

TEN

The Creature from the Marsh

I

T<small>HE ONLY THING</small> strange about me is my profession. I happen to be one of those few persons who pursue the farther history of man on the planet earth, what Darwin once called "the great subject." But my business is not with the formal art of the history books. Take, for example, today.

Today I have been walking in the ruins of the city. The city still moves, it is true, the air drills ring against iron, and I am aware of laughter and of feet hurrying by at the noon hour. Nevertheless the city is in ruins. This is what the trained eye makes of it. It stands here in the morning sun while rust flakes the steel rails and the leaves of innumerable autumns blow mistily through the ribs of skyscrapers and over the fallen brick work lies a tangle of morning glories. I have seen this before in the dead cities of Mexico—the long centuries wavering past with the curious distortion of things seen through deep sea water. Even the black snake gliding down the steps of the cathedral seems a repetition, past and future being equally resolvable in the curious perspective of the archaeological eye.

But it was not for this that I hurried out to walk in the streets of the city. I wanted to find a symbol, something that would stand for us when the time came, something that might be proud after there was no stone upon another—some work of art, perhaps, or a gay conceit that the rains had not tarnished, something that would tell our story to whatever strange minds might come groping there.

I think I must have walked miles in those ruins. I studied a hundred shop windows. I weighed with a quarter-century of digging experience the lasting qualities of metal, stone, and glass. I hesitated over the noble inscriptions upon public buildings, while the rain dissolved in the locked containers the files of treaties and the betrayal of all human trust. I passed by the signs of coruscating heat and the wilted metal of huge guns. I found the china head of a doll in the metal of a baby carriage that my mind took hold of and considered carefully, though later I realized I stood by a living nursemaid in the Park.

I looked at tools, and at flowers in the windows of tenements. (They will creep out and grow, I thought.) And I heard a dog howl in the waste streets for the comforting hand of man. I saw the vacant, ashy leaves of books blow by, but I did not pick them up. There were dead television screens and the curious detached loneliness of telephone receivers whose broken wires still thrummed in the winds over the Sierra.

It will never do, I thought; there was more to us than this—with all the evil, with all the cruelty. I remembered an inscribed gold ring in a pawnshop window. "From Tom to Mary," it had read, "for always." It is back there, I thought; that might be it—there was love in us, things we spoke to each other in the evening or on deathbeds, the eyes frank at last. It might be there. I turned and hurried back.

But the little shop was gone, and finally I came up short at a place where stones had tumbled in a peculiar way, half sheltering a fallen window. There were bones there, and at first this made no sense because bones in exposed places do not last well, and there are so many of them finally that the meaning escapes you.

There was a broken sign LED——— Fi——— Av———. And among the bits of glass, a little cluster of feathers, and under a shattered pane, the delicate bones of a woman's hand that, dying, had reached wistfully out, caught there, when the time came.

Why not? I mused. The human hand, the hand is the story. I touched one of the long, graceful bones. It had come the evolutionary way up from far eons and watery abysses only to perish here.

There was a little restless stirring beside me.

So it died after all that effort, I thought on. Five hundred million years expended in order that the shining thread of life could die reaching after a little creation of feathers in the window of a shop. And why not? Even my antique reptilian eye had a feel for something—a kind of beauty here . . .

The tugging at my sleeve continued. The slow, affectionate voice of my wife said to me, "Wait here. I want to go in."

"Of course," I said, and took and squeezed her slender hand as I returned from some far place. "It will look becoming," I said. "I will stand here and watch."

"The gloves to match it are of green lizard skin," she exulted.

"That will look just right," I ventured, and did not quite know what I meant. "I will stand here and watch."

A swinging light like a warning at a railroad crossing began flashing in the darkness below consciousness. A bell began jangling. Then it subsided. My wife was pointing, for the benefit of an attentive clerk, at a little cluster of feathers in the window. I came forward then and beckoned hastily. "But why—?" she said. "Another day," I said, wiping my forehead. It was just that— "Another time," I promised and urged her quickly away.

It is the nature, you see, of the profession—the terrible *déjà vu* of the archaeologist, the memory that scans before and after. For instance, take the case of the black skull, a retreat, some might say, in another direction.

II

The skull was black when they brought it to me. It was black from the irons and acids and mineral replacements of ice-age gravels. It was polished and worn and gleaming from the alterations of unnumbered years. It had made strange journeys after the death of its occupant; it had moved with glacial slowness in the beds of rivers; it had been tumbled by floods and, becoming an object of grisly beauty, had been picked up and passed from hand to hand by men the individual had never seen in life.

Finally it was brought to me.

It was my duty to tell them about the skull.

It was my professional duty to clothe these bones once more with the faint essence of a personality, to speak of a man or a woman, young or old, as the bones might tell the story. It was my task to read the racial features in a forgotten face, stare deep into the hollow sockets through which had once passed in endless procession the days and seasons and the shed tears of long ago.

The woman had been young. I could tell them that. I could tell them she had once fallen or been struck and that after a long time the bone had mended and she had recovered—how, it was difficult to say, for it had been a dangerous and compound fracture. Today such a wound would mean months of immobilization in a hospital. This woman had survived without medical attention through the endless marchings and journeyings of the hunters' world. Even the broken orbit of the left eye had dropped by a quarter of an inch—a serious disfigurement. Nevertheless she had endured and lived on toward some doom that had come fast upon her but was not written in the bones. It was, in all likelihood, a death by violence. Her skull had not been drawn from a grave. It had come from beneath the restless waters of a giant river that is known to keep its secrets well.

They asked me for the time of these events, and again, obediently, I went down that frail ladder which stretches below us into the night of time. I went slowly, by groping deductions and the hesitant intuitions of long experience that only scholars know. I passed through ages where water was wearing away the shapes of river pebbles into crystalline sand and the only sound in the autumn thickets was the gathering of south-flying birds. Somewhere in the neighborhood of the five thousandth millennium—I could place it no closer than that—the ladder failed me. The river was still there but larger—an enormous rolling waste of water and marshes out of which rose a vast October moon.

They interrupted me then, querulously, asking if archaeologists could do no better than this, and was it not true that there were new and clever methods by which physicists could call the year in the century and mark the passage of time by the tick of atoms in the substance of things. And I said, yes, within limits it was true, but that the methods were not always usable, and that the subtle contaminations possible among radioactive objects sometimes defeated our attempts.

At this point they shook their heads unwillingly, for, as I quickly saw, they had the passion of modern men for the precision of machines and disliked vagueness of any sort. But the skull lay there on the table between us, and over it one man lingered, fascinated in spite of himself. I knew what he was thinking: Where am I going? When shall I become like this?

I heard this in his mind for just an instant while I stared across at him from among my boxes of teeth and flint arrow-heads that had grown chalky and dull with the passage of long centuries in the ground.

"Thank you," the visitor said finally, moving after his party to the door. He was, I saw, unsure for what it was he thanked me.

"You are quite welcome," I said, still returning slowly from that waste of forgotten water over which the birds of another century cried dolefully, so that I could hear them keening in my head. Like the man who asks a medium to bring back some whimpering memoryless ghost and make it speak out of a living mouth for the amusement of a group of curiosity seekers, he may have felt remorse. At any rate, he nodded uncertainly and fled.

I was the instrument. I had made this journey a hundred times for students who scrawled their initials on my skulls, a hundred times for reporters who wanted sensational accounts of monkey-men, a hundred times for people who came up at the end of lectures and asked, "How much money are the bones worth, doctor? Are they easy to find?"

In spite of this I have continued to make these journeys. It is old habit now. I go back into the past alone. I would do so if I fled my job and sought safety in some obscure room. My sense of time is so heightened that I can feel the frost at work in stones, the first creeping advance of grass in a deserted street. I have stood by the carved sarcophagi of dead knights in a Euro-pean cathedral, men seven hundred years away from us with their steel and their ladies, and from that point striven to hurl the mind still backward into the wilderness where man coughs bestially and vanishes into the shape of beasts.

I cannot say I am a student of the dates in the history books. My life is mostly occupied with caves filled up and drifted over with the leaves of ten thousand thousand autumns. My special-ity is the time when man was changing into man. But, like a river that twists, evades, hesitates through slow miles, and then leaps violently down over a succession of cataracts, man can be called a crisis animal. Crisis is the most powerful element in his definition. Of his entire history, this he understands the least. Only man has continued to turn his own definition around upon his tongue until, in the end, he has looked outside of nature to something invisible to any eye but his own. Long ago, this emotion was well expressed in the Old Testament. "Oh Lord," exclaimed the prophet Jeremiah, "I know that the way of man is not in himself." Therefore, I would add, as a modern evolutionist, "the way" only lies through man and has to be sought beyond him. It was this that led to a very remarkable experience.

III

"The greatest prize of all," once confessed the British plant explorer F. Kingdon Ward, "is the skull of primitive man." Ward forgot one thing: there are other clues to primitive men than those confined to skulls. The bones of fossil men are few because the earth tolerated them in scant numbers. We call them missing links on the road to ourselves. A little less tooth here, a little more brain there, and you can see them changing toward ourselves in that long historyless time when the great continental ice sheets ebbed and flowed across the northern continents. Like all the students of that age, I wanted to find a missing link in human history. That is what this record is about, for I stumbled on the track of one.

Some men would maintain that a vague thing called atmosphere accounts for such an episode as I am about to relate, that there are houses that demand a murder and wait patiently until the murderer and his victim arrive, that there are great cliffs that draw the potential suicide from afar or mountains of so austere a nature that they write their message on the face of a man who looks up at them. This all may be. I do not deny it. But when I encountered the footprint in the mud of that remote place I think the thing that terrified me most was the fact that I knew to whom it belonged and yet I did not want to know him. He was a stranger to me and remains so to this day. Because of a certain knowledge I had, however, he succeeded in impressing himself upon me in a most insidious manner. I have never been the same since the event took place, and often

at night I start up sweating and think uncannily that the creature is there with me in the dark. If the sense of his presence
grows, I switch on the light, but I never look into the mirror.
This is a matter of old habit with me.

First off, though, we must get straight what we mean by a
missing link.

A missing link is a day in the life of a species that is changing
its form and habits, just as, on a smaller scale, one's appearance
and behavior at the age of five are a link in one's development
to an adult man or woman. The individual person may have
changed and grown, but still the boy or girl of many years ago
is linked to the present by a long series of steps. And if one is
really alive and not already a living fossil, one will go on changing till the end of one's life and perhaps be the better for it.
The term "missing link" was coined because some of the
physical links in the history of man as a species are lost, and
those people who, like myself, are curious about the past look
for them.

My album is the earth, and the pictures in it are faded and
badly torn and have to be pieced together by detective work. If
one thinks of oneself at five years of age, one may get a thin
wisp of disconnected memory pictures. By contrast, the past of
a living species is without memory except as that past has
written its physical record in vestigial organs like the appendix
or a certain pattern on our molar teeth. To eke out what those
physical stigmata tell us, we have to go grubbing about in
caves and gravel for the bones of very ancient men. If one can
conceive of the trouble an archaeologist might have in locating
one's remains a half-million years from now, supposing they
still existed, one will get an idea of the difficulties involved in
finding traces of man before his bones were crowded together
in cities and cemeteries.

I was wandering inland along a sunken shore when the thing
happened—the thing I had dreamed of so long. In other
words, I got a clue to man. The beaches on that coast I had
come to visit are treacherous and sandy and the tides are always
shifting things about among the mangrove roots. It is not a
place to which I would willingly return and you will get no
bearings from me. Anyway, what it was I found there could be
discovered on any man's coast if he looked sharp for it. I had

come to that place with other things in mind, and a notion of being alone. I was tired. I wanted to lie in the sun or clamber about like an animal in the swamps and the forest. To secure such rest from the turmoil of a modern city is the most difficult thing in the world to accomplish and I have only achieved it twice: once in one of the most absolute deserts in the world and again in this tropical marsh.

By day and night strange forms of life scuttled and gurgled underfoot or oozed wetly along outthrust branches; luminous tropical insects blundered by in the dark like the lamps of hesitant burglars. Overhead, on higher ground, another life shrieked distantly or was expectantly still in the treetops. Somehow, alone as I was, I got to listening as if all that world were listening, waiting for something to happen. The trees drooped a little lower listening, the tide lurked and hesitated on the beach, and even a tree snake dropped a loop and hung with his face behind a spider web, immobile in the still air.

A world like that is not really natural, or (the thought strikes one later) perhaps it really is, only more so. Parts of it are neither land nor sea and so everything is moving from one element to another, wearing uneasily the queer transitional bodies that life adopts in such places. Fish, some of them, come out and breathe air and sit about watching you. Plants take to eating insects, mammals go back to the water and grow elongate like fish, crabs climb trees. Nothing stays put where it began because everything is constantly climbing in, or climbing out, of its unstable environment.

Along drowned coasts of this variety you only see, in a sort of speeded-up way, what is true of the whole world and everything upon it: the Darwinian world of passage, of missing links, of beetles with soldered, flightless wings, of snakes with vestigial feet dragging slowly through the underbrush. Everything is marred and maimed and slightly out of focus—everything in the world. As for man, he is no different from the rest. His back aches, he ruptures easily, his women have difficulties in childbirth—all because he has struggled up upon his hind legs without having achieved a perfect adjustment to his new posture.

On this particular afternoon, I came upon a swamp full of

huge waterlilies where I had once before ventured. The wind had begun to rise and rain was falling at intervals. As far as I could see, giant green leaves velvetly impervious to water were rolling and twisting in the wind. It was a species of lily in which part of the leaves projected on stalks for a short distance above the water, and as they rolled and tossed the whole swamp flashed and quivered from the innumerable water drops that were rolling around and around like quicksilver in the great cupped leaves. Everything seemed flickering and changing as if in some gigantic illusion, but so soft was the green light and so delicate the brushing of the leaves against each other that the whole effect was quite restful, as though one could be assured that nothing was actually tangible or real and no one in his senses would want it to be, as long as he could sway and nod and roll reflecting water drops about over the surface of his brain.

Just as I finally turned away to climb a little ridge I found the first footprint. It was in a patch of damp, exposed mud and was pointed away from the water as though the creature had emerged directly out of the swamp and was heading up the shore toward the interior. I had thought I was alone, and in that place it was wise to know one's neighbors. Worst of all, as I stood studying the footprint, and then another, still heading up the little rise, it struck me that though undoubtedly human the prints were different in some indefinable way. I will tell you once more that this happened on the coast of another country in a place where form itself is an illusion and no shape of man or beast is totally impossible. I crouched anxiously in the mud while all about the great leaves continued to rotate on their stems and to flash their endlessly rolling jewels.

But there were these footprints. They did not disappear. As I fixed the lowermost footprint with every iota of scientific attention I could muster, it became increasingly apparent that I was dealing with some transitional form of man. The arch, as revealed in the soft mud, was low and flat and implied to the skilled eye an inadequate adjustment to the upright posture. This, in its turn, suggested certain things about the spine and the nature of the skull. It was only then, I think, that the full import of my discovery came to me.

Good Lord, I thought consciously for the first time, the thing is alive. I had spent so many years analyzing the bones of past ages or brooding over lizard tracks turned to stone in remote epochs that I had never contemplated this possibility before. The thing was alive and it was human. I looked uneasily about before settling down into the mud once more. One could make out that the prints were big but what drew my fascinated eye from the first was the nature of the second toe. It was longer than the big toe, and as I crawled excitedly back and forth between the two wet prints in the open mud, I saw that there was a remaining hint of prehensile flexibility about them.

Most decidedly, as a means of ground locomotion this foot was transitional and imperfect. Its loose, splayed aspect suggested inadequate protection against sprains. That second toe was unnecessarily long for life on the ground, although the little toe was already approximating the rudimentary condition so characteristic of modern man. Could it be that I was dealing with an unreported living fossil, an archaic ancestral survival? What else could be walking the mangrove jungle with a foot that betrayed clearly the marks of ancient intimacy with the arboreal attic, an intimacy so long continued that now, after hundreds of thousands of years of ground life, the creature had squiggled his unnecessarily long toes about in the mud as though an opportunity to clutch at something had delighted his secret soul.

I crouched by the footprint and thought. I remembered that comparisons with the living fauna, whenever available, are good scientific procedure and a great aid to precise taxonomy. I sat down and took off my shoes.

I had never had much occasion to look critically at my own feet before. In modern man they are generally encased in shoes—something that still suggests a slight imperfection in our adaptations. After all, we don't normally find it necessary to go about with our hands constantly enclosed in gloves. As I sat contemplating and comparing my feet with the footprints, a faintly disturbing memory floated hazily across my mind. It had involved a swimming party many years before at the home of one of the most distinguished comparative anatomists in the world. As we had sat on the bench alongside his pool, I had glanced up suddenly and caught him staring with what had seemed unnecessary fascination at my feet. I remembered now that he had blushed a deep pink under his white hair and had diverted my inquiring glance deftly to the scenery about us.

Why I should have remembered the incident at all was unclear to me. I thought of the possibility of getting plaster casts of a footprint, and I also debated whether I should attempt to trail the creature farther up the slope toward which he appeared to have been headed. It was no moment for hesitation. Still, I did hesitate. The uneasy memory grew stronger, and a thought finally struck me. A little sheepishly and with a glance around to see that I was not observed, I lowered my own muddy foot into the footprint. It fitted.

I stood there contemplatively clutching, but this time consciously, the mud in my naked toes. I was the dark being on that island shore whose body carried the marks of its strange passage. I was my own dogging Man Friday, the beast from the past who had come with weapons through the marsh. The wind had died and the great green leaves with their rolling jewels were still. The mistake I had made was the mistake of all of us.

The story of man was not all there behind us in the caves of remote epochs. Even our physical bodies gave evidence that the change was not completed. As for our minds, they were still odd compounds of beast and saint. But it was not by turning back toward the marsh out of which we had come that the truly human kingdom was to be possessed and entered— that kingdom dreamed of in many religions and spoken of in many barbarous tongues. A philosopher once said in my presence, "The universe is a series of leaping sparks—everything else is interpretation." But what, I hesitated, was man's interpretation to be?

I drew a foot out of the little steaming swamp that sucked at it. The air hung heavily about me. I listened as the first beast might have listened who came from the water up the shore and did not return again to his old element. Everything about me listened in turn and seemed to be waiting for some decision on my part. I swayed a moment on my unstable footing.

Then, warily, I stepped higher up the shore and let the water and the silt fill in that footprint to make it, a hundred million years away, a fossil sign of an unknown creature slipping from the shadows of a marsh toward something else that awaited him. I had found the missing link. He walked on misshapen feet. The stones hurt him and his belly sagged. There were dreams like Christmas ornaments in his head, intermingled with an ancient malevolent viciousness. I knew because I was the missing link, but for the first time I sensed where I was going.

I have said I never look into the mirror. It is a matter of old habit now. If that other presence grows too oppressive I light the light and read.

ELEVEN

One Night's Dying

T HERE IS ALWAYS a soft radiance beyond the bedroom door from a night-light behind my chair. I have lived this way for many years now. I sleep or I do not sleep, and the light makes no difference except if I wake. Then, as I awaken, the dim forms of objects sustain my grip on reality. The familiar chair, the walls of the book-lined study reassert my own existence.

I do not lie and toss with doubt any longer, as I did in earlier years. I get up and write, as I am writing now, or I read in the old chair that is as worn as I am. I read philosophy, metaphysics, difficult works that sometime, soon or late, draw a veil over my eyes so that I drowse in my chair.

It is not that I fail to learn from these midnight examinations of the world. It is merely that I choose that examination to remain as remote and abstruse as possible. Even so, I cannot always prophesy the result. An obscure line may whirl me into a wide-awake, ferocious concentration in which ideas like animals leap at me out of the dark, in which sudden odd trains of thought drive me inexorably to my desk and paper. I am, in short, a victim of insomnia—sporadic, wearing, violent, and melancholic. In the words of Shakespeare, for me the world "does murder sleep." It has been so since my twentieth year.

In that year my father died—a man well loved, the mainstay of our small afflicted family. He died slowly in severe bodily torture. My mother was stone-deaf. I, his son, saw and heard him die. We lived in a place and time not free with the pain-alleviating drugs of later decades. When the episode of many weeks' duration was over, a curious thing happened: I could no longer bear the ticking of the alarm clock in my own bedroom.

At first I smothered it with an extra blanket in a box beside my cot, but the ticking persisted as though it came from my own head. I used to lie for hours staring into the dark of the sleeping house, feeling the loneliness that only the sleepless know when the queer feeling comes that it is the sleeping who are alive and those awake are disembodied ghosts. Finally, in desperation, I gave up the attempt to sleep and turned to reading, though it was difficult to concentrate.

It was then that human help appeared. My grandmother saw the light burning through the curtains of my door and came to sit with me. A few years later, when I touched her hair in farewell at the beginning of a journey from which I would not return to see her alive, I knew she had saved my sanity. Into that lonely room at midnight she had come, abandoning her own sleep, in order to sit with one in trouble. We had not talked much, but we had sat together by the lamp, reasserting our common humanity before the great empty dark that is the universe.

Grandmother knew nothing of psychiatry. She had not reestablished my sleep patterns, but she had done something more important. She had brought me out of a dark room and retied my thread of life to the living world. Henceforward, by night or day, though I have been subject to the moods of depression or gaiety which are a part of the lives of all of us, I have been able not merely to endure but to make the best of what many regard as an unbearable affliction.

It is true that as an educational administrator I can occasionally be caught nodding in lengthy committee meetings, but so, I have observed, can men who come from sound nights on their pillows. Strangely, I, who frequently grow round-eyed and alert as an owl at the stroke of midnight, find it pleasant to nap in daylight among friends. I can roll up on a couch and sleep peacefully while my wife and chatting friends who know my peculiarities keep the daytime universe safely under control. Or so it seems. For, deep-seated in my subconscious, is perhaps the idea that the black bedroom door is the gateway to the tomb.

I try in that bedroom to sleep high on two pillows, to have ears and eyes alert. Something shadowy has to be held in place and controlled. At night one has to sustain reality without help. One has to hear lest hearing be lost, see lest sight not return to follow moonbeams across the floor, touch lest the sense of objects vanish. Oh, sleeping, soundlessly sleeping ones, do you ever think who knits your universe together safely from one day's memory to the next? It is the insomniac, not the night policeman on his beat.

Many will challenge this point of view. They will say that electric power does the trick, that many a roisterer stumbles down the long street at dawn, after having served his purpose of holding the links of the mad world together. There are parts of the nighttime world, men say to me, that it is just as well I do not know. Go home and sleep, man. Others will keep your giddy world together. Let the thief pass quickly in the shadow, he is awake. Let the juvenile gangs which sidle like bands of evil crabs up from the dark waters of poverty into prosperous streets pass without finding you at midnight.

The advice is good, but in the city or the country small things important to our lives have no reporter except as he who does not sleep may observe them. And that man must be disencumbered of reality. He must have no commitments to the dark, as do the murderer and thief. Only he must see, though what he sees may come from the night side of the planet that no man knows well. For even in the early dawn, while men lie unstirring in their sleep or stumble sleepy-eyed to work, some single episode may turn the whole world for a moment into the place of marvel that it is but that we grow too day-worn to accept.

For example, I call the place where I am writing now the bay of broken things. In the February storms, spume wraiths climb the hundred-foot cliff to fight and fall like bitter rain in the moonlight upon the cabin roof. The earth shakes from the drum roll of the surf. I lie awake and watch through the window beyond my bed. This is no ticking in my brain; this is the elemental night of chaos. This is the sea chewing its million-year way into the heart of the continent.

The caves beneath the cliff resound with thunder. Again those warring wraiths shoot high over the house. Impelled as though I were a part of all those leaping ghosts, I dress in the dark and come forth. With my back against the door, like an ancient necromancer, I hurl my mind into the white spray and try to summon back, among those leaping forms, the faces and features of the dead I know. The shapes rise endlessly, but they pass inland before the wind, indifferent to my mortal voice.

I walk a half mile to a pathway that descends upon a little beach. Below me is a stretch of white sand. No shell is ever found unbroken, even on quiet days, upon that shore. Everything comes over the rocks to seaward. Wood is riven into splinters; the bones of seamen and of sea lions are pounded equally into white and shining sand. Throughout the night the long black rollers, like lines of frothing cavalry, form ranks, drum towering forward, and fall, fall till the mind is dizzy with the spume that fills it. I wait in the shelter of a rock for daybreak. At last the sea eases a trifle. The tide is going out.

I stroll shivering along the shore, and there, exposed in inescapable nakedness, I see the elemental cruelty of the natural world. A broken-winged gull, hurled by the wind against the cliff, runs before me wearily along the beach. It will starve or, mercifully, the dogs will find it. I try not to hurry it, and walk on. A little later in a quieter bend of the shore, I see ahead of me a bleeding, bedraggled blot on the edge of the white surf. As I approach, it starts warily to its feet. We look at each other. It is a wild duck, also with a shattered wing. It does not run ahead of me like the longer-limbed gull. Before I can cut off its retreat it waddles painfully from its brief refuge into the water.

The sea continues to fall heavily. The duck dives awkwardly, but with long knowledge and instinctive skill, under the fall of the first two inshore waves. I see its head working seaward. A long green roller, far taller than my head, rises and crashes forward. The black head of the waterlogged duck disappears. This is the way wild things die, without question, without knowledge of mercy in the universe, knowing only themselves and their own pathway to the end. I wonder, walking farther up the beach, if the man who shot that bird will die as well.

This is the chaos before man came, before sages imbued with pity walked the earth. Indeed it is true, and in my faraway study my hands have often touched with affection the backs of the volumes which line my shelves. Nevertheless, I have endured the nights and mornings of the city. I have seen old homeless men who have slept for hours sitting upright on ledges along the outer hallway of one of the great Eastern stations straighten stiffly in the dawn and limp away with feigned businesslike aloofness before the approach of the policeman on his rounds. I know that on these cold winter mornings sometimes a man, like the pigeons I have seen roosting as closely as possible over warm hotel air vents, will fall stiffly and not awaken. It is true that there are shelters for the homeless, but some men, like their ice-age forebears, prefer their independence to the end.

The loneliness of the city was brought home to me one early sleepless morning, not by men like me tossing in lonely rooms, not by poverty and degradation, not by old men trying with desperate futility to be out among others in the great roaring hive, but by a single one of those same pigeons which I had seen from my hotel window, looking down at midnight upon the smoking air vents and chimneys.

The pigeon, *Columba livia*, is the city bird *par excellence*. He is a descendant of the rock pigeon that in the Old World lived among the cliffs and crevices above the caves that early man inhabited. He has been with us since our beginning and has adapted as readily as ourselves to the artificial cliffs of man's first cities. He has known the Roman palaces and the cities of Byzantium. His little flat feet, suited to high and precarious walking, have sauntered in the temples of vanished gods as readily as in New York's old Pennsylvania Station. In my dim morning strolls, waiting for the restaurants to open, I have seen him march quickly into the back end of a delivery truck while the driver was inside a store engaged in his orders with the proprietor. Yet for all its apparent tolerance of these highly adapted and often comic birds, New York also has a beach of broken things more merciless than the reefs and rollers of the ocean shore.

One morning, strolling sleepless as usual toward early breakfast time in Manhattan, I saw a sick pigeon huddled at an uncomfortable slant against a building wall on a street corner. I felt sorry for the bird, but I had no box, no instrument of help, and had learned long ago that pursuing wounded birds on city streets is a hopeless, dangerous activity. Pigeons, like men, die in scores every day in New York. As I hesitantly walked on, however, I wondered why the doomed bird was assuming such a desperately contorted position under the cornice that projected slightly over it.

At this moment I grew aware of something I had heard more loudly in European streets as the factory whistles blew, but never in such intensity as here, even though American shoes are built of softer materials. All around me the march of people was intensifying. It was New York on the way to work.

Space was shrinking before my eyes. The tread of innumerable feet passed from an echo to the steady murmuring of a stream, then to a drumming. A dreadful robot rhythm began to rack my head, a sound like the boots of Nazis in their heyday of power. I was carried along in an irresistible surge of bodies.

A block away, jamming myself between a waste-disposal basket and a lightpost, I managed to look back. No one hesitated at that corner. The human tide pressed on, jostling and pushing. My bird had vanished under that crunching, multifooted current as remorselessly as the wounded duck under the indifferent combers of the sea. I watched this human ocean, of which I was an unwilling droplet, rolling past, its individual faces like whitecaps passing on a night of storm, fixed, merciless, indifferent; man in the mass marching like the machinery of which he is already a replaceable part, toward desks, computers, missiles, and machines, marching like the waves toward his own death with a conscious ruthlessness no watery shore could ever duplicate. I have never returned to search in that particular street for the face of humanity. I prefer the endlessly rolling pebbles of the tide, the moonstones polished by the pulling moon.

And yet, plunged as I am in dire memories and midnight reading, I have said that it is the sufferer from insomnia who knits the torn edges of men's dreams together in the hour before dawn. It is he who from his hidden, winter vantage point sees the desperate high-hearted bird fly through the doorway of the grand hotel while the sleepy doorman nods, a deed equivalent in human terms to that of some starving wretch evading Peter at heaven's gate, and an act, I think, very likely to be forgiven.

It is a night more mystical, however, that haunts my memory. Around me I see again the parchment of old books and remember how, on one rare evening, I sat in the shadows while a firefly flew from volume to volume lighting its small flame, as if in literate curiosity. Choosing the last title it had illuminated, I came immediately upon these words from St. Paul: "Beareth all things, believeth all things, hopeth all things, endureth all things." In this final episode I shall ask you to bear with me and also to believe.

I sat, once more in the late hours of darkness, in the airport of a foreign city. I was tired as only both the sufferer from insomnia and the traveler can be tired. I had missed a plane and had almost a whole night's wait before me. I could not sleep. The long corridor was deserted. Even the cleaning women had passed by.

In that white efficient glare I grew ever more depressed and weary. I was tired of the endless comings and goings of my profession; I was tired of customs officers and police. I was lonely for home. My eyes hurt. I was, unconsciously perhaps, looking for that warm stone, that hawthorn leaf, where, in the words of the poet, man trades in at last his wife and friend. I had an ocean to cross; the effort seemed unbearable. I rested my aching head upon my hand.

Later, beginning at the far end of that desolate corridor, I saw a man moving slowly toward me. In a small corner of my eye I merely noted him. He limped, painfully and grotesquely, upon a heavy cane. He was far away, and it was no matter to me. I shifted the unpleasant mote out of my eye.

But, after a time, I could still feel him approaching, and in one of those white moments of penetration which are so dreadful, my eyes were drawn back to him as he came on. With an anatomist's eye I saw this amazing conglomeration of sticks and broken, misshapen pulleys which make up the body of man. Here was an apt subject, and I flew to a raging mental dissection. How could anyone, I contended, trapped in this mechanical thing of joints and sliding wires expect the acts it performed to go other than awry?

The man limped on, relentlessly.

How, oh God, I entreated, did we become trapped within this substance out of which we stare so hopelessly upon our own eventual dissolution? How for a single minute could we dream or imagine that thought would save us, children deliver us, from the body of this death? Not in time, my mind rang with my despair; not in mortal time, not in this place, not anywhere in the world would blood be stanched, or the dark wrong be forever righted, or the parted be rejoined. Not in this time, not mortal time. The substance was too gross, our utopias bought with too much pain.

The man was almost upon me, breathing heavily, lunging and shuffling upon his cane. Though an odor emanated from him, I did not draw back. I had lived with death too many years. And then this strange thing happened, which I do not mean physically and cannot explain. The man entered me. From that moment I saw him no more. For a moment I was contorted within his shape, and then out of his body—our bodies, rather—there arose some inexplicable sweetness of union, some understanding between spirit and body which I had never before experienced. Was it I, the joints and pulleys only, who desired this peace so much?

I limped with growing age as I gathered up my luggage. Something of that terrible passer lingered in my bones, yet I was released, the very room had dilated. As I went toward my plane the words the firefly had found for me came automatically to my lips. "Beareth all things," believe, believe. It is thus that one day and the next are welded together, that one night's dying becomes tomorrow's birth. I, who do not sleep, can tell you this.

TWELVE

Obituary of a Bone Hunter

THE PAPERS and the magazines reprint the stories endlessly these days—of Sybaris the sin city, or, even further back, that skull at Tepexpan. One's ears are filled with chatter about assorted magnetometers and how they are used to pick up the traces of buried objects and no one has to guess at all. They unearth the city, or find the buried skull and bring it home. Then everyone concerned is famous overnight.

I'm the man who didn't find the skull. I'm the man who'd just been looking twenty years for something like it. This isn't sour grapes. It's their skull and welcome to it. What made me sigh was the geophysics equipment. The greatest gambling game in the world—the greatest wit-sharpener—and now they do it with amplifiers and electronically mapped grids. An effete age, gentlemen, and the fun gone out of it.

There are really two kinds of bone hunters—the big bone hunters and the little bone hunters. The little bone hunters may hunt big bones, but they're little bone hunters just the same. They are the consistent losers in the most difficult game of chance that men can play: the search for human origins. Eugène Dubois, the discoverer of Pithecanthropus, hit the jackpot in a gamble with such stupendous odds that the most devoted numbers enthusiast would have had better sense than to stake his life on them.

I am a little bone hunter. I've played this game for a twenty-year losing streak. I used to think it all lay in the odds—that it was luck that made the difference between the big and little bone hunters. Now I'm not so sure any longer. Maybe it's something else.

Maybe sometimes an uncanny clairvoyance is involved, and if it comes you must act or the time goes by. Anyhow I've thought about it a lot in these later years. You think that way as you begin to get grayer and you see pretty plainly that the game is not going to end as you planned.

With me I think now that there were three chances: the cave of spiders, the matter of the owl's egg, and the old man out of the Golden Age. I muffed them all. And maybe the old man just came to show me I'd sat in the big game for the last time.

II

In that first incident of the spiders, I was playing a hunch, a long one, but a good one still. I wanted to find Neanderthal man, or any kind of ice-age man, in America. One or two important authorities were willing to admit he *might* have got in before the last ice sheet; that he *might* have crossed Bering Strait with the mammoth. He might have, they said, but it wasn't likely. And if he had, it would be like looking for hummingbirds in the Bronx to find him.

Well, the odds were only a hundred to one against me, so I figured I'd look. That was how I landed in the cave of spiders. It was somewhere west out of Carlsbad, New Mexico, in the Guadalupe country. Dry. With sunlight that would blister cactus. We were cavehunting with a dynamiter and a young Harvard assistant. The dynamiter was to blow boulders away from fallen entrances so we could dig what lay underneath.

We found the cave up a side canyon, the entrance blocked with fallen boulders. Even to my youthful eyes it looked old, incredibly old. The waters and the frosts of centuries had eaten at the boulders and gnawed the cave roof. Down by the vanished stream bed a little gleam of worked flints caught our eye.

We stayed there for days, digging where we could and leaving the blasting till the last. We got the Basket Maker remains we had come to get—the earliest people that the scientists of that time would concede had lived in the Southwest. Was there anything more? We tamped a charge under one huge stone that blocked the wall of the cave and scrambled for the outside. A dull boom echoed down the canyon and the smoke and dust slowly blew away.

Inside the cave mouth the shattered boulder revealed a crack behind it. An opening that ran off beyond our spot lights. The hackles on my neck crawled. This might be the road to— something earlier? There was room for only one man to worm his way in. The dynamiter was busy with his tools. "It's probably nothing," I said to the assistant. "I'll just take a quick look."

As I crawled down that passage on my belly I thought once or twice about rattlesnakes and what it might be like to meet one on its own level where it could look you in the eye. But after all I had met snakes before in this country, and besides I had the feeling that there was something worth getting to beyond.

I had it strong—too strong to turn back. I twisted on and suddenly dropped into a little chamber. My light shot across it. It was low and close, and this was the end of the cave. But there was earth on the floor beneath me, the soft earth that must be dug, that might hold something more ancient than the cave entrance. I couldn't stand up; the roof was too low. I would have to dig on hands and knees. I set the light beside me and started to probe the floor with a trench shovel. It was just then that the fear got me.

The light lay beside me shining on the ceiling—a dull, velvety-looking ceiling, different from the stone around. I don't know when I first sensed that something was wrong, that the ceiling was moving, that waves were passing over it like the wind in a stand of wheat. But suddenly I did; suddenly I dropped the shovel and thrust the light closer against the roof. Things began to detach themselves and drop wherever the light touched them. Things with legs. I could hear them plop on the soft earth around me.

I shut off the light. The plopping ceased. I sat on my knees in the darkness, listening. My mind was centered on just one thing—escape. I knew what that wavering velvet wall was. Millions upon millions of daddy-long-legs—packed in until they hung in layers. Daddy-long-legs, the most innocent and familiar of all the spider family. I wish I could say I had seen black widows there among them. It would help now, in telling this.

But I didn't. I didn't really see anything. If I turned on the light that hideous dropping and stirring would commence again. The light woke them. They disliked it.

If I could have stood up it would have been different. If they had not been overhead it would have been different. But they had me on my knees and they were above and all around. Millions upon millions. How they got there I don't know. All I know is that up out of the instinctive well of my being flowed some ancient, primal fear of the crawler, the walker by night. One clambered over my hand. And above they dangled, dangled. . . . What if they all began to drop at once?

I did not light the light. I had seen enough. I buttoned my jacket close, and my sleeves. I plunged blindly back up the passage down which I had wriggled and which, luckily, was free of them.

Outside the crew looked at me. I was sweating, and a little queer. "Close air," I gasped; "a small hole, nothing there."

We went away then in our trucks. I suppose in due time the dust settled, and the fox found his way in. Probably all that horrible fecund mass eventually crept, in its single individualities, back into the desert where it frightened no one. What it was doing there, what evil unknown to mankind it was plotting, I do not know to this day. The evil and the horror, I think now, welled out of my own mind, but somehow that multitude of ancient life in a little low dark chamber touched it off. It did not pass away until I could stand upright again. It was a fear out of the old, four-footed world that sleeps within us still.

Neanderthal man? He might have been there. But I was young and that was only a first chance gone. Yes, there were things I might have done, but I didn't do them. You don't tell your chief dynamiter that you ran from a daddy-long-legs. Not in that country. But do you see, it wasn't *one* daddy-long-legs. That's what I can't seem to make clear to anyone. It wasn't just one daddy-long-legs. It was millions of them. Enough to bury you. And have you ever thought of being buried under spiders? I thought not. You begin to get the idea?

III

I had a second chance and again it was in a cave I found. This time I was alone, tramping up a canyon watching for bones, and I just happened to glance upward in the one place where the cave could be seen. I studied it a long time—until I could feel the chill crawling down my back. This might be it; this might be the place. . . . This time I would know. This time there would be no spiders.

Through the glasses I could make out a fire-blackened roof, a projecting ledge above the cave mouth, and another one below. It was a small, strange hide-out, difficult to reach, but it commanded the valley on which the canyon opened. And there was the ancient soot-impregnated cave roof. Ancient men had been there.

I made that climb. Don't ask me how I did it. Probably there had been an easier route ages ago. But I came up a naked chimney of rock down which I lost my knapsack and finally the geologist's pick that had helped me hack out a foothold in the softening rock.

When I flung myself over the ledge where the cave mouth opened, I was shaking from the exhausting muscle tension and fear. No one, I was sure, had come that way for a thousand years, and no one after me would come again. I did not know how I would get down. It was enough momentarily to be safe. In front of me the cave mouth ran away darkly into the mountain.

I took the flashlight from my belt and loosened my sheath knife. I began to crawl downward and forward, wedging my-self over sticks and fallen boulders. It was a clean cave and something was there, I was sure of it. Only, the walls were small and tight. . . .

They were tighter when the voice and the eyes came. I re-member the eyes best. I caught them in my flashlight the same instant that I rammed my nose into the dirt and covered my head. They were big eyes and coming my way.

I never thought at all. I just lay there dazed while a great roaring buffeting thing beat its way out over my body and went away.

It went out into the silence beyond the cave mouth. A half minute afterward, I peered through my fingers and rolled weakly over. Enough is enough. But this time I wasn't going back empty-handed. Not I. Not on account of a mere bird. Not if I *had* thought it was a mountain lion, which it could just as well have been. No owl was going to stop me, not even if it was the biggest owl in the Rocky Mountains.

I twitched my ripped shirt into my pants and crawled on. It wasn't much farther. Over the heap of debris down which the great owl had charged at me, I found the last low chamber, the place I was seeking. And there in a pile of sticks lay an egg, an impressive egg, glimmering palely in the cavernous gloom, full of potentialities, and fraught, if I may say so, with destiny.

I affected at first to ignore it. I was after the buried treasures that lay beneath its nest in the cave floor. The egg was simply going to have to look after itself. Its parent had gone, and in a pretty rude fashion, too. I was no vandal, but I was going to be firm. If an owl's egg stood in the path of science— But suddenly the egg seemed very helpless, very much alone. I probed in the earth around the nest. The nest got in the way. This was a time for decision.

I know a primatologist who will lift a rifle and shoot a baby monkey out of its mother's arms for the sake of science. He is a good man, too, and goes home nights to his wife. I tried to focus on this thought as I faced the egg.

I knew it was a rare egg. The race of its great and lonely mother was growing scant in these mountains and would soon be gone. Under it might lie a treasure that would make me famed in the capitals of science, but suppose there was nothing under the nest after all and I destroyed it? Suppose . . .

Here in this high, sterile silence with the wind crying over frightful precipices, myself and that egg were the only living things. That seemed to me to mean something. At last and quietly I backed out of the cave and slipped down into the chasm out of which I had come. By luck I did not fall.

Sometimes in these later years I think perhaps the skull was there, the skull that could have made me famous. It is not so bad, however, when I think that the egg became an owl. I had had charge of it in the universe's sight for a single hour, and I had done well by life.

It is not the loss of the skull that torments me sometimes on winter evenings. Suppose the big, unutterably frightened bird never came back to its egg? A feeling of vast loss and desolation sweeps over me then. I begin to perceive what it is to doubt.

IV

It was years later that I met the old man. He was waiting in my office when I came in. It was obvious from the timid glances of my secretary that he had been passed from hand to hand and that he had outwitted everybody. Someone in the background made a twisting motion at his forehead.

The old man sat, a colossal ruin, in the reception chair. The squirrel-like twitterings of the office people did not disturb him.

As I came forward he fished in a ragged wallet and produced a clipping. "You made this speech?" he asked.

"Why, yes," I said.

"You said men came here late? A few thousand years ago?"

"Yes, you see—"

"Young man," he interrupted, "you are frightfully wrong."

I was aware that his eyes were contracted to pin points and seemed in some danger of protruding on stalks.

"You have ignored," he rumbled, "the matter of the Miocene period—the Golden Age. A great civilization existed then, far more splendid than this—degenerate time." He struck the floor fiercely with his cane.

"But," I protested, "that period is twenty million years ago. Man wasn't even in existence. Geology shows—"

"Nothing!" said the massive relic. "Geology has nothing to do with it. Sit down. I know all about the Golden Age. I will prove to you that you are wrong."

I collapsed doubtfully into a chair. He told me that he was from some little town in Missouri, but I never believed it for a moment. He smelled bad, and it was obvious that if he brought news of the Golden Age, as he claimed, he had come by devious and dreadful ways from that far era.

"I have here," he said, thrusting his head forward and breathing heavily into my face, "a human jaw. I will unwrap it a little and you can see. It is from a cave I found."

"It is embedded in stalactite drippings," I murmured, hypnotized against my will. "That might represent considerable age. Where did you find it?"

He raised a protesting hand. "Later, son, later. You admit then—?"

I strained forward. "Those teeth," I said, "they are large—they look primitive." The feeling I had had at the mouth of the owl's cave came to me again overpoweringly. "Let me see a little more of the jaw. If the mental eminence should be lacking, you may have something important. Just let me handle it a moment."

With the scuttling alacrity of a crab, the old man drew back and popped the papers over his find. "You admit, then, that it is important? That it proves the Golden Age was real?"

Baffled, I looked at him. He eyed me with an equal wariness.

"Where did you find it?" I asked. "In this light it seemed—it might be—a fossil man. We have been looking a long time. If you would only let me see—"

"I found it in a cave in Missouri," he droned in a rote fashion. "You can never find the cave alone. If you will make a statement to the papers that the Golden Age is true, I will go with you. You have seen the evidence."

Once more I started to protest. "But this has nothing to do with the Golden Age. You may have a rare human fossil there. You are denying science—"

"Science," said the old man with frightening dignity, "is illusion." He arose. "I will not come back. You must make a choice."

For one long moment we looked at each other across the fantastic barriers of our individual minds. Then, on his heavy oakwood cane, he hobbled to the door and was gone. I watched through the window as he crossed the street in a patch of autumn sunlight as phantasmal and unreal as he. Leaves fell raggedly around him until, a tatter among tatters, he passed from sight.

I rubbed a hand over my eyes, and it seemed the secretary looked at me strangely. How was it that I had failed this time? By unbelief? But the man was mad. I could not possibly have made such a statement as he wanted.

Was it pride that cost me that strange jaw bone? Was it academic dignity? Should I have followed him? Found where he lived? Importuned his relatives? Stolen if necessary, that remarkable fragment?

Of course I should! I know that now. Of course I should.

Thirty years have passed since the old man came to see me. I have crawled in many caverns, stooped with infinite aching patience over the bones of many men. I have made no great discoveries.

I think now that in some strange way that old man out of the autumn leaf-fall was the last test of the inscrutable gods. There will be no further chances. The egg and the spiders and the madman—in them is the obituary of a life dedicated to the folly of doubt, the life of a small bone hunter.

THIRTEEN

The Mind as Nature

WHEN I was a small boy I lived, more than most children, in two worlds. One was dark, hidden, and self-examining, though in its own way not without compensations. The other world in which I somehow also managed to exist was external, boisterous, and what I suppose the average parent would call normal or extroverted. These two worlds simultaneously existing in one growing brain had in them something of the dichotomy present in the actual universe, where one finds, behind the ridiculous, wonderful tent-show of woodpeckers, giraffes, and hoptoads, some kind of dark, brooding, but creative void out of which these things emerge—some antimatter universe, some web of dark tensions running beneath and creating the superficial show of form that so delights us. If I develop this little story of a personal experience as a kind of parable, it is because I believe that in one way or another we mirror in ourselves the universe with all its dark vacuity and also its simultaneous urge to create anew, in each generation, the beauty and the terror of our mortal existence.

In my own case, through the accidents of fortune, the disparity between these two worlds was vastly heightened. How I managed to exist in both I do not know. Children under such disharmony often grow sick, retire inward, choose to return to the void. I have known such cases. I am not unaware that I paid a certain price for my survival and indeed have been paying for it ever since. Yet the curious thing is that I survived and, looking back, I have a growing feeling that the experience was good for me. I think I learned something from it even while I passed through certain humiliations and an utter and profound loneliness. I was living in a primitive world at the same time that I was inhabiting the modern world as it existed in the second decade of this century. I am not talking now about the tree-house, cave-building activities of normal boys. I am talking about the minds of the first dawning human consciousness —about a kind of mental ice age, and of how a light came in from outside until, as I have indicated, two worlds existed in which a boy, still a single unsplit personality, walked readily from one world to the other by day and by night without anyone observing the invisible boundaries he passed.

I

To begin this story I have to strip myself of certain conventions, but since all the major figures of my childhood are dead I can harm no one but myself. I think, if we are to find our way into the nature of creativeness, into those multitudinous universes that inhabit the minds of men, such case histories—though I hate this demeaning term—have a certain value. Perhaps if we were franker on these matters, we might reach out and occasionally touch, with a passing radiance, some other star in the night.

I was born in the first decade of this century, conceived in and part of the rolling yellow cloud that occasionally raises up a rainy silver eye to look upon itself before subsiding into dust again. That cloud has been blowing in my part of the Middle West since the ice age. Only a few months ago, flying across the continent, I knew we were passing over it in its customary place. It was still there and its taste was still upon my tongue.

In those days I lived, like most American boys of that section, in a small house where the uncemented cellar occasionally filled with water and the parlor was kept shuttered in a perpetual cool darkness. We never had visitors. No minister ever called on us, so the curtains were never raised. We were, in a sense, social outcasts. We were not bad people nor did we belong to a racial minority. We were simply shunned as unimportant and odd.

The neighbors were justified in this view. As I have mentioned, my mother was stone-deaf; my father worked the long hours of a time when labor was still labor. I was growing up alone in a house whose dead silence was broken only by the harsh discordant jangling of a voice that could not hear itself. My mother had lost her hearing as a young girl. I never learned what had attracted my father to her. I never learned by what fantastic chance I had come to exist at all. Only the cloud would know, I sometimes think to this day, only the yellow loess cloud rolling, impenetrable as it was when our ancestors first emerged from it on the ice-age steppes of Europe, or when they followed the bison into its heart on the wide American plains.

I turned over the bricks of our front sidewalk and watched ants with a vague interest. There grew up between my mother and myself an improvised system of communication, consisting of hand signals, stampings on the floor to create vibrations, exaggerated lip movements vaguely reminiscent of an anthropoid society. We did not consciously work at this; we were far too ignorant. Certain acts were merely found useful and came to be repeated and to take on symbolic value. It was something of the kind of communication which may have been conducted by the man-apes of the early ice age. One might say we were at the speech threshold—not much more.

I did not go to church, and since the family was not agreed upon any mode of worship, I merely wondered as I grew older how it was that things came to be. In short, I would have been diagnosed today by social workers as a person suffering from societal deprivation and headed for trouble.

There was another curious aspect of this family which involved my father. He was a good man who bore the asperities of my afflicted mother with dignity and restraint. He had been a strolling itinerant actor in his younger years, a largely self-trained member of one of those little troupes who played *East Lynne* and declaimed raw Shakespearean melodrama to unsophisticated audiences in the little Midwestern "opera houses." He had a beautiful resonant speaking voice. Although we owned no books, and although when I knew him in middle age a harsh life had dimmed every hunger except that for rest, he could still declaim long rolling Elizabethan passages that caused shivers to run up my back.

> "Give me my Robe, put on my Crowne; I have
>   Immortall longings in me."

Like many failures of his time he used to speak wistfully of cheap land in Arkansas and send for catalogues, searching for something that was permanently lost. It was the last of the dream that had finally perished under the yellow cloud. I use the word "failure" in a worldly sense only. He was not a failure as a man. He reared a son, the product of an unfortunate marriage from which he might easily have fled, leaving me inarticulate. He was kind and thoughtful with an innate courtesy that no school in that rough land had taught him. Although he was intensely sensitive, I saw him weep but twice. The first time, when I was young and very ill, I looked up in astonishment as a tear splashed on my hand. The second time, long years afterward, belongs to him and his life alone. I will merely say he had had a great genius for love and that his luck was very bad. He was not fitted for life under the yellow cloud. He knew it, yet played out his role there to the end. So poor were we, it took me twenty years to put a monument upon his grave.

II

We come now to the two worlds of which I spoke, the two worlds making up the mind and heart of this curiously deprived and solitary child—a child whose mother's speech was negligible and disordered and which left him for the greater part of his early childhood involved with only rudimentary communication and the conscious rebuffs of neighbors, or another ill-understood world of haunting grandiloquent words at which his playmates laughed; or even a third world where he sat, the last little boy allowed in the street, and watched the green night moths beat past under the street lamps.

Into what well of being does one then descend for strength? How does one choose one's life? Or does one just go on without guidance, as in the dark town of my youth, from one spot of light to the next? "Uncertainty," as John Dewey has well said, "is primarily a practical matter. It signifies uncertainty of the *issue* of present experiences. These are fraught with future peril. . . . The intrinsic troublesome and uncertain quality of situations lies in the fact that they hold outcomes in suspense; they move to evil or to good fortune. The natural tendency of man is to do something at once, there is impatience with suspense, and lust for immediate action."

It is here, amid a chaos of complexities, that the teacher, frequently with blindness, with uncertainties of his own, must fight with circumstance for the developing mind—perhaps even for the very survival of the child. The issue cannot be long delayed, because, as Dewey observes, man—and far more, the child—has a lust for the immediate, for action. Yet the teacher is fighting for an oncoming future, for something that has not emerged, which may, in fact, never emerge. His lot is worse than that of the sculptors in snow, which Sidney Hook once described us as being. Rather, the teacher is a sculptor of the intangible future. There is no more dangerous occupation on the planet, for what we conceive as our masterpiece may appear out of time to mock us—a horrible caricature of ourselves.

The teacher must ever walk warily between the necessity of inducing those conformities which in every generation reaffirm our rebellious humanity, and of allowing for the free play of the creative spirit. It is not only for the sake of the future that the true educator fights, it is for the justification of himself, his profession, and the state of his own soul. He, too, amid contingencies and weariness, without mental antennae, and with tests that fail him, is a savior of souls. He is giving shapes to time, and the shapes themselves, driven by their own inner violence, wrench free of his control—must, if they are truly sculptured, surge like released genii from the classroom, or tragically shrink to something less than bottle size.

The teacher cannot create, any more than can the sculptor, the stone upon which he exercises his talents; he cannot, it is true, promote gene changes and substitutions in the bodies with which he works. But here again Dewey has words of peculiar pertinence to us—words which remove from the genes something of the utter determinacy in which the geneticist sometimes revels. "In the continuous ongoing of life," he contends, "objects part with something of their final character and become conditions of subsequent experiences. There is regulation of the change in which . . . a causal character is rendered preparatory and instrumental."

The boy under the street lamp may become fascinated by night-flying moths or the delinquent whisperings of companions. Or he may lie awake in the moonlight of his room, quaking with the insecurity of a divided household and the terrors of approaching adulthood. He may quietly continue some lost part of childhood by playing gentle and abstract games with toys he would not dare to introduce among his raucous companions of the street. He wanders forlornly through a museum and is impressed by a kindly scientist engrossed in studying some huge bones.

Objects do indeed "part with something of their final character," and so do those who teach. There are subjects in which I have remained dwarfed all of my adult life because of the ill-considered blow of someone nursing pent-up aggressions, or because of words more violent in their end effects than blows. There are other subjects for which I have more than ordinary affection because they are associated in my mind with kindly and understanding men or women—sculptors who left even upon such impliant clay as mine the delicate chiseling of refined genius, who gave unwittingly something of their final character to most unpromising material. Sculptors reaching blindly forward into time, they struck out their creation, scarce living to view the result.

Now, for many years an educator, I often feel the need to seek out a quiet park bench to survey mentally that vast and nameless river of students which has poured under my hands. In pain I have meditated: "This man is dead—a suicide. Was it I, all unknowingly, who directed, in some black hour, his hand upon the gun? This man is a liar and a cheat. Where did my stroke go wrong?" Or there comes to memory the man who, after long endeavors, returned happily to the farm from which he had come. Did I serve him, if not in the world's eye, well? Or the richly endowed young poet whom I sheltered from his father's wrath—was I pampering or defending—and at the right or the wrong moment in his life? Contingency, contingency, and each day by word or deed the chisel falling true or blind upon the future of some boy or girl.

Ours is an ill-paid profession and we have our share of fools. We, too, like the generation before us, are the cracked, the battered, the malformed products of remoter chisels shaping the most obstinate substance in the universe—the substance of man. Someone has to do it, but perhaps it might be done more kindly, more precisely, to the extent that we are consciously aware of what we do—even if that thought sometimes congeals our hearts with terror. Or, if we were more conscious of our task, would our hands shake or grow immobilized upon the chisel?

I do not know. I know only that in these late faint-hearted years I sometimes pause with my hand upon the knob before I go forth into the classroom. I am afflicted in this fashion because I have come to follow Dewey in his statement that "nature is seen to be marked by histories." As an evolutionist I am familiar with that vast sprawling emergent, the universe, and its even more fantastic shadow, life. Stranger still, however, is the record of the artist who creates the symbols by which we live. As Dewey has again anticipated, "No mechanically exact science of an individual is possible. An individual is a history unique in character. But"—he adds—"constituents of an individual are known when they are regarded not as qualitative, but as statistical constants derived from a series of operations."

I should like to survey briefly a few such constants from the lives of certain great literary figures with whose works I happen to be reasonably familiar. I choose to do so because creativity —that enigma to which the modern student of educational psychology is devoting more and more attention—is particularly "set in the invisible." "The tangible," Dewey insisted in *Experience and Nature*, "rests precariously upon the untouched and ungrasped." Dewey abhorred the inculcation of fixed conclusions at the expense of man's originality. Though my case histories are neither numerous nor similar, they contain certain constants which, without revealing the total nature of genius, throw light upon the odd landscapes and interiors that have nurtured it.

III

"A person too early cut off from the common interests of men," Jean Rostand, the French biologist, once remarked, "is exposed to inner impoverishment. Like those islands which are lacking in some whole class of mammals." Naturally there are degrees of such isolation, but I would venture the observation that this eminent observer has overlooked one thing: sometimes on such desert islands there has been a great evolutionary proliferation amongst the flora and fauna that remain. Strange shapes, exotic growths, which, on the mainland, would have been quickly strangled by formidable enemies, here spring up readily. Sometimes the rare, the beautiful, can only emerge or survive in isolation. In a similar manner, some degree of withdrawal serves to nurture man's creative powers. The artist and the scientist bring out of the dark void, like the mysterious universe itself, the unique, the strange, the unexpected. Numerous observers have testified upon the loneliness of the process.

"The whole of my pleasure," wrote Charles Darwin of his travels with illiterate companions on the high Andean uplands, "was derived from what passed in my mind." The mind, in other words, has a latent, lurking fertility, not unrelated to the universe from which it sprang. Even in want and in jail it will labor, and if it does not produce a physical escape, it will appear, assuming the motivational drive be great enough, in *Pilgrim's Progress*, that still enormously moving account of one who walked through the wilderness of this world and laid him down and dreamed a dream. That dream, as John Bunyan himself wrote, "will make a Traveller of Thee." It was the account of the journey from the City of Destruction to the City of God. It was written by a man in homespun not even aware of being a conscious artist, nor interested in personal fame. Three centuries later he is better known than many kings.

The fact is that many of us who walk to and fro upon our usual tasks are prisoners drawing mental maps of escape. I once knew a brilliant and discerning philosopher who spent many hours each week alone in movie houses, watching indifferently pictures of a quality far below his actual intellectual tastes. I knew him as an able, friendly, and normal person. Somewhere behind this sunny mask, however, he was in flight, from what, I never knew. Was it job, home, family—or was it rather something lost that he was seeking? Whatever it was, the pictures that passed before his eyes, the sounds, only half-heard, could have meant little except for an occasional face, a voice, a fading bar of music. The darkness and the isolation were what he wanted, something in the deep night of himself that called him home.

The silver screen was only a doorway to a land he had entered long ago. It was weirdly like hashish or opium. He taught well; he was far better read than many who climbed to national reputation upon fewer abilities than he possessed. His kindness to others was proverbial, his advice the sanest a friend could give. Only the pen was denied to him and so he passed toward his end, leaving behind the quick streak of a falling star that slips from sight. A genius in personal relationships, he was voiceless—somewhere a door had been softly, courteously, but inexorably closed within his brain. It would never open again within his lifetime.

I knew another man of similar capacities—a scholar who had shifted in his last graduate days from the field of the classics to the intricacies of zoology. A scintillating piece of research had rocked his profession, and he had marched steadily to the leadership of a great department. He was a graying, handsome man, with the world at his feet. He did not fail in health; his students loved him, and he loved them. The research died. This happens to other men. His problem was more serious: he could not answer letters. His best pupils could not depend upon him even to recommend them for posts or scholarships. Airmail letters lay unopened in his box. It was not that he was cruel. Anything a man could do by word of mouth he would do for his students, even the assumption of unpleasant tasks. Firm, upright, with a grave old-fashioned gallantry, in him also a door had closed forever. One never heard him speak of his family. Somewhere behind that door was a landscape we were never permitted to enter.

One can also read case histories. There is, for example, the brilliant child who had lost a parent and then a guardian abroad. Here, in some strange transmutation, arose a keen cartographer, a map maker, seeking a way back to the lost—a student of continents, time tables, odd towns, and fading roads. This is a juvenile case and the end therefore uncertain. One wonders whether there will come a breaking point where, as eventually they must, the trails within dissolve to waving grass and the crossroad signs lie twisted and askew on rotting posts. Where, then, will the wanderer turn? Will the last sign guide him safely home at last—or will he become one of the dawdlers, the evaders, unconscious fighters of some cruel inner master? It is of great interest in this connection that Herman Melville, who had lost his father under painful circumstances in his youth, describes in *Redburn*, in fictional guise, what must have lain close to his own heart: the following of a thirty-year-old map which traced the wanderings of Redburn's father in Liverpool. These passages are handled with the kind of imaginative power that indicates Melville's deep personal involvement in this aspect of Redburn's story. He speaks of running in the hope of overtaking the lost father at the next street; then the map fails him, just as the father himself had gone where no son's search could find him. Again in *Moby Dick* he cries out, "Where is the foundling's father hidden?" I think of my own slow journey homeward along those arc lights in a city whose name now comes with difficulty to my tongue.

In some of us a child—lost, strayed off the beaten path—goes wandering to the end of time while we, in another garb, grow up, marry or seduce, have children, hold jobs, or sit in movies, and refuse to answer our mail. Or, by contrast, we haunt our mailboxes, impelled by some strange anticipation of a message that will never come. "A man," Thoreau has commented, "needs only to be turned around once with his eyes shut in this world to be lost."

But now an odd thing happens. Some of the men with maps in their heads do not remain mute. Instead, they develop the power to draw the outside world within and lose us there. Or, as scientists, after some deep inner colloquy, they venture even to remake reality.

What would the modern chronicler of the lives of Hollywood celebrities feel if he were told he had to produce a great autobiography out of a year spent in a shack by a little pond, seeing scarcely anyone? Yet Thoreau did just that, and in entering what was essentially an inner forest, he influenced the lives of thousands of people all over the world and, it would appear, through succeeding generations.

There was another man, Nathaniel Hawthorne, who, as he put it, "sat down by the wayside of life, like a man under enchantment." For over a decade he wrote in a room in Salem, subsisting on a small income, and scarcely going out before evening. "I am surrounding myself with shadows," he wrote, "which bewilder me by aping the realities of life." He found in the human heart "a terrible gloom, and monsters of diverse kinds . . . but deeper still . . . the eternal beauty." This region of guttering candles, ungainly night birds, "fragments of a world," are an interior geography through which even the modern callous reader ventures with awe.

One could run through other great creative landscapes in the literary field. One could move with Hudson over the vast Patagonian landscape which haunted him even in his long English exile. One could, in fact, devise an anthology in which, out of the same natural background, under the same stars, beneath the same forests, or upon the same seas, each man would evoke such smoky figures from his own heart, such individual sunlight and shadow as would be his alone. Antoine St. Exupéry had his own flyer's vision of the little South American towns, or of the Andes, when flying was still young. Or Herman Melville, whose Pacific was "a Potters Field of all four continents . . . mixed shades and shadows, drowned dreams, somnambulisms, reveries."

Or, to change scene into the city-world, there is the vision of London in Arthur Machen's *Hill of Dreams*: "one grey temple of an awful rite, ring within ring of wizard stones circled about some central place, every circle was an initiation, every initiation eternal loss." Perhaps it should not go unnoticed that in this tale of solitude in a great city Machen dwells upon the hatred of the average man for the artist, "a deep instinctive dread of all that was strange, uncanny, alien to his nature." Julian, Machen's hero, "could not gain the art of letters and he had lost the art of humanity." He was turning fatally inward as surely as those men whose stories I have recounted.

Perhaps there is a moral here which should not go unobserved, and which makes the artist's problem greater. It also extends to the scientist, particularly as in the case of Darwin or Freud, or, in earlier centuries, such men as Giordano Bruno or Francis Bacon. "Humanity is not, as was once thought," says John Dewey, "the end for which all things were formed; it is but a slight and feeble thing, perhaps an episodic one, in the vast stretch of the universe. But for man, man is the center of interest and the measure of importance."

IV

It is frequently the tragedy of the great artist, as it is of the great scientist, that he frightens the ordinary man. If he is more than a popular story-teller it may take humanity a generation to absorb and grow accustomed to the new geography

with which the scientist or artist presents us. Even then, perhaps only the more imaginative and literate may accept him.
Subconsciously the genius is feared as an image breaker; frequently he does not accept the opinions of the mass, or man's
opinion of himself. He has voiced through the ages, in one
form or another, this very loneliness and detachment which
Dewey saw so clearly at the outcome of our extending knowledge. The custom-bound, uneducated, intolerant man projects
his fear and hatred upon the seer. The artist is frequently a
human mirror. If what we see there displeases us, if we see all
too clearly our own insignificance and vanity, we tend to revolt, not against ourselves, but in order to martyrize the unfortunate soul who forced us into self-examination.

In short, like the herd animals we are, we sniff warily at the
strange one among us. If he is fortunate enough finally to be
accepted, it is likely to be after a trial of ridicule and after the
sting has been removed from his work by long familiarization
and bowdlerizing, when the alien quality of his thought has
been mitigated or removed. Cecil Schneer recounts that Einstein made so little impression on his superiors, it was with
difficulty that he obtained even a junior clerkship in the Swiss
Patents office at Bern, after having failed of consideration as a
scholar of promise. Not surprisingly, theoretical physicists fa-
vored his views before the experimentalists capitulated. As
Schneer remarks: "It was not easy to have a twenty-six-year-old
clerk in the Swiss Patents office explain the meaning of experiments on which one had labored for years." Implacable hatred,
as well as praise, was to be Einstein's lot.

To an anthropologist, the social reception of invention re-
minds one of the manner in which a strange young male is first
repulsed, then tolerated, upon the fringes of a group of howler
monkeys he wishes to join. Finally, since the memories of the
animals are short, he becomes familiar, is accepted, and fades
into the mass. In a similar way, discoveries made by Darwin
and Wallace were at first castigated and then by degrees absorbed. In the process both men experienced forms of loneliness and isolation, not simply as a necessity for discovery but as
a penalty for having dared to redraw the map of our outer,
rather than inner, cosmos.

This fear of the upheld mirror in the hand of genius extends to the teaching profession and perhaps to the primary and secondary school teacher most of all. The teacher occupies, as we shall see a little further on, a particularly anomalous and exposed position in a society subject to rapid change or threatened by exterior enemies. Society is never totally sure of what it wants of its educators. It wants, first of all, the inculcation of custom, tradition, and all that socializes the child into the good citizen. In the lower grades the demand for conformity is likely to be intense. The child himself, as well as the teacher, is frequently under the surveillance of critical, if not opinionated, parents. Secondly, however, society wants the child to absorb new learning which will simultaneously benefit that society and enhance the individual's prospects of success.

Thus the teacher, in some degree, stands as interpreter and disseminator of the cultural mutations introduced by the individual genius into society. Some of the fear, the projected guilt feelings, of those who do not wish to look into the mirrors held up to them by men of the Hawthorne stamp of genius, falls upon us. Moving among innovators of ideas as we do, sifting and judging them daily, something of the suspicion with which the mass of mankind still tends to regard its own cultural creators falls upon the teacher who plays a role of great significance in this process of cultural diffusion. He is, to a degree, placed in a paradoxical position. He is expected both to be the guardian of stability and the exponent of societal change. Since all persons do not accept new ideas at the same rate, it is impossible for the educator to please the entire society even if he remains abjectly servile. This is particularly true in a dynamic and rapidly changing era like the present.

Moreover, the true teacher has another allegiance than that to parents alone. More than any other class in society, teachers mold the future in the minds of the young. They transmit to them the aspirations of great thinkers of which their parents may have only the faintest notions. The teacher is often the first to discover the talented and unusual scholar. How he handles and encourages, or discourages, such a child may make all the difference in the world to that child's future—and to the world. Perhaps he can induce in stubborn parents the conviction that their child is unusual and should be encouraged in his studies. If the teacher is sufficiently judicious, he may even be able to help a child over the teetering planks of a broken home and a bad neighborhood. Like a responsible doctor, he knows that he will fail in many instances—that circumstances will destroy, or genes prove defective beyond hope. There is a limit, furthermore, to the energy of one particular man or woman in dealing individually with a growing mass of students.

It is just here, however—in our search for what we might call the able, all-purpose, success-modeled student—that I feel it so necessary not to lose sight of those darker, more uncertain, late-maturing, sometimes painfully abstracted youths who may represent the Darwins, Thoreaus, and Hawthornes of the next generation. As Dr. Carroll Newsom emphasized in his admirable book, *A University President Speaks Out*: real college education is not a four-year process; it should be lifelong. Men, moreover, mature in many ways and fashions. It is uncertain what Darwin's or Wallace's chances of passing a modern college board examination might have been.

I believe it useful, and not demeaning to the teaching profession, to remember Melville's words in 1850, at a time when he was fighting horribly with the materials of what was to become his greatest book. The words, besides being prophetic in his case, bespeak the philosopher who looks beyond man as he is. He said: "I somehow cling to the wondrous fancy, that in all men hiddenly reside certain wondrous, occult properties—as in some plants and minerals—which by some happy but very rare accident . . . may chance to be called forth here on earth."

As a teacher I know little about how these remarkable events come about, but I have seen them happen. I believe in them. I believe they are more likely to happen late in those whose background has been one of long deprivation. I believe that the good teacher should never grow indifferent to their possibility —not, at least, if there is evidence, even in the face of failure in some subjects, of high motivation and intelligence in some specific field.

At the height of his creative powers, Thoreau wrote that "we should treat our minds as innocent and ingenuous children whose guardians we are—be careful what objects and what subjects we thrust on their attention. Even the facts of science may dust the mind by their dryness, unless they are in a sense effaced each morning, or rather rendered fertile by the dews of fresh and living truth. Every thought that passes through the mind helps to wear and tear it, and to deepen the ruts, which, as in the streets of Pompeii, evince how much it has been used. How many things there are concerning which we might well deliberate whether we had better know them!"

V

Educators responsible to society will appreciate that certain of these ancient institutions by which men live are, however involved with human imperfection, the supporting bones of the societal body. Without them, without a certain degree of conformity and habit, society would literally cease to exist. The problem lies in sustaining the airy flight of the superior intellect above the necessary ruts it is forced to travel. As Thoreau comments, the heel of the true author wears out no road. "A thinker's weight is in his thought, not in his tread."

A direct analogy is evident in the biological domain, where uncontrolled diversification at the species level would make for maladaptation to the environment. Yet without the emergence of superior or differently adapted individuals—beneficial mutations, in other words—the doorways to prolonged survival of the species would, under changing conditions, be closed. Similarly, if society sinks into the absolute rut of custom, if it refuses to accept beneficial mutations in the cultural realm or to tolerate, if not promote, the life of genius, then its unwieldy slumbers may be its last. Worse is the fact that all we know of beauty and the delights of free untrammeled thought may sink to a few concealed sparks glimmering warily behind the foreheads of men no longer in a position to transfer these miraculous mutations to the society which gave them birth.

Such repression is equivalent to placing an animal with a remarkable genetic heritage alone on a desert island. His strain will perish without issue. And so it is, in analogous ways, in oppressive societies or even in societies not consciously oppressive. It is all too easy to exist in an atmosphere of supposed free speech and yet bring such pressures upon the individual that he is afraid to speak openly. I do not refer to political matters alone. There was a time in the American world when

Thoreau's advice to "catalogue stars, those thoughts whose orbits are as rarely calculated as comets" could be set down without undergoing the scrutiny of twenty editors or publishers' readers. The observer of the fields was free as the astronomer to watch the aspect of his own interior heavens. He could even say boldly that he was "not concerned to express that kind of truth which Nature has expressed. Who knows but I may suggest some things to her?"

Our faith in science has become so great that, though the open-ended and novelty-producing aspect of nature is scientifically recognized in the physics and biology of our time, there is often a reluctance to give voice to it in other than professional jargon. It has been my own experience among students, laymen, and scholars that to express even wonder about the universe—in other words, to benefit from some humble consideration of what we do not know, as well as marching to the constant drumbeat of what we call the age of technology—is regarded askance in some quarters. I have had the vague word "mystic," applied to me because I have not been able to shut out wonder occasionally, when I have looked at the world. I have been lectured by at least one member of my profession who advised me to "explain myself"—words which sound for all the world like a humorless request for the self-accusations so popular in Communist lands. Although I have never disturbed the journals in my field with my off-hour compositions, there seemed to be a feeling on the part of this eminent colleague that something vaguely heretical about the state of my interior heavens demanded exposure or "confession" in a scientific journal. This man was unaware, in his tough laboratory attitude, that there was another world of pure reverie that is of at least equal importance to the human soul. Ironically, only a decade ago Robert Hofstadter, Nobel Prize winner in physics, revealed a humility which would greatly become the lesser men of our age. "Man," he said wistfully, "will never find the end of the trail."

VI

Directly stated, the evolution of the entire universe—stars, elements, life, man—is a process of drawing something out of nothing, out of the utter void of nonbeing. The creative element in the mind of man—that latency which can conceive gods, carve statues, move the heart with the symbols of great poetry, or devise the formulas of modern physics—emerges in as mysterious a fashion as those elementary particles which leap into momentary existence in great cyclotrons, only to vanish again like infinitesimal ghosts. The reality we know in our limited lifetimes is dwarfed by the unseen potential of the abyss where science stops. In a similar way the smaller universe of the individual human brain has its lonely cometary passages or flares suddenly like a supernova, only to subside in death while the waves of energy it has released roll on through unnumbered generations.

As the astrophysicist gazes upon the rare releases of power capable of devastating an entire solar system, so does the student of the behavioral sciences wonder at the manifestations of creative genius and consider whether the dark mechanisms that control the doorways of the human mind might be tripped open at more frequent intervals. Does genius emerge from the genes alone? Does the largely unknown chemistry of the brain contain at least part of the secret? Or is the number of potential cell connections involved? Or do we ordinary men carry it irretrievably locked within our subconscious minds?

That the *manifestations* of genius are culturally controlled we are well aware. The urban world, in all its diversity, provides a background, a cultural base, without which—whatever may be hidden in great minds—creativity would have had to seek other and more ephemeral expression or remain mute. Yet no development in art or scientific theory from the upper Stone Age onward seems to have demanded any further development in the brain of man. Mathematical theory, science, the glories of art lurked hidden as the potential seeds of the universe itself, in the minds of children rocked to sleep by cave fires in ice-age Europe.

If genius is a purely biological phenomenon one must assume that the chances of its appearance should increase with the size of populations. Yet it is plain that, as with toadstools which spring up in the night in fairy rings and then vanish, there is some delicate soil which nurtures genius—the cultural circumstance and the play of minds must meet. It is not a matter of population statistics alone, else there would not have been so surprising an efflorescence of genius in fifth-century Greece—a thing we still marvel at in our vastly expanded world. Darwin, committed to biological explanations alone, was left fumbling uncertainly with a problem that was essentially not reducible to a simplistic biological explanation. Without ignoring the importance of biology as one aspect of an infinitely complicated subject, therefore, the modern researcher favors the view that the intensive examination of the creative mind and its environment may offer some hope of stimulating the sources from which it springs, or, at the very least, of nurturing it more carefully.

I have touched upon loneliness, the dweller in the forest as represented by Thoreau, the isolated man in the room who was Hawthorne, and such wandering recluse scientists as Darwin and Wallace. This loneliness, in the case of literary men, frequently leads to an intense self-examination. "Who placed us with eyes between a microscopic and telescopic world?" questions Thoreau. "I have the habit of attention to such excess that my senses get no rest, but suffer from a constant strain."

Thoreau here expresses the intense self-awareness which is both the burden and delight of the true artist. It is not merely the fact that such men create their universe as surely as shipwrecked bits of life run riot and transform themselves on oceanic islands. It is that in this supremely heightened consciousness of genius the mind insists on expression. The spirit literally cannot remain within itself. It will talk if it talks on paper only to itself, as did Thoreau.

Anxiety, the disease which many psychiatrists seek to excise completely from the human psyche, is here carried to painful but enormously creative heights. The freedom of genius, its passage beyond the bonds of culture which control the behavior of the average man, in itself demands the creation of new modes of being. William Butler Yeats comments with penetration:

"Man's life is thought and he despite his terror cannot cease
Ravening through century after century . . .
That he may come
Into the desolation of reality."

Within that desolation, whether he be scientist or poet,

man—for this is the nature of his inmost being—will build ever anew. It is not in his nature to do otherwise.

Thoreau, however, presents in his writing an interesting paradox. In his reference to the excessive strain of heightened attention, one might get the impression that creativity was to him a highly conscious exercise that had wriggled into his very fingertips. That he was an intensely perceptive observer there can be no question. Yet he wrote, in those pre-Freudian, pre-Jungian days of 1852:

> "I catch myself philosophizing most abstractly when first returning to consciousness in the night or morning. I make the truest observations and distinctions then, when the will is yet wholly asleep and the mind works like a machine without friction. I am conscious of having, in my sleep, transcended the limits of the individual, and made observations and carried on conversations which in my waking hours I can neither recall nor appreciate. As if in sleep our individual fell into the infinite mind, and at the moment of awakening we found ourselves on the confines of the latter."

"It is," he confides in another place, "the material of all things loose and set afloat that makes my sea."

The psychiatrist Lawrence Kubie has speculated that "the creative person is one who in some manner, which today is still accidental, has retained his capacity to use his pre-conscious functions more freely than is true of others who may potentially be equally gifted." While I do not believe that the time will ever come when each man can release his own Shakespeare, I do not doubt that the freedom to create is somehow linked with facility of access to those obscure regions below the conscious mind.

There is, perhaps, a wonderful analogy here between the potential fecundity of life in the universe and those novelties which natural selection in a given era permits to break through the living screen, the biosphere, into reality. Organic opportunity has thus placed sharp limits upon a far greater life potential than is ever permitted to enter the actual world. This other hidden world, a world of possible but nonexistent futures, is a constant accompaniment, a real but wholly latent twin, of the nature in which we have our being. In a strangely similar manner the mental censor of a too rigidly blocked or distorted unconscious may interfere, not alone with genius, but even with what might be called ordinary productivity.

Just as, in a given situation, the living biological screen may prevent the emergence of a higher form of life or precipitate its destruction, so in that dark, soundless area of the brain, which parallels the similarly pregnant void of space, much may be barred from creation that exists only as a potentiality. Here again, culturally imposed forms and individual experiences may open or keep permanently closed the doorways of life. The role of purely genetic expression becomes frightfully obscured by the environmental complexities which surround the birth and development of the individual. There is no doubt that clinical studies devoted to creativity, including private interviews with cooperating and contemporary men of genius, offer the prospect of gaining greater insight into those dark alleys and byways out of which stumble at infrequent intervals the Shelleys, the Shakespeares, the Newtons, and the Darwins of our world.

Sometimes they are starved by poverty, sometimes self-schooled, sometimes they have known wealth, sometimes they have appeared like comets across an age of violence. Or they have been selfless, they have been beautiful or unlovely of body, they have been rake or puritan. One thing alone they have had in common: thought, music, art, transmissible but unique.

VII

If we, as ordinary educators, have the task which Dewey envisioned of transmitting from these enrichers of life their wisdom to the unformed turbulent future, of transforming reflection into action consonant with their thought, then some of their luminosity must encompass our minds; their passion must, in some degree, break through our opaque thoughts and descend to us. Whether we will it so or not, we play, in another form, the part of the biological screen in the natural world, or the psychological censor to the individual human mind.

As educators, we play this role in our own culture. In innumerable small ways, if we are rigid, dogmatic, arrogant, we shall be laying stone upon stone, an ugly thing. We shall, for such is our power, give the semblance of necessary reality to a future that need never have been permitted. The educator can be the withholder as well as the giver of life.

I began by offering a case history for examination, the case history of an obscure educator who, within the course of a single lifetime has passed from a world of almost primordial illiteracy and isolation to one which permits wandering at will among such towering minds as I have been discussing. In the end, their loneliness has been my loneliness; their poverty I have endured; their wasted days have been my own. Even their desolate islands, their deserts, and their forests have been mine to tread.

Unlike them I cannot speak with tongues, unlike them I cannot even adequately describe my wanderings. Yet for a brief interval as a teacher and lecturer I have been allowed to act as their interpreter and because no man knows what vibrations he sets in motion in his lifetime, I am content. Has not Saint Paul said that there are many kinds of voices in the world and none is without signification?

At a university's opening exercises, in this era of carefully directed advising, in this day of grueling college board exam-inations and aptitude tests, I have been permitted just once to cry out to our herded youngsters: "Wait, forget the Dean of Admissions who, if I came today in youth before him might not have permitted me to register; be wary of our dubious advice. Freshmen, sophomores, with the gift of youth upon you, do not be prematurely withered up by us. Are you uncer-tain about your destiny? Take heart, in middle age I am still seeking my true calling. I was born a stranger. Perhaps some of you are strangers too." I said this, and much more besides, and was blushing for my impulsive folly, when students I did not know began to invade my office or come up to speak to me on the campus.

I learned in that moment just how much we have lost in our inability to communicate across the generations. I learned how deep, how wrapped in mummy cloth, repressed, long buried, lies in our minds the darkest wound that time has given us. The men of my profession speak frequently of the physical scars of evolution. They mean by this that we carry in our bodies evidence of the long way we have traveled. There is written even in our bones the many passages at arms upon the road. To the student of the past we are as scarred and ragged as old battle flags. We drag with us into the future the tatters of defeat as well as of victory: impulses of deep-buried animal aggression, unconscious midbrain rivalries that hurl us into senseless accidents upon the road, even as nations, which, after all, are but a few men magnified, similarly destroy themselves upon the even more dangerous roads of history.

From the standpoint of the hungry spirit, however, humanity has suffered an even greater wound—the ability of the mind to extend itself across a duration greater than the capacity of mortal flesh to endure. It is part of the burden of all these hungry creative voices that assail us, as it is part of the unsophisticated but equally hungry students who followed me to my office, not because of me but because of some chord in their minds which I must have momentarily and unconsciously plucked. It is, in brief, the wound of time—that genius of man, which, as Emerson long ago remarked, "is a continuation of the power that made him and that has not done making him."

When man acquired "otherness," when he left the safe confines of the instinctive world of animals, he became conscious of his own mortality. Locked in this evolution-scarred and wounded body was a mind which needed only the stimulus of knowledge to reach across the eons of the past or hurl itself upon the future. Has it occurred to us sufficiently that it is part of the continuing growth of this mind that it may desire to be lost—lost among whalebones in the farthest seas, in that great book of which its author, Melville wrote, "a polar wind blows through it," lost among the instinctive villainies of insects in the parched field where Henri Fabre labored, lost with St. Exupéry amidst the crevasses and thin air of the high Andes, down which the first airmail pilots drifted to their deaths? Are these great visions and insights matter for one lifetime that we must needs compress them at twenty, safely controlled, into the little rivulet of a single profession—the powers exorcised, the magnificent torrent siphoned into the safe container of a single life—this mind which, mortal and encased in flesh, would contain the past and seek to devour eternity? "As the dead man is spiritualized . . . ," remarks Thoreau, "so the imagination requires a long range. It is the faculty of the poet to see present things as if also past and future; as if distant or universally significant." The evolutionary wound we bear has been the creation of a thing abstracted out of time yet trapped within it: the mind, by chance distorted, locked into a white-ribbed cage which effervesces into air the moment it approaches wisdom.

These are our students and ourselves, sticks of fragile calcium, little sacs of watery humors that dry away in too much heat or turn crystalline with cold; no great thing, really, if the thermometer is to be our gauge, or the small clerk stool of a single profession is to measure life. Rather, "What shall be grand in thee," wrote Melville searching his far-flung waters of memory, "must needs be plucked at from the skies, and dived for in the deep, and featured in the unbodied air."

VIII

Not all men possess such stamina, not all sustain such flights of intellect, but each of us should be aware that they exist. It is important that creature comfort does not dull the mind to somnolence. "The more an organism learns," John Dewey wrote, "the more it has to learn in order to keep itself going." This acceleration, so well documented in the history of civilization, is perceived by Dewey to characterize inflexibly the life of the individual. As an archaeologist, however, gazing from the air upon the faint outlines of the neolithic hill forts still visible upon the English downs or, similarly, upon the monoliths of Stonehenge, I am aware of something other than the geometric extension of power, whether in a civilization or a man.

There comes a time when the thistles spring up over man's ruins with a sense of relief. It is as though the wasting away of power through time had brought with it the retreat of something shadowy and not untouched with evil. The tiny incremental thoughts of men tend to congeal in vast fabrics, from gladiatorial coliseums to skyscrapers, and then mutely demand release. In the end the mind rejects the hewn stone and rusting iron it has used as the visible expression of its inner dream. Instead it asks release for new casts at eternity, new opportunities to confine the uncapturable and elusive gods.

It is one of the true functions of education to teach, in just this figurative way, the pure recompense of observing sunlight and the nodding weed wash over our own individual years and ruins. Joy was there and lingers in the grasses, the black wrong lies forever buried, and the tortured mind may seek its peace. Here all is open to the sun. The youthful rebel lies down with his double, the successful man. Or the reverse—who is to judge?

I, who endured the solitude of an ice age in my youth, re-member now the yellow buttercups of the only picnic I was ever taken on in kindergarten. There are other truths than those contained in laboratory burners, on blackboards, or in test tubes. With the careful suppressions of age the buttercups grow clearer in my memory year by year. I am trying to write honestly from my own experience. I am trying to say that buttercups, a mastodon tooth, a giant snail, and a rolling Eliz-abethan line are a part of my own ruins over which the weeds grow tall.

I have not paused there in many years, but the light grows long in retrospect, and I have peace because I am released from pain. I know not how, yet I know also that I have been in some degree created by those lost objects in the grass. We are in truth sculptors in snow, we educators, but, thank God, we are sometimes aided by that wild fitfulness which is called "hazard," "contingency," and the indeterminacy which Dewey labeled "thinking." If the mind is indigenous and integral to nature itself in its unfolding, and operates in nature's ways and under nature's laws, we must seek to understand this creative aspect of nature in its implications for the human mind. I have tried, therefore, to point out that the natural laws of the mind include an emergent novelty with which education has to cope and elaborate for its best and fullest realization.

In Bimini, on the old Spanish Main, a black girl once said to me, "Those as hunts treasure must go alone, at night, and when they find it they have to leave a little of their blood be-hind them."

I have never heard a finer, cleaner estimate of the price of wisdom. I wrote it down at once under a sea lamp, like the belated pirate I was, for the girl had given me unknowingly the latitude and longitude of a treasure—a treasure more valuable than all the aptitude tests of this age.

FOURTEEN

The Brown Wasps

THERE IS A CORNER in the waiting room of one of the great
Eastern stations where women never sit. It is always in the
shadow and overhung by rows of lockers. It is, however, always frequented—not so much by genuine travelers as by the
dying. It is here that a certain element of the abandoned poor
seeks a refuge out of the weather, clinging for a few hours longer to the city that has fathered them. In a precisely similar
manner I have seen, on a sunny day in midwinter, a few old
brown wasps creep slowly over an abandoned wasp nest in a
thicket. Numbed and forgetful and frost-blackened, the hum
of the spring hive still resounded faintly in their sodden tissues.
Then the temperature would fall and they would drop away
into the white oblivion of the snow. Here in the station it is in
no way different save that the city is busy in its snows. But the
old ones cling to their seats as though these were symbolic and
could not be given up. Now and then they sleep, their gray old
heads resting with painful awkwardness on the backs of the
benches.

Also they are not at rest. For an hour they may sleep in the
gasping exhaustion of the ill-nourished and aged who have to
walk in the night. Then a policeman comes by on his round
and nudges them upright.

"You can't sleep here," he growls.

A strange ritual then begins. An old man is difficult to
waken. After a muttered conversation the policeman presses a
coin into his hand and passes fiercely along the benches prod-
ding and gesturing toward the door. In his wake, like birds
rising and settling behind the passage of a farmer through a
cornfield, the men totter up, move a few paces, and subside
once more upon the benches.

One man, after a slight, apologetic lurch, does not move at
all. Tubercularly thin, he sleeps on steadily. The policeman
does not look back. To him, too, this has become a ritual. He
will not have to notice it again officially for another hour.

Once in a while one of the sleepers will not awake. Like the brown wasps, he will have had his wish to die in the great droning center of the hive rather than in some lonely room. It is not so bad here with the shuffle of footsteps and the knowledge that there are others who share the bad luck of the world. There are also the whistles and the sounds of everyone, everyone in the world, starting on journeys. Amidst so many journeys somebody is bound to come out all right. Somebody.

Maybe it was on a like thought that the brown wasps fell away from the old paper nest in the thicket. You hold till the last, even if it is only to a public seat in a railroad station. You want your place in the hive more than you want a room or a place where the aged can be eased gently out of the way. It is the place that matters, the place at the heart of things. It is life that you want, that bruises your gray old head with the hard chairs; a man has a right to his place.

But sometimes the place is lost in the years behind us. Or sometimes it is a thing of air, a kind of vaporous distortion above a heap of rubble. We cling to a time and a place because without them man is lost, not only man but life. This is why the voices, real or unreal, which speak from the floating trumpets at spiritualist seances are so unnerving. They are voices out of nowhere whose only reality lies in their ability to stir the memory of a living person with some fragment of the past. Before the medium's cabinet both the dead and the living revolve endlessly about an episode, a place, an event that has already been engulfed by time.

This feeling runs deep in life; it brings stray cats running over endless miles, and birds homing from the ends of the earth. It is as though all living creatures, and particularly the more intelligent, can survive only by fixing or transforming a bit of time into space or by securing a bit of space with its objects immortalized and made permanent in time. For example, I once saw, on a flower pot in my own living room, the efforts of a field mouse to build a remembered field. I have lived to see this episode repeated in a thousand guises, and since I have spent a large portion of my life in the shade of a nonexistent tree I think I am entitled to speak for the field mouse.

One day as I cut across the field which at that time extended on one side of our suburban shopping center, I found a giant slug feeding from a runnel of pink ice cream in an abandoned Dixie cup. I could see his eyes telescope and protrude in a kind of dim uncertain ecstasy as his dark body bunched and elongated in the curve of the cup. Then, as I stood there at the edge of the concrete, contemplating the slug, I began to realize it was like standing on a shore where a different type of life creeps up and fumbles tentatively among the rocks and sea wrack. It knows its place and will only creep so far until something changes. Little by little as I stood there I began to see more of this shore that surrounds the place of man. I looked with sudden care and attention at things I had been running over thoughtlessly for years. I even waded out a short way into the grass and the wild-rose thickets to see more. A huge black-belted bee went droning by and there were some indistinct scurryings in the underbrush.

Then I came to a sign which informed me that this field was to be the site of a new Wanamaker suburban store. Thousands of obscure lives were about to perish, the spores of puffballs would go smoking off to new fields, and the bodies of little white-footed mice would be crunched under the inexorable wheels of the bulldozers. Life disappears or modifies its appearances so fast that everything takes on an aspect of illusion —a momentary fizzing and boiling with smoke rings, like pouring dissident chemicals into a retort. Here man was advancing, but in a few years his plaster and bricks would be disappearing once more into the insatiable maw of the clover. Being of an archaeological cast of mind, I thought of this fact with an obscure sense of satisfaction and waded back through the rose thickets to the concrete parking lot. As I did so, a mouse scurried ahead of me, frightened of my steps if not of that ominous Wanamaker sign. I saw him vanish in the general direction of my apartment house, his little body quivering with fear in the great open sun on the blazing concrete. Blinded and confused, he was running straight away from his field. In another week scores would follow him.

I forgot the episode then and went home to the quiet of my living room. It was not until a week later, letting myself into the apartment, that I realized I had a visitor. I am fond of plants and had several ferns standing on the floor in pots to avoid the noon glare by the south window.

As I snapped on the light and glanced carelessly around the room, I saw a little heap of earth on the carpet and a scrabble of pebbles that had been kicked merrily over the edge of one of the flower pots. To my astonishment I discovered a full-fledged burrow delving downward among the fern roots. I waited silently. The creature who had made the burrow did not appear. I remembered the wild field then, and the flight of the mice. No house mouse, no *Mus domesticus*, had kicked up this little heap of earth or sought refuge under a fern root in a flower pot. I thought of the desperate little creature I had seen fleeing from the wild-rose thicket. Through intricacies of pipes and attics, he, or one of his fellows, had climbed to this high green solitary room. I could visualize what had occurred. He had an image in his head, a world of seed pods and quiet, of green sheltering leaves in the dim light among the weed stems. It was the only world he knew and it was gone.

Somehow in his flight he had found his way to this room with drawn shades where no one would come till nightfall. And here he had smelled green leaves and run quickly up the flower pot to dabble his paws in common earth. He had even struggled half the afternoon to carry his burrow deeper and had failed. I examined the hole, but no whiskered twitching face appeared. He was gone. I gathered up the earth and re-filled the burrow. I did not expect to find traces of him again.

Yet for three nights thereafter I came home to the darkened room and my ferns to find the dirt kicked gaily about the rug and the burrow reopened, though I was never able to catch the field mouse within it. I dropped a little food about the mouth of the burrow, but it was never touched. I looked under beds or sat reading with one ear cocked for rustlings in the ferns. It was all in vain; I never saw him. Probably he ended in a trap in some other tenant's room.

But before he disappeared I had come to look hopefully for his evening burrow. About my ferns there had begun to linger the insubstantial vapor of an autumn field, the distilled essence, as it were, of a mouse brain in exile from its home. It was a small dream, like our dreams, carried a long and weary journey along pipes and through spider webs, past holes over which loomed the shadows of waiting cats, and finally, desperately, into this room where he had played in the shuttered daylight for an hour among the green ferns on the floor. Every day these invisible dreams pass us on the street, or rise from beneath our feet, or look out upon us from beneath a bush.

Some years ago the old elevated railway in Philadelphia was torn down and replaced by a subway system. This ancient El with its barnlike stations containing nut-vending machines and scattered food scraps has, for generations, been the favorite feeding ground of flocks of pigeons, generally one flock to a station along the route of the El. Hundreds of pigeons were dependent upon the system. They flapped in and out of its stanchions and steel work or gathered in watchful little audiences about the feet of anyone who rattled the peanut-vending machines. They even watched people who jingled change in their hands, and prospected for food under the feet of the crowds who gathered between trains. Probably very few among the waiting people who tossed a crumb to an eager pigeon realized that this El was like a food-bearing river, and that the life which haunted its banks was dependent upon the running of the trains with their human freight.

I saw the river stop.

The time came when the underground tubes were ready; the traffic was transferred to a realm unreachable by pigeons. It was like a great river subsiding suddenly into desert sands. For a day, for two days, pigeons continued to circle over the El or stand close to the red vending machines. They were patient birds, and surely this great river which had flowed through the lives of unnumbered generations was merely suffering from some momentary drought.

They listened for the familiar vibrations that had always heralded an approaching train; they flapped hopefully about the head of an occasional workman walking along the steel runways. They passed from one empty station to another, all the while growing hungrier. Finally they flew away.

I thought I had seen the last of them about the El, but there was a revival and it provided a curious instance of the memory of living things for a way of life or a locality that has long been cherished. Some weeks after the El was abandoned workmen began to tear it down. I went to work every morning by one particular station, and the time came when the demolition crews reached this spot. Acetylene torches showered passers-by with sparks, pneumatic drills hammered at the base of the structure, and a blind man who, like the pigeons, had clung with his cup to a stairway leading to the change booth, was forced to give up his place.

It was then, strangely, momentarily, one morning that I witnessed the return of a little band of the familiar pigeons. I even recognized one or two members of the flock that had lived around this particular station before they were dispersed into the streets. They flew bravely in and out among the sparks and the hammers and the shouting workmen. They had returned— and they had returned because the hubbub of the wreckers had convinced them that the river was about to flow once more. For several hours they flapped in and out through the empty windows, nodding their heads and watching the fall of girders with attentive little eyes. By the following morning the station was reduced to some burned-off stanchions in the street. My bird friends had gone. It was plain, however, that they retained a memory for an insubstantial structure now compounded of air and time. Even the blind man clung to it. Someone had provided him with a chair, and he sat at the same corner staring sightlessly at an invisible stairway where, so far as he was concerned, the crowds were still ascending to the trains.

I have said my life has been passed in the shade of a nonexistent tree, so that such sights do not offend me. Prematurely I am one of the brown wasps and I often sit with them in the great droning hive of the station, dreaming sometimes of a certain tree. It was planted sixty years ago by a boy with a bucket and a toy spade in a little Nebraska town. That boy was myself. It was a cottonwood sapling and the boy remembered it because of some words spoken by his father and because everyone died or moved away who was supposed to wait and grow old under its shade. The boy was passed from hand to hand, but the tree for some intangible reason had taken root in his mind. It was under its branches that he sheltered; it was from this tree that his memories, which are my memories, led away into the world.

After sixty years the mood of the brown wasps grows heavier upon one. During a long inward struggle I thought it would do me good to go and look upon that actual tree. I found a

rational excuse in which to clothe this madness. I purchased a
ticket and at the end of two thousand miles I walked another
mile to an address that was still the same. The house had not
been altered.

I came close to the white picket fence and reluctantly, with
great effort, looked down the long vista of the yard. There was
nothing there to see. For sixty years that cottonwood had been
growing in my mind. Season by season its seeds had been floating farther on the hot prairie winds. We had planted it lovingly
there, my father and I, because he had a great hunger for soil
and live things growing, and because none of these things had
long been ours to protect. We had planted the little sapling
and watered it faithfully, and I remembered that I had run out
with my small bucket to drench its roots the day we moved
away. And all the years since it had been growing in my mind,
a huge tree that somehow stood for my father and the love I
bore him. I took a grasp on the picket fence and forced myself
to look again.

A boy with the hard bird eye of youth pedaled a tricycle
slowly up beside me.

"What'cha lookin' at?" he asked curiously.

"A tree," I said.

"What for?" he said.

"It isn't there," I said, to myself mostly, and began to walk
away at a pace just slow enough not to seem to be running.

"What isn't there?" the boy asked. I didn't answer. It was
obvious I was attached by a thread to a thing that had never
been there, or certainly not for long. Something that had to be
held in the air, or sustained in the mind, because it was part of
my orientation in the universe and I could not survive without
it. There was more than an animal's attachment to a place.
There was something else, the attachment of the spirit to a
grouping of events in time; it was part of our mortality.

So I had come home at last, driven by a memory in the brain
as surely as the field mouse who had delved long ago into my
flower pot or the pigeons flying forever amidst the rattle of
nut-vending machines. These, the burrow under the greenery
in my living room and the red-bellied bowls of peanuts now
hovering in midair in the minds of pigeons, were all part of an
elusive world that existed nowhere and yet everywhere. I

looked once at the real world about me while the persistent boy pedaled at my heels.

It was without meaning, though my feet took a remembered path. In sixty years the house and street had rotted out of my mind. But the tree, the tree that no longer was, that had perished in its first season, bloomed on in my individual mind, unblemished as my father's words. "We'll plant a tree here, son, and we're not going to move any more. And when you're an old, old man you can sit under it and think how we planted it here, you and me, together."

I began to outpace the boy on the tricycle.

"Do you live here, Mister?" he shouted after me suspiciously. I took a firm grasp on airy nothing—to be precise, on the bole of a great tree. "I do," I said. I spoke for myself, one field mouse, and several pigeons. We were all out of touch but somehow permanent. It was the world that had changed.

Bibliography

Aiken, Conrad. *Collected Poems*. New York: Oxford University Press, 1953.

Arvin, Newton. *Herman Melville: A Critical Biography*. New York: Viking, Compass Books, 1957.

Brandon, S. G. F. *History, Time and Deity*. New York: Barnes and Noble, 1965.

Cottrell, Leonard. *Lost Cities*. New York: Rinehart, 1957.

Cruttwell, Patrick. *The Shakespearian Moment*. New York: Random House, 1959.

De La Mare, Walter. *Ding Dong Bell*. London: Faber & Faber, 1924.

Dewey, John. *The Quest for Certainty*. New York: Putnam, Capricorn Edition, 1960.

———. *Experience and Nature*. 2nd ed. New York: Dover Publications, 1958.

Dodds, E. R. *Pagan and Christian in an Age of Anxiety*. New York: Cambridge University Press, 1965.

Eiseley, Loren. *Darwin's Century*. New York: Doubleday, 1958.

Flaceliere, Robert. *Greek Oracles*. London: Elek Books, 1965.

Goneim, M. Zakaria. *The Lost Pyramid*. New York: Rinehart, 1956.

Heim, Karl. *The Transformation of the Scientific World View*. London: SCM Press Ltd., 1953.

Jarrell, Randall. "On Preparing to Read Kipling," *American Scholar*, 31 (1962), 220–235.

Kubie, Lawrence. *Neurotic Distortion of the Creative Process*. Lawrence: University of Kansas Press, 1958.

Machen, Arthur. *The Hill of Dreams*. New York: Alfred A. Knopf, 1927.

Manuel, Frank E. *Shapes of Philosophical History*. Stanford, Calif.: Stanford University Press, 1965.

Matthews, Kenneth B., Jr. *Cities in the Sand: Leptis Magna and Sabratha in Roman Africa*. Philadelphia: University of Pennsylvania Press, 1957.

Medawar, P. B. *The Future of Man*. New York: Mentor Books, 1961.

Newsom, Carroll. *A University President Speaks Out*. New York: Harper & Row, 1961.

Nicolson, Marjorie Hope. *Mountain Gloom and Mountain Glory*. New York: W. W. Norton, The Norton Library, 1963.

Olson, Charles. *Call Me Ishmael: A Study of Melville*. New York: Grove Press, 1947.

Petitclerc, Denne. Interview with Robert Hofstadter, *San Francisco Chronicle*, November 4, 1961, p. 4.

Rieff, Philip. *Freud: The Mind of the Moralist*. New York: Doubleday Anchor Books, 1961.

Robertson, John (ed.). *The Philosophical Works of Francis Bacon*. 1 vol. ed. London: George Routledge & Sons, 1905.

Rokeach, Milton. "The Pursuit of the Creative Process," in *The Creative Organization*, Gary A. Steiner (ed.), Glencoe, Ill.: The Free Press, 1962.

Rostand, Jean. *The Substance of Man*. New York: Doubleday, 1962.

Schneer, Cecil J. *The Search for Order*. New York: Harper & Row, 1960.

Thompson, Ruth D'Arcy. *D'Arcy Wentworth Thompson: The Scholar-Naturalist, 1860–1948*. London: Oxford University Press, 1958.

Thoreau, Henry David. *A Writer's Journal*. Edited by Laurence Stapleton. New York: Dover Publications, 1960.

Van Doren, Mark. *Nathaniel Hawthorne: A Critical Biography*. New York: Viking, Compass Books, 1957.

Ward, F. Kingdon. *Modern Exploration*. London: Jonathan Cape, 1945.

Weigal, Arthur E. P. *Travels in the Upper Egyptian Deserts*. London: William Blackwood and Sons, 1913.

Williams, Charles. *Bacon*. New York: Harper & Brothers [n.d.].

Essays from

THE STAR THROWER

The Long Loneliness

THERE IS NOTHING more alone in the universe than man. He is alone because he has the intellectual capacity to know that he is separated by a vast gulf of social memory and experiment from the lives of his animal associates. He has entered into the strange world of history, of social and intellectual change, while his brothers of the field and forest remain subject to the invisible laws of biological evolution. Animals are molded by natural forces they do not comprehend. To their minds there is no past and no future. There is only the everlasting present of a single generation—its trails in the forest, its hidden pathways of the air and in the sea.

Man, by contrast, is alone with the knowledge of his history until the day of his death. When we were children we wanted to talk to animals and struggled to understand why this was impossible. Slowly we gave up the attempt as we grew into the solitary world of human adulthood; the rabbit was left on the lawn, the dog was relegated to his kennel. Only in acts of inarticulate compassion, in rare and hidden moments of communion with nature, does man briefly escape his solitary destiny. Frequently in science fiction he dreams of worlds with creatures whose communicative power is the equivalent of his own.

It is with a feeling of startlement, therefore, and eager interest touching the lost child in every one of us, that the public has received the recent accounts of naval research upon the intelligence of one of our brother mammals—the sea-dwelling bottle-nosed porpoise or dolphin.

These small whales who left the land millions of years ago to return to the great mother element of life, the sea, are now being regarded by researchers as perhaps the most intelligent form of life on our planet next to man. Dr. John Lilly of the Communications Research Institute in the Virgin Islands reports that the brain of the porpoise is 40 per cent larger than man's and is just as complex in its functional units. Amazed by the rapidity with which captive porpoises solved problems that even monkeys found difficult, Dr. Lilly is quoted as expressing the view that "man's position at the top of the hierarchy [of intelligence] begins to be questioned."

Dr. Lilly found that his captives communicated in a series of underwater whistles and that, in addition, they showed an amazing "verbalizing" ability in copying certain sounds heard in the laboratory. The experimental animal obviously hoped to elicit by this means a reproduction of the pleasurable sensations he had been made to experience under laboratory conditions. It is reported that in spite of living in a medium different from the one that man inhabits, and therefore having quite a different throat structure, one of the porpoises even uttered in a Donald-Duckish voice a short number series it had heard spoken by one of the laboratory investigators.

The import of these discoveries is tremendous and may not be adequately known for a long time. An animal from a little-explored medium, which places great barriers in the way of the psychologist, has been found to have not only a strong social organization but to show a degree of initiative in experimental communicative activity unmatched by man's closest relatives, the great apes. The porpoises reveal, moreover, a touching altruism and friendliness in their attempts to aid injured companions. Can it be, one inevitably wonders, that man is so locked in his own type of intelligence—an intelligence that is linked to a prehensile, grasping hand giving him power over his environment—that he is unable to comprehend the intellectual life of a highly endowed creature from another domain such as the sea?

Perhaps the water barrier has shut us away from a potentially communicative and jolly companion. Perhaps we have some things still to learn from the natural world around us before we turn to the far shores of space and whatever creatures may await us there. After all, the porpoise is a mammal. He shares with us an ancient way of birth and affectionate motherhood. His blood is warm, he breathes air as we do. We both bear in our bodies the remnants of a common skeleton torn asunder for divergent purposes far back in the dim dawn of mammalian life. The porpoise has been superficially streamlined like a fish.

His are not, however, the cold-blooded ways of the true fishes. Far higher on the tree of life than fishes, the dolphin's paddles are made-over paws, rather than fins. He is an ever-constant reminder of the versatility of life and its willingness to pass through strange dimensions of experience. There are environmental worlds on earth every bit as weird as what we may imagine to revolve by far-off suns. It is our superficial familiarity with this planet that inhibits our appreciation of the unknown until a porpoise, rearing from a tank to say Three-Two-Three, re-creates for us the utter wonder of childhood.

Unless we are specialists in the study of communication and its relation to intelligence, however, we are apt to oversimplify or define poorly what intelligence is, what communication and language are, and thus confuse and mystify both ourselves and others. The mysteries surrounding the behavior of the bottle-nosed porpoise, and even of man himself, are not things to be probed simply by the dissector's scalpel. They lie deeper. They involve the whole nature of the mind and its role in the universe.

We are forced to ask ourselves whether native intelligence in another form than man's might be as high as or even higher than his own, yet be marked by no such material monuments as man has placed upon the earth. At first glance we are alien to this idea, because man is particularly a creature who has turned the tables on his environment so that he is now engrossed in shaping it, rather than being shaped by it. Man expresses himself upon his environment through the use of tools. We therefore tend to equate the use of tools in a one-to-one relationship with intelligence.

The question we must now ask ourselves, however, is whether this involves an unconsciously man-centered way of looking at intelligence. Let us try for a moment to enter the dolphin's kingdom and the dolphin's body, retaining, at the same time, our human intelligence. In this imaginative act, it may be possible to divest ourselves of certain human preconceptions about our kind of intelligence and at the same time to see more clearly why mind, even advanced mind, may have manifestations other than the tools and railroad tracks and laboratories that we regard as evidence of intellect. If we are particularly adept in escaping from our own bodies, we may even learn to discount a little the kind of world of rockets and death that our type of busy human curiosity, linked to a hand noted for its ability to open assorted Pandora's boxes, has succeeded in foisting upon the world as a symbol of universal intelligence.

We have now sacrificed, in our imagination, our hands for flippers and our familiar land environment for the ocean. We will go down into the deep waters as naked of possessions as when we entered life itself. We will take with us one thing alone that exists among porpoises as among men: an ingrained biological gregariousness—a sociality that in our new world will permit us to run in schools, just as early man ran in the packs that were his ancient anthropoid heritage. We will assume in the light of Dr. Lilly's researches that our native intelligence, as distinguished from our culturally transmitted habits, is very high. The waters have closed finally over us, our paws have been sacrificed for the necessary flippers with which to navigate.

The result is immediately evident and quite clear: No matter how well we communicate with our fellows through the water medium we will never build drowned empires in the coral; we will never inscribe on palace walls the victorious boasts of porpoise kings. We will know only water and the wastes of water beyond the power of man to describe. We will be secret visitors in hidden canyons beneath the mouths of torrential rivers. We will survey in innocent astonishment the flotsam that pours from the veins of continents—dead men, great serpents, giant trees—or perhaps the little toy boat of a child loosed far upstream will come floating past. Bottles with winking green lights will plunge by us into the all-embracing ooze. Meaningless appearances and disappearances will comprise our philosophies. We will hear the earth's heart ticking in its thin granitic shell. Volcanic fires will growl ominously in steam-filled crevices. Vapor, bird cries, and sea wrack will compose our memories. We will see death in many forms and, on occasion, the slow majestic fall of battleships through the green light that comes from beyond our domain.

Over all that region of wondrous beauty we will exercise no more control than the simplest mollusk. Even the octopus with flexible arms will build little shelters that we cannot imitate. Without hands we will have only the freedom to follow the untrammeled sea winds across the planet.

Perhaps if those whistling sounds that porpoises make are truly symbolic and capable of manipulation in our brains, we will wonder about the world in which we find ourselves—but it will be a world not susceptible to experiment. At best we may nuzzle in curiosity a passing shipbottom and be harpooned for our pains. Our thoughts, in other words, will be as limited as those of the first men who roved in little bands in the times before fire and the writing that was to open to man the great doorway of his past.

Man without writing cannot long retain his history in his head. His intelligence permits him to grasp some kind of succession of generations; but without writing, the tale of the past rapidly degenerates into fumbling myth and fable. Man's greatest epic, his four long battles with the advancing ice of the great continental glaciers, has vanished from human memory without a trace. Our illiterate fathers disappeared and with them, in a few scant generations, died one of the great stories of all time. This episode has nothing to do with the biological quality of a brain as between then and now. It has to do instead with a device, an invention made possible by the hand. That invention came too late in time to record eyewitness accounts of the years of the Giant Frost.

Primitives of our own species, even today, are historically shallow in their knowledge of the past. Only the poet who writes speaks his message across the millennia to other hearts. Only in writing can the cry from the great cross on Golgotha still be heard in the minds of men. The thinker of perceptive insight, even if we allow him for the moment to be a porpoise rather than a man, has only his individual glimpse of the universe until such time as he can impose that insight upon unnumbered generations. In centuries of pondering, man has come upon but one answer to this problem: speech translated into writing that passes beyond human mortality.

Writing, and later printing, is the product of our adaptable many-purposed hands. It is thus, through writing, with no increase in genetic, inborn capacity since the last ice advance, that modern man carries in his mind the intellectual triumphs of all his predecessors who were able to inscribe their thoughts for posterity.

All animals which man has reason to believe are more than usually intelligent—our relatives the great apes, the elephant, the raccoon, the wolverine, among others—are problem solvers, and in at least a small way manipulators of their environment.

Save for the instinctive calls of their species, however, they cannot communicate except by direct imitation. They cannot invent words for new situations nor get their fellows to use such words. No matter how high the individual intelligence, its private world remains a private possession locked forever within a single, perishable brain. It is this fact that finally balks our hunger to communicate even with the sensitive dog who shares our fireside.

Dr. Lilly insists, however, that the porpoises communicate in high-pitched, underwater whistles that seem to transmit their wishes and problems. The question then becomes one of ascertaining whether these sounds represent true language—in the sense of symbolic meanings, additive, learned elements—or whether they are simply the instinctive signals of a pack animal. To this there is as yet no clear answer, but the eagerness with which laboratory sounds and voices were copied by captive porpoises suggests a vocalizing ability extending perhaps to or beyond the threshold of speech.

Most of the intelligent land animals have prehensile, grasping organs for exploring their environment—hands in man and his anthropoid relatives, the sensitive inquiring trunk in the elephant. One of the surprising things about the porpoise is that his superior brain is unaccompanied by any type of manipulative organ. He has, however, a remarkable range-finding ability involving some sort of echo-sounding. Perhaps this acute sense—far more accurate than any man has been able to devise artificially—brings him greater knowledge of his watery surroundings than might at first seem possible. Human beings think of intelligence as geared to things. The hand and the tool are to us the unconscious symbols of our intellectual achievement. It is difficult for us to visualize another kind of lonely, almost disembodied intelligence floating in the wavering green fairyland of the sea—an intelligence possibly near or comparable to our own but without hands to build, to transmit knowledge by writing, or to alter by one hairsbreadth the planet's surface. Yet at the same time there are indications that this is a warm, friendly and eager intelligence quite capable of coming to the assistance of injured companions and striving to rescue them from drowning. Porpoises left the land when mammalian brains were still small and primitive. Without the stimulus provided by agile exploring fingers, these great sea mammals have yet taken a divergent road toward intelligence of a high order. Hidden in their sleek bodies is an impressively elaborated instrument, the reason for whose appearance is a complete enigma. It is as though both man and porpoise were each part of some great eye which yearned to look both outward on eternity and inward to the sea's heart—that fertile entity so like the mind in its swarming and grotesque life.

Perhaps man has something to learn after all from fellow creatures without the ability to drive harpoons through living flesh, or poison with strontium the planetary winds. One is reminded of those watery blue vaults in which, as in some idyllic eternity, Herman Melville once saw the sperm whales nurse their young. And as Melville wrote of the sperm whale, so we might now paraphrase his words in speaking of the porpoise. "Genius in the porpoise? Has the porpoise ever written a book, spoken a speech? No, his great genius is declared in his doing nothing particular to prove it. It is declared in his pyramidical silence." If man had sacrificed his hands for flukes, the moral might run, he would still be a philosopher, but there would have been taken from him the devastating power to wreak his thought upon the body of the world. Instead he would have lived and wandered, like the porpoise, homeless across currents and winds and oceans, intelligent, but forever the lonely and curious observer of unknown wreckage falling through the blue light of eternity. This role would now be a deserved penitence for man. Perhaps such a transformation would bring him once more into that mood of childhood innocence in which he talked successfully to all things living but had no power and no urge to harm. It is worth at least a wistful thought that someday the porpoise may talk to us and we to him. It would break, perhaps, the long loneliness that has made man a frequent terror and abomination even to himself.

Man the Firemaker

MAN, IT IS WELL to remember, is the discoverer but not the inventor of fire. Long before this meddling little Prometheus took to experimenting with flints, then matches, and finally (we hope not too finally) hydrogen bombs, fires had burned on this planet. Volcanoes had belched molten lava, lightning had struck in dry grass, winds had rubbed dead branches against each other until they burst into flame. There are evidences of fire in ancient fossil beds that lie deep below the time of man.

Man did not invent fire but he did make it one of the giant powers on the earth. He began this experiment long ago in the red morning of the human mind. Today he continues it in the midst of coruscating heat that is capable of rending the very fabric of his universe. Man's long adventure with knowledge has, to a very marked degree, been a climb up the heat ladder, for heat alone enables man to mold metals and glassware, to create his great chemical industries, to drive his swift machines. It is my intention here to trace man's manipulation of this force far back into its ice-age beginnings and to observe the part that fire has played in the human journey across the planet. The torch has been carried smoking through the ages of glacial advance. As we follow man on this journey, we shall learn another aspect of his nature: that he is himself a consuming fire.

At just what level in his intellectual development man mastered the art of making fire is still unknown. Neanderthal man of 50,000 years ago certainly knew the art. Traces of the use of fire have turned up in a cave of Peking man, the primitive human being of at least 250,000 years ago who had a brain only about two-thirds the size of modern man's. And in 1947 Raymond Dart of Witwatersrand University announced the discovery in South Africa of *Australopithecus prometheus*, a man-ape cranium recovered from deposits which he believed showed traces of burned bone.

This startling announcement of the possible use of fire by a

subhuman creature raised a considerable storm in anthropological circles. The chemical identifications purporting to indicate
evidence of fire are now considered highly questionable. It has
also been intimated that the evidence may represent only traces
of a natural brush fire. Certainly, so long as the South African
man-apes have not been clearly shown to be tool users, wide
doubts about their use of fire will remain. There are later sites
of tool-using human beings which do not show traces of fire.

Until there is proof to the contrary, it would seem wise to
date the earliest use of fire to Peking man—*Sinanthropus.*
Other human sites of the same antiquity have not yielded evidence of ash, but this is not surprising, for as a new discovery
the use of fire would have taken time to diffuse from one group
to another. Whether it was discovered once or several times we
have no way of knowing. The fact that fire was in worldwide
use at the beginning of man's civilized history enables us to
infer that it is an old human culture trait—doubtless one of the
earliest. Furthermore, it is likely that man used fire long before
he became sophisticated enough to produce it himself.

In 1865 Sir John Lubbock, a British banker who made a hobby
of popular writing on science, observed: "There can be no
doubt that man originally crept over the earth's surface, little by
little, year by year, just, for instance, as the weeds of Europe
are now gradually but surely creeping over the surface of Australia." This remark was, in its time, a very shrewd and sensible
observation. We know today, however, that there have been
times when man suddenly made great strides across the face of
the earth. I want to review one of those startling expansions—a
lost episode in which fire played a tremendous part. To make
its outlines clear we shall have to review the human drama in
three acts.

The earliest humanlike animals we can discern are the man-
apes of South Africa. Perhaps walking upright on two feet, this
creature seems to have been roaming the East African grasslands about one million years ago. Our ancestor, proto-man,
probably emerged from the tropics and diffused over the region
of warm climate in Eurasia and North Africa. He must have
been dependent upon small game, insects, wild seeds, and
fruits. His life was hard, his search for food incessant, his numbers were small.

The second stage in human history is represented by the first true men. Paleoanthropic man is clearly a tool user, a worker in stone and bone, but there is still something of the isolated tinkerer and fumbler about him. His numbers are still sparse, judging from the paucity of skeletal remains. Short, stocky, and powerful, he spread over the most temperate portions of the Afro-Eurasiatic land mass but never attempted the passage through the high Arctic to America. Through scores of millennia he drifted with the seasons, seemingly content with his troglodyte existence, making little serious change in his array of flint tools. It is quite clear that some of these men knew the use of fire, but many may not have.

The third act begins some 15,000 or 20,000 years ago. The last great ice sheet still lies across northern Europe and North America. Roving on the open tundra and grasslands below those ice sheets is the best-fed and most varied assemblage of grass-eating animals the world has ever seen. Giant long-horned bison, the huge wild cattle of the Pleistocene, graze on both continents. Mammoth and mastodon wander about in such numbers that their bones are later to astonish the first American colonists. Suddenly, into this late paradise of game, there erupts our own species of man—*Homo sapiens.* Just where he came from we do not know. Tall, lithe, long-limbed, he is destined to overrun the continents in the blink of a geological eye. He has an excellent projectile weapon in the shape of the spear thrower. His flint work is meticulous and sharp. And the most aggressive carnivore the world has ever seen comes at a time made for his success: the grasslands are alive with seemingly inexhaustible herds of game.

Yet fire as much as flesh was the magic that opened the way for the supremacy of *Homo sapiens.* We know that he was already the master of fire, for the track of it runs from camp to buried camp: the blackened bones of the animals he killed, mute testimony to the relentless step of man across the continents, lie in hundreds of sites in the Old and the New Worlds. Meat, more precious than the gold for which men later struggled, supplied the energy that carried man across the world. Had it not been for fire, however, all that enormous source of life would have been denied to him: he would have gone on drinking the blood from small kills, chewing wearily at uncooked bone ends or masticating the crackling bodies of grasshoppers.

Fire shortens the digestive process. It breaks down tough masses of flesh into food that the human stomach can easily assimilate. Fire made the difference that enabled man to expand his numbers rapidly and to press on from hunting to more advanced cultures. Yet we take fire so much for granted that this first great upswing in human numbers, this first real gain in the seizure of vast quantities of free energy, has to a remarkable degree eluded our attention.

With fire primitive man did more than cook his meat. He extended the pasture for grazing herds. A considerable school of thought, represented by such men as the geographer Carl Sauer and the anthropologist Omer Stewart, believes that the early use of fire by the aborigines of the New World greatly expanded the grassland areas. Stewart says: "The number of tribes reported using fire leads one to the conclusion that burning of vegetation was a universal culture pattern among the Indians of the U.S. Furthermore, the amount of burning leads to the deduction that nearly all vegetation in America at the time of discovery and exploration was what ecologists would call fire vegetation. That is to say, fire was a major factor, along with soil, moisture, temperature, wind, animals, and so forth, in determining the types of plants occurring in any region. It follows then, that the vegetation of the Great Plains was a fire vegetation." In short, the so-called primeval wilderness which awed our forefathers had already felt the fire of the Indian hunter. Here, as in many other regions, man's fire altered the ecology of the earth.

It had its effect not only on the flora but also on the fauna. Of the great herds of grazing animals that flourished in America in the last Ice Age, not a single trace remains—the American elephants, camels, long-horned bison are all gone. Not all of them were struck down by the hunters' weapons. Sauer argues that a major explanation of the extinction of the great American mammals may be fire. He says that the aborigines used fire drives to stampede game, and he contends that this weapon would have worked with peculiar effectiveness to exterminate such lumbering creatures as the mammoth. I have stood in a gully in western Kansas and seen outlined in the earth the fragmented black bones of scores of bison who had perished in what was probably a man-made conflagration. If, at the end of Pleistocene times, vast ecological changes occurred, if climates shifted, if lakes dried and in other places forests sprang up, and if, in this uncertain and unsteady time, man came with flint and fire upon the animal world about him, he may well have triggered a catastrophic decline and extinction. Five thousand years of man and his smoking weapon rolling down the wind may have finished the story for many a slow-witted animal species. In the great scale of geological time this act of destruction amounts to but one brief hunt.

Man, as I have said, is himself a flame. He has burned through the animal world and appropriated its vast stores of protein for his own. When the great herds failed over many areas, he had to devise new ways to feed his increase or drop back himself into a precarious balance with nature. Here and there on the world's margins there have survived into modern times men who were forced into just such local adjustments. Simple hunters and collectors of small game in impoverished areas, they maintain themselves with difficulty. Their numbers remain the same through generations. Their economy permits no bursts of energy beyond what is necessary for the simple age-old struggle with nature. Perhaps, as we view the looming shadow of atomic disaster, this way of life takes on a certain dignity today.

Nevertheless there is no road back; the primitive way is no longer our way. We are the inheritors of an aggressive culture which, when the great herds disappeared, turned to agriculture. Here again the magic of fire fed the great human wave and built up man's numbers and civilization.

Man's first chemical experiment involving the use of heat was to make foods digestible. He had cooked his meat; now he used fire to crack his grain. In the process of adopting the agricultural way of life he made his second chemical experiment with heat: baking pottery. Ceramics may have sprung in part from the need for storage vessels to protect harvested grain from the incursions of rats and mice and moisture. At any rate, the potter's art spread with the revolutionary shift in food production in early Neolithic times.

People who have only played with mud pies or made little

sundried vessels of clay are apt to think of ceramics as a simple
art. Actually it is not. The sundried vessels of our childhood
experiments would melt in the first rain that struck them. To
produce true pottery one must destroy the elasticity of clay
through a chemical process which can be induced only by
subjecting the clay to an intense baking at a temperature of at
least 400 or 500 degrees centigrade. The baking drives out the
so-called water of constitution from the aluminum silicate in
the clay. Thereafter the clay will no longer dissolve in water; a
truly fired vessel will survive in the ground for centuries. This
is why pottery is so important to the archaeologist. It is impervious to the decay that overtakes many other substances, and,
since it was manufactured in quantity, it may tell tales of the
past when other clues fail us.

Pottery can be hardened in an open campfire, but the results
can never be so excellent as those achieved in a kiln. At some
point the early potter must have learned that he could concentrate and conserve heat by covering his fire—perhaps making it
in a hole or trench. From this it was a step to the true closed
kiln, in which there was a lower chamber for the fire and an
upper one for the pottery. Most of the earthenware of simple
cultures was fired at temperatures around 500 degrees centigrade, but really thorough firing demands temperatures in the
neighborhood of 900 degrees.

After man had learned to change the chemical nature of clay,
he began to use fire to transform other raw materials—ores
into metals, for instance. One measure of civilization is the
number of materials manipulated. The savage contents himself
with a few raw materials which can be shaped without the application of high temperatures. Civilized man uses fire to ex-
tract, alter, or synthesize a multitude of substances.

By the time metals came into extended use, the precious
flame no longer burned in the open campfire, radiating its heat
away into the dark or flickering on the bronzed faces of the
hunters. Instead it roared in confined furnaces and was fed
oxygen through crude bellows. One of the by-products of
more intensified experiments with heat was glass—the strange,
impassive substance which, in the form of the chemist's flask,
the astronomer's telescope, the biologist's microscope, and the

mirror, has contributed so vastly to our knowledge of ourselves and the universe.

We hear a good deal about the Iron Age, or age of metals, as a great jump forward in man's history; actually the metals themselves played a comparatively small part in the rise of the first great civilizations. While men learned to use bronze, which demands little more heat than is necessary to produce good ceramics, and later iron, for tools and ornaments, the use of metal did not make a really massive change in civilization for well over 1,500 years. It was what Leslie White of the University of Michigan calls the "Fuel Revolution" that brought the metals into their own. Coal, oil, and gas, new sources of energy, combined with the invention of the steam and combustion engines, ushered in the new age. It was not metals as tools, but metals combined with heat in new furnaces and power machinery that took human society off its thousand-year plateau and made possible another enormous upswing in human numbers, with all the social repercussions.

Today the flames grow hotter in the furnaces. Man has come far up the heat ladder. The creature that crept furred through the glitter of blue glacial nights lives surrounded by the hiss of steam, the roar of engines, and the bubbling of vats. Like a long-armed crab, he manipulates the tongs in dangerous atomic furnaces. In asbestos suits he plunges into the flaming debris of hideous accidents. With intricate heat-measuring instruments he investigates the secrets of the stars, and he has already found heat-resistant alloys that have enabled him to hurl himself into space.

How far will he go? Three hundred years of the scientific method have built the great sky-touching buildings and nourished the incalculable fertility of the human species. But man is also *Homo duplex*, as they knew in the darker ages. He partakes of evil and of good, of god and of man. Both struggle in him perpetually. And he is himself a flame—a great, roaring, wasteful furnace devouring irreplaceable substances of the earth. Before this century is out, either *Homo duplex* must learn that knowledge without greatness of spirit is not enough for man, or there will remain only his calcined cities and the little charcoal of his bones.

The Ghostly Guardian

THERE IS AN ANIMAL that is followed everywhere by a ghost. The ghost floats uncannily a little above and just back of the animal's head. Whether the creature is clambering up the bars of a cage in the zoo or ascending trees in the dense rain forests of the Amazon or the Orinoco, the ghost is always there, sensitive, exploring, shrinking back or protruding forward as the occasion may demand.

The animal is a skinny creature in a funereal black dress. Its legs, arms, and tail are so elongated in proportion to its body that, seen obscurely through a curtain of leaves, it often appears like a huge and repulsive spider as it sidles about in the forest. As a result it has earned the name of "spider monkey," and the ghost that accompanies it is its tremendously lengthy tail. This tail is one of the most remarkable organs to be found anywhere among a group of animals, the monkeys, noted for their addiction to quite varied styles in posterior adornment.

The spider monkey's tail is prehensile—that is, it is capable of grasping objects and transferring them to its owner's mouth, or of holding him safe while he is seated on a waving branch, or its almost preternatural grasping power may save his life as he hurtles headlong from some lofty spot in the forest attic. The muscles and nerves of such tails are of so extraordinary a character that even in death they may hold their owner to his aerial refuge after he has been slain by rifles from the ground.

The spider monkey's tail, however, is not the strangest part of the story. He lives in an area where a great variety of animals of quite unrelated ancestry have tails of this general type, although the spider monkey's is perhaps the most clever and appears to have almost literally taken on an independent life of its own. It is, in very truth, a guardian ghost, but the tail itself is not nearly so mysterious as the way in which numerous diverse creatures in one particular area of forest have acquired these hovering appendages.

These contemporaneous animals, all of one time and place but of widely different origins and habits, share in common one thing: the prehensile grasping tail. In a living world regarded by many scientists as the creation of chance mutations in animals acted upon by selective forces in the environment, they present a strange spectacle. These dangling creatures with their uncanny third hand suggest, inescapably, that nature, somewhere in the vast intricacies with which she complicates the world of life, has here been playing with loaded dice.

On the basis of pure chance it is hard to see why the mysterious evolutionary forces behind the universe should have bestowed this gift with such incredible profusion in only one area of the world, the Amazon basin. It is as though all human beings had spun coins at a particular moment but only those who tossed in San Francisco came up with "tails." This is an obvious simplification, but it makes the issue clear. We would regard such an event as outside the known laws of mathematics. We would feel a hair-raising chill down our spines and demand an explanation.

That is the way one feels when one looks at the ghostly guardian hovering behind a spider monkey, particularly when one walks to the next cage of the mammal house and sees *Coendou*, the prehensile-tailed porcupine of South America; or *Disactylus*, the pygmy anteater, "note the strongly prehensile tail." At such times, though I am supposed to be a naturalist and adjusted to such matters, I get the chill to which I referred. I slink out, almost as if I were followed by a tail myself, and go and sit on a bench to meditate upon the inscrutable way of tails in nature. Of six whole genera—not species, but genera—of South American monkeys, only one lacks a grasping tail.

If you are not a student of these matters you may think indifferently, So what? All monkeys, you see cartoons of them everywhere, hang by their tails. It's the way of monkeys. But the "way of things" is a cover-up for our ignorance. As a matter of simple fact, there isn't a single solitary monkey on the continent of Africa who possesses a true prehensile tail. Yet many monkeys live lives in the high green attic of Africa that, so far as we can tell, are the precise equivalent of those lived by their relatives in the South American forests. But the monkeys of Africa climb without the help of their adorning tails, and as for other African animals only an obscure lizard and a peculiar scaled anteater, the pangolin, seem to be endowed with this helpful little secondary personality.

No, there is something else to be seen here than the "way of things," or, at least, the way of things turns out to be unaccountably marvelous whenever we turn sharply and begin to scrutinize it instead of using the phrase as an opiate. Involved in this matter of tails are some strange factors of inheritance and change which still trouble the modern scientist and over which philosophers have puzzled in vain. Only by climbing ourselves into that lofty world hung on the unstable rafters of strangling vines or by peering down at the wandering and uncertain waters that drench the forest floor in the season of floods can we expect to learn anything of the dark forces that created the grotesque little appendage which, in the spider monkey, may affectionately caress his fellows or faithfully sustain his broken body beyond the hour of death.

The basins of the Amazon and Orinoco rivers contain the largest untouched tropical rain forest left anywhere in the world. Here is the world of the past as it existed over much of the globe before the crawling continent-wide glaciers of the Age of Ice met and forced back, in million-year-long battles, the forces of the green. Now, harried by the fires and axes of man, forests are everywhere in the process of disappearance. Their last strongholds lie in the damp hot lands along the equator in Africa, in South America, in a few South Sea Islands, and in the remoter confines of Asia. In these areas, and particularly in the state of Amazonas in Brazil, man can move only along the waterways that pour under the great archways of the forest. He is still a puny shadow who, a few feet from the protecting river, may vanish without a trace. If he flies above that endless green expanse and falls, he will be as soundlessly engulfed as though he had gone down into the sea itself. For this world *is* a sea, a sea over whose swaying green billows pass the wind and the birds. Below, the depths are still. Nevertheless, there is life there, scampering wildly through orchid gardens high in the dim green forest garret, just as, below the waves of the sea, there is also life among the forests of the coral.

The forest floor of the great jungles can almost be compared to the abyssal depths of the sea. It is barren and dark and it receives the dead that fall from above. Few higher forms of life haunt its dim recesses. The real life of the rain forest lives among the rafters of a thousand-mile attic more than one hundred feet above the ground. Here, safely elevated above the floods that swirl over hundreds of miles of forest during the rainy season, the monkeys pass on their trembling pathways, serpents creep through networks of vines, brilliantly colored parrots shriek discordantly, and even frogs are born in aerial ponds and never go beneath their floor of leaves. In short, this is a world whose life needs the help of a friendly ghost, and among all the animals who creep and climb on rickety stairs and crumbling balustrades, tails are much in evidence. Not just any old tail, but tails that creep along behind and hold one safe. Tails with naked sensitive tips furrowed like fingers with nonskid whorls of skin.

In a place where death lurks frequently on either side of some quivering patchwork of cables it is no great marvel that everything from porcupines to arboreal anteaters should have a second life in their tails. The real mystery, to the student of tails, lies in another direction. It is the problem of how so many totally unrelated animals have acquired these tails in the Amazonian basin while in the great forests along the Congo scarcely a single aerial performer makes use of a prehensile tail. This is an uncanny situation. The dice of chance have fallen in a certain way all in one place.

The student of animal evolution is accustomed to the assumption that a series of chance changes in the heredity of an animal—"mutations," the biologist calls them—happen to promote the animal's welfare in a given way of life. Slowly, by infinitesimal degrees, such advantageous mutations may be followed by others until an animal is equipped with some highly improved organ such as a grasping tail or other useful character. All of this, as far as present-day experimentation can tell us, is the result of chance adaptations which are very infrequent compared with poor or disadvantageous variations that lead to an animal's destruction. Chance adaptation, in other words, might have readily explained the appearance of one species of monkey with a prehensile tail. It might even explain the occurrence of a grasping tail in two closely related forms, because closely related forms, possibly because of similarities in their body chemistry, may show a tendency to produce similar mutations and hence to evolve along the same pathway, particularly if their environment is such as to similarly select them.

But here, in a single great jungle, numerous quite unrelated animals have all chosen to pull out of the grab bag of chance a single useful organ: the grasping tail. Within a reasonably limited period of geological time, they have drawn from the mysterious little packets of the genes, the hereditary substance, an identical solution to a similar problem, and supposedly it has been done by chance variation and selection.

"Easy," runs one solution. "Tails all tend to vary in this direction. It's one of the few useful ways they can vary. Nothing to it."

If one says this, one is forced to look at the arboreal life of Africa. The minute one does so the ease of this solution slips away. There are many tails; among the African mammals there are no prehensile ones save that of the scaled and ancient pangolin. If the grasping tail has been so easy to produce by chance, why is it so scarce in Africa and so common in South America?

One writer, confining himself discreetly to monkeys, argues that the South American forms like the spider monkey are thumbless; as a consequence they have developed the prehensile tail as a compensation for this absence of a digit. Very well, but the colobus monkeys of Africa are thumbless and they do not possess a grasping tail. The correlation does not seem to hold, and in any case it does not explain why certain arboreal opossums, kinkajous, and other queer Amazonian beasts with paws and no missing digits should go in for grasping tails.

Another suggestion is that the Amazonian lowlands, being frequently inundated by floods, have stimulated the selection of prehensile tails more vigorously than the African forests. In this connection one must consider the Uakari monkey. He is a queer little fellow with an absolutely bald head and a choleric, equally bald face. Uncannily human in facial appearance, he looks, in his coat of moth-eaten brown fur, as though he were a human being who had been playing an orangutang at a masked ball and had just removed his head mask. He has a nonprehensile, stumpy, rudimentary, and almost useless tail, but he continues to flourish happily among the Amazonian floods.

In any case it is to be noticed that while these varied theories attempt to explain the selection of mutations leading to the development of grasping tails, none of them really comes to grips with the more formidable problem of why these apt variations were all conveniently at hand in this one place. The difficulty is enhanced by the presence of such a diverse array of creatures all showing the same characteristic, which comprises several lesser adaptations such as suitably flattened vertebrae, muscular enhancement in the tail and lower back, a certain way of rolling up or manipulating the organ, and its control by the brain.

Darwin, always troubled by such problems, used to speak of the mysterious and unknown laws governing these matters. Unlike many of his followers he had no illusions that he had solved all the mysteries of life and evolution. Reading his works one is often made aware of how that great mind hesitated painfully over much that his followers take for granted.

Once he expressed himself to the effect that the independent duplication of a single animal form, if proven for two separate areas of the world, might force him to entertain the possibility of some other explanation for evolution than that offered by chance mutations acted upon by natural selection. Strangely enough, these monkeys of South America offer an evolutionary problem very close to, though admittedly not identical with, the hypothetical case proposed by Darwin. These monkeys, while not totally identical anatomically with those of the Old World, are remarkably similar to them. In fact, only the sophisticated observer can recognize the differences. Yet the separation of these two monkey groups from each other is ancient, and no ancestral monkey remains connecting the Old World with the New are known by fossil hunters.

Instead, it is believed by many authorities that the New World and Old World monkeys have arisen as parallel developments from older pre-monkey forms, the lemurs, which were once spread over the whole region from Asia to South America. In this case, creatures of very similar brain, face, habits, and general appearance would have come into being in separate parts of the world from ancestors far below the monkey level. Such a development might suggest latent evolutionary powers not entirely the simple product of what we, in our ignorance of a better word, call chance. We are not in a position as yet to verify absolutely this interpretation of the separate origins of the Old and New World monkeys, but it is the most reasonable theory that we possess.

There is no doubt that this mystery would have intrigued the brooding mind of Darwin, particularly since one of the South American monkeys involved has a brain which in proportion to its body weight is larger than man's. Perhaps fortunately for us, however, Chrysothrix, the squirrel monkey, stayed small and remained in the trees so that his large and elflike cranium presents no threat to us. Yet withal, these strange parallelisms leave one with an odd feeling about what the biologist means by chance.

Elusive, tantalizing, and remote in the green attic of the elder world one catches an occasional glimpse of faces hauntingly like one's own and yet different, as though one peered into a charmed mirror. And slowly, through the shifting endless greenery, crawl the multitudinous things with tails, the tails that curl and hold, or wave forward like slender ghosts. There is something, particularly in a spider monkey's tail, that is too bold and purposeful to be easily called the product of simple chance. It floats there like a complete little personality. At least it may cause the true philosopher to pause hesitantly and ponder before he dismisses the universe as totally a world of chance.

The Fire Apes

I WAS THE ONLY man in the world who saw him do it. Every-body else was hurrying. Everybody else around that hospi-tal was busy, or flat on his back and beyond seeing. I had a smashed ankle and was using a crutch, so I couldn't hurry. That was the only reason I was on the grounds and allowed to sit on a bench. If it hadn't been for that I would have missed it. I saw what it meant, too. I had the perspective, you see, and the time to think about it. In the end I hardly knew whether to be glad or sorry, but it was a frightening experience, perhaps not so much frightening as weird because I suddenly and pre-ternaturally saw very close to the end—the end of all of us—and it happened because of that squirrel.

The bird-feeding station stood on the lawn before my bench. Whoever had erected it was a bird-lover, not a squirrel enthusiast, that much was certain. It was on top of a section of thin pipe stuck upright in the ground, and over the end of the pipe half of a bread can had been inverted. The thin, smooth pipe and the bread can were to keep squirrels from the little wooden platform and roof where the birds congregated to feed. The feeding platform was attached just above the tin shield that protected it from the squirrels. I could see that considerable thought had gone into the production of this apparatus and that it was carefully placed so that no squirrel could spring across from a nearby tree.

In the space of the morning I watched five squirrels lope easily across the lawn and try their wits on the puzzle. It was clear that they knew the bread was there—the problem was to reach it. Five squirrels in succession clawed their way up the thin pipe only to discover they were foiled by the tin umbrella around which they could not pass. Each squirrel in turn slid slowly and protestingly back to earth, flinched at my distant chuckle, and went away with a careful appearance of total dis-interest that preserved his dignity.

There was a sixth squirrel that came after a time, but I was bored by then, and only half watching. God knows how many things a man misses by becoming smug and assuming that matters will take their natural course. I almost drowsed enough to miss it, and if I had, I might have gone away from there still believing in the fixity of species, or the inviolability of the human plane of existence. I might even have died believing some crass anthropocentric dogma about the uniqueness of the human brain.

As it was, I had just one sleepy eye half open, and it was through that that I saw the end of humanity. It was really a very little episode, and if it hadn't been for the squirrel I wouldn't have seen it at all. The thing was: he stopped to think. He stopped right there at the bottom of the pole and looked up and I knew he was thinking. Then he went up.

He went up with a bound that swayed the thin pipe slightly and teetered the loose shield. In practically the next second he had caught the tilted rim of the shield with an outstretched paw, flicked his body on to and over it, and was sitting on the platform where only birds were supposed to be. He dined well there and daintily, and went away in due time in the neat quick fashion by which he had arrived. I clucked at him and he stopped a moment in his leisurely sweep over the grass, holding up one paw and looking at me with the small shrewd glance of the wood people. There are times now when I think it was a momentous meeting and that for just a second in that sunlit glade, the present and the future measured each other, half conscious in some strange way of their destinies. Then he was loping away with the autumn sunlight flickering on his fur, to a tree where I could not follow him. I turned away and limped back to the shadow of my bench.

He's a smart squirrel, all right, I tried to reassure myself. He's a super-smart squirrel, but just the same he's only a squirrel. Besides, there are monkeys that can solve better problems than that. A nice bit of natural history, an insight into a one-ounce brain at its best, but what's the significance of—

It was just then I got it. The chill that had been slowly crawling up my back as I faced that squirrel. You have to remember what I said about perspective. I have been steeped in geological eras; my mind is filled with the osseous debris of a hundred graveyards. Up till now I had dealt with the past. I was one of the planet's undisputed masters. But that squirrel had busy fingers. He was loping away from me into the future.

The chill came with the pictures, and those pictures rose dim and vast, as though evoked from my subconscious memory by that small uplifted paw. They were not pleasant pictures. They had to do with times far off and alien. There was one, I remember, of gasping amphibian heads on the shores of marshes, with all about them the birdless silence of a land into which no vertebrate life had ever penetrated because it could not leave the water. There was another in which great brainless monsters bellowed in the steaming hollows of a fern forest, while tiny wraithlike mammals eyed them from the underbrush. There was a vast lonely stretch of air, through which occasionally skittered the ill-aimed flight of lizardlike birds. And finally there was a small gibbonlike primate teetering along through a great open parkland, upright on his two hind feet. Once he turned, and I seemed to see something familiar about him, but he passed into the shade.

There were more pictures, but always they seemed to depict great empty corridors, corridors in the sense of a planet's spaces, first empty and then filled with life. Always along those corridors as they filled were eager watchers, watching from the leaves, watching from the grasses, watching from the woods' edge. Sometimes the watchers ventured out a little way and retreated. Sometimes they emerged and strange changes overtook the corridor.

It was somewhere there at the last on the edge of a dying city that I thought I recognized my squirrel. He was farther out of the woods now, bolder, and a bit more insolent, but he was still a squirrel. The city was dying, that was plain, but the cause was undiscernible. I saw with a slight shock that nothing seemed very important about it. It was dying slowly, in the length of centuries, and all about it the little eyes under the leaves were closing in. It was then that I understood, finally, and no longer felt particularly glad or sorry. The city was forfeit to those little shining brains at the woods' edge. I knew how long they had waited. And we, too, had been at the woods' edge in our time. We could afford to go now. Our vast intellectual corridor might stretch empty for a million years. It did not matter. My squirrel would attend to it. And if not he, then the wood rats. They were all there waiting under the leaves. . . .

Easter: The Isle of Faces

EASTER ISLAND, twelve miles long by seven broad, juts up from the Pacific waters like the triconodont tooth of some primeval mammal. Volcanic cones like cusps mark the corners of the island. On the slopes that fall toward the sea lie great stone platforms cut and fitted from volcanic rock. Standing here and there on the slopes or lying unmoved in the old quarries are the huge stone faces which, like the island itself, hint of a distant and unimaginable past. Once the isle must have been forested by whatever trees could be wafted over the desolate mid-Pacific waters before man came. Today it is clothed sparsely by bushes, grazed by cattle, and supports a few hundred people. It has been known since the Dutch navigator Roggeveen discovered it on Easter day in 1772. Other great Pacific captains such as James Cook and Jean François de la Perouse touched there a few years later and recorded a flourishing population of 2000 to 3000 people. The story of how they came there is lost.

We know only that here man engaged in an incalculable frenzy of creative efforts which apparently ended as suddenly as it began. More than 600 figures lie in the quarries or on the slopes—figures weighing many tons. On the heads of the erect completed torsos were placed huge hats made of reddish stone from separate quarries. Dropped tools and unfinished efforts suggest a hasty and perhaps tragic ending to an unknown religious cult whose purpose is lost. Did a wearied, god-enslaved populace destroy its priesthood? There are suggestions of this in island legends, but these are as misty as modern ideas of how the great statues were transported from their quarries and erected. Even small populations imbued with religious fervor can achieve remarkable feats. The story which we shall never read and which will forever intrigue mankind lies in the meaning of this fantastic performance. It is as though some mad sculptor had bent an entire people to his will—as though a collective will to expend all one's days in the creation of men in stone had spread throughout Easter society. There was an intent, judging from the remains in the quarries, to populate the

entire island with gods—if gods they were. Even today the remains are so inspiring in terms of the effort they represent that
one can be sure of just one thing: only a tremendous religious
stimulus could have incited such labors. Only man, the seer of
visions, could have produced such work.

The distances in Polynesia are enormous. Beyond Easter Island the waves run sheer for 2000 miles to the South American
shore. In this waste of waters the venturesome canoes that had
always found land in the low archipelagoes must have drifted
into oblivion with their parched occupants. Easter was the last
isle, the outpost on the edge of desolation. Easter was the end.
When primitive man found Easter he had finally encompassed
the world. He had proved that courage, a stone ax, and an
outrigger canoe could take man anywhere on the planet.

The great Pacific triangle known as Polynesia is drawn on
the maps through Hawaii, Easter, and New Zealand. It was the
last livable area in the world to be settled by man and in some
ways the strangest: mostly it was water, but the little groups of
islands scattered in the endless seas were fertile. For 3000 years
before the dawn of the modern era man had been learning the
keys to that vast region. Floating, drifting, paddling, through
typhoons and unknown breakers, he had passed from one isle
to another farther into the illimitable reaches of the Pacific.
After a time, as from a living brain beneath the water, strange
psychological worlds had thrust up. Cannibalism on one isle,
great mysterious faces on another. Here man lived alone, as he
has lived in no other place. Here in the great waters he was free
to indulge his cultural fantasies without the visible contradictions presented by too-close neighbors. As in Robert Louis
Stevenson's story of "The Isle of Voices" with its unseen population of sorcerers and little fires, so the Polynesian until the
eighteenth century remained invisible to the rest of mankind.
The water dwellers from the mid-Pacific are, in their history,
bodiless and without substance. They make maps of strangely
assembled sticks, they steer by the stars and know the secrets of
clouds that hover over islands. Their legends are all of lands
dragged up from the primeval waters or of lands sunken again
into the abyss. Like all history-less people they attract history.
The sea vapors that have concealed their passage seem to hint
of sunken continents, lost cities, and forgotten arts. Who are

these people? From whence did they come? And why at Easter Island, at the very edge of nowhere, did they carve the great sad faces whose eyes look into distances too remote for man? Romantics have linked them to the lost cities of Asia, the cities lying beneath desert sands or crumbling in the green depths of tropical jungles. Others have reversed these wanderings of the Polynesians and seen them as drifting originally westward from Peru. Many of these tales and carefully spun theories are fantastic, but so also is the life of man. Easter Island since the early nineteenth century has become a symbol in Western literature, a symbol of human aspiration to an alien people—ourselves. These seaward-gazing stone faces, transmuted out of their time and place, have played an undying role in a great human drama whose end it is perhaps only now possible to express.

In 1885 Stevenson dreamed a strange dream of transformation which resulted in a story known the world over as "Dr. Jekyll and Mr. Hyde." In it the ill-fated Dr. Jekyll made the horrifying discovery that man is not one but two; that the human consciousness is racked by opposing forces. Stevenson was epitomizing one of the great preoccupations of the Victorians—the struggle between good and evil. The philosophy of this subject is as old as man. What marked the Victorian approach to the problem was the emergence of a new conception of man's mind which had arisen along with the Darwinian discovery of the evolution of man's body.

Darwin and his followers had seen man as arising from the struggle of beast against beast and as therefore inevitably dragging with him out of his midnight experiences a vestigial ferocity only slowly eroding away under conditions of civilized enlightenment. The intricacies of the human mind had not yet been touched by the penetrating scalpel of Freud, but it was becoming plain that, if man's way had come upward from the beast, then the myth of the deathless Garden and the moral fall which occurred there would have to be abandoned. Ascending ape or fallen angel—man would have to make his choice. Oddly enough, it was by the circumstances of this theological quarrel in England that the giant faces upon Easter Island reentered history and have occupied scientific attention ever since.

Legend has it that in some remote castle on the Continent two intertwined stairs run upward in a tower. So clever was the architect, so remarkable was his design of the stairway, that although the steps twine and intertwine in their ascent, a man ascending gets no glimpse at any point of his counterpart coming down. Both are private pathways. In a similar fashion the moral imperfections of man can be seen as a corrupt descent from a state of perfection and absolute innocence, or the flaws of a creature wearily dragging with him up a dark stairwell the imperfections of the slime from which he rose. The two themes circle about each other but are not one.

This was the dilemma that confronted western Europe after the rise of the evolutionary philosophy. The ancestors of genteel Englishmen appeared in the Ice Age past to have chewed knucklebones by open fires and to have shaped and battered common stones for tools. For a proud people at the height of world empire this was a distasteful notion. As always in such cases a school of thought arose to combat the heresy. It was in this connection that men's thoughts returned to Easter Island. Man has never accepted his low-born state. He dreams still. All over the world he dreams, amid the wreckage of great palaces, about the road he has come down. Only in the west where life was young had he devised this other dream—the belief that he had crept outward into the sunlight long ago from the boughs of a dark forest. In the England of Darwin's day these two ideas met and mingled. To many intellectuals, after the publication of the *Origin of Species* and the finding of shaped implements in the earth, the enigma of man could be represented by the allegory of the two stairs. Either man had slowly and painfully made his way upward through the ages while his mind and his body changed, or, on the other hand, the crude remnants of early cultures found in the earth were those of a creature fallen from a state of grace, fallen from divine inspiration—a creature possessing no memory of his great past and dwelling barbarically amid the fallen monuments of his predecessors. Heated controversy echoed in the halls of scientific societies, and learned men debated the question with vigor. Primitive tools alone could not prove human evolution. This only fossils could achieve, and acceptable human "missing links" were not to be discovered for several decades. The "degenerationists," so far as man was concerned, had successfully inverted the archaeological argument for evolution. Man had not arisen from barbarism, they argued; instead he had sunk to it. "No simple tool," wrote one savant of the period, "of and by itself, would be sufficient to prove that there was a condition of mankind lying near that of animals." The foes of human evolution began to search for the evidences of man's coming down in the world.

Albert Mott, the president of the Philosophical Society of Liverpool, addressing that body in 1873 on "The Origin of Savage Life," publicized Easter Island and made much of the inferiority of the living inhabitants to their ancestors. Mott dwelt upon the great size and weight of the stone images scattered around the island, the terraces of fitted stone, the utter isolation of this little pinnacle thrust upward in the vast waste of waters. "If," Mott argued, "this island was first peopled by the accidental drifting of a canoe, it is incredible that the art of making these images and terraces should have developed there. To suppose that savages, under such circumstances, would spend their time and strength upon such labors is altogether past belief."

Instead of accepting a local origin for the sculptures, Mott contended that the islanders must have been an outpost colony of some maritime power long since vanished. Their descendants "though nearly white in colour have degenerated into ordinary savages." Turning from a consideration of Easter Island, Mott hinted that similar ancient ruins would be found in other unexplored islands of the Pacific. Passing to other parts of the world he brought forward additional arguments to prove his point that "savage life is the result of decay and degradation." Such examples as Easter Island, Mott concluded, make it unnecessary to believe "that the earth has been peopled through vast periods of time by the most brutish forms of humanity alone."

The degenerationist-evolutionist controversy, which I have allegorized in the legend of the two stairways, echoed for thirty years in the literature of America and Europe. It faded in importance only with the discovery of genuine human fossils and a better understanding of the archaeological past. The degenerationists, though mistaken in their main theme, contributed many astute observations upon human culture. Mott, for example, profoundly influenced Darwin's great compatriot and fellow scholar Alfred Russel Wallace. Easter Island, lifted out of its Pacific obscurity and attached to a great epoch in human thinking, became known all over the civilized world as a place of mystery. Just as the shadowy continent of Atlantis seemed for a time to haunt the mists of the Atlantic, so Easter Island contributed to romantic theories which involved a similar continental submergence in the mid-Pacific.

Long after the argument of the degenerationists had been forgotten, Easter Island and its inscrutable stone faces continued to be a source of literary dissension which has persisted right down to the present day. Hardly a year passes that some scholar, eccentric or otherwise, does not publish a volume or a special theory about Easter Island. Mott's published lecture, it is now clear, struck some universal chord in the make-up of the human mind. It touched the love of mystery that lies in us all. We are caught in the spell of the dreaming faces and of the three-coned volcanic isle without history. The whole vast sweep of Polynesia and its people take on significance in many minds merely as a clue to the unknown tragedy of which Easter is the heart. Has a continent vanished, has a high civilization foundered and gone down forever into the depths? Was Polynesia populated from Asia or from America? And what of the Polynesians themselves—those near-whites of whom Mott spoke? Of what race are they and from where? What is the lesson of their wanderings? Eternally these questions repeat themselves.

The story even in careful scientific terms is an enthralling one. It has to be pieced together from such debris as litters the shores of a hundred sea-pounded isles. Here it may be a green-stone adze blade dropped on a coral strand which contains no such stone. There it may be an outrigger canoe rotting on an uninhabited beach. From the cargo of the canoe a few coconut palms have sprouted and wave in the light-filled Pacific air. The bleached skeleton of a man is subsiding into the sand a little way up the shore. A few rats that survived the journey have scuttled away into the underbrush. Or one learns of an atoll upon which marooned men without their women survived for years, scanning, until they grew old, the endless sea-glint of the far horizon. One by one they died and were buried until the last survivor went mad with loneliness among the graves. Or men and women on genial trading visits to nearby islands vanished behind a screen of rain never to be seen again, unless, if they were fortunate, some uninhabited island, hundreds of miles removed, received them. If not, they vanished like the insubstantial vapors of the sea itself. Mostly there was no way back from these adventures; no return for wandering Odysseus. The seas were too vast, the isles of home far too elusive. Yet the old chronicles tell that rescued men have been known, in nostalgic desperation, to launch their frail vessels once more in the hope of finding home or death in the waves of earth's greatest sea.

For us, the continent dwellers, this Pacific world will never be totally understandable. We look upon it from too great a height of security. We have mapped in the course of three centuries the entire world. The compass and chronometer and sextant take us through dark nights unerringly across thousands of miles of sea to a speck of land not much larger than a single farm. By contrast, only two peoples in the world have known what it is to be alone: the polar Eskimos of the nineteenth century who thought they were the only men in the world and that the explorers who came to them were ghosts, and those inhabitants of the more remote Pacific islets who learned in astonishment that theirs was not the only land above the water that stretched, as they thought, to infinity.

The people, the sturdy golden-brown people of whom Mott spoke as nearly white, are certainly from Asia, not America. They have ridden with the east wind rather than the west. Their domestic animals, such as the pig, are of the east. It is true that one or two plants in the islands, such as the sweet potato, come from America, but the cultural connection cannot have been sustained. At best it represents the single passage of some great lost voyager. The Polynesian racial strain contains, beside Mongoloid elements, forgotten Asiatic white components which were later drowned in the rising yellow tide. In Fiji in the far south an admixture of darker Melanesian blood is evident. The islands themselves have doubtless exerted a selective influence upon the people. Small islands promote inbreeding. In 3000 years a comparatively small number of families can multiply and spread over a great area; such a development may well have added to the homogeneity of the original population of Polynesia. It is not necessary to postulate great planned migrations. The steady drift of an increasing population into regions of uninhabited land is sufficient to explain the peopling of the mid-Pacific. Except for the poles, it was the last open region of the earth. Each isle was a small paradise until crowding produced war. Local differences of culture and language arose, but these are minor. The basic linguistic stock is the same over the whole vast Pacific triangle.

These archipelagoes in which even the gods traveled in coconut shells or on the backs of fishes contain no trace or rumor of a vanished continent. The islands reveal no signs of animals or plants which could be called continental in type. The vegetation, the animal life, is all such as the sea in long ages brings by chance to island shores: the seeds that survive immersion in salt water, the plant and animal life that travels with primitive man. Yet man's pigs, dogs, and chickens, as well as his food plants, reveal clearly that Polynesia has seen many undirected comings and goings. The vagaries of voyaging without compass have dropped humanly transported plants and animals upon islands where man did not survive, or by contrast men have survived in circumstances where the full complement of his cultural items was not present.

Men forgot, in the island world, much they had learned on the continents, but they learned new things as well. The low coral islands limited man. They constricted his use of stone and forced him into greater reliance upon shells for tools. The high volcanic isles like Easter offered another environment. The superficial similarities of an island existence conceal cultural differences. In an archipelago of closely related islets, voyaging and sea craft may be encouraged. On the distant edge of nowhere, as at lonely Easter Island, sea skills may lapse because the venturesome disappear. The whole of Polynesia, containing perhaps a few hundred thousand souls at the time of discovery, has been populated at an enormous cost in human lives over something like 3000 years in time. Skilled seamen though the islanders were, the lack of navigating instruments must have led to innumerable tragedies at the same time that it resulted in movement farther and farther into the central Pacific.

Great voyaging canoes carrying scores of people and their chattels were used in some regions, and there is evidence that mounting population pressure led to searches for new land.

Nevertheless there can be no doubt that these deliberate efforts have been exaggerated by later writers. Captain Cook, who saw the natives in their own untouched environment, felt that chance rather than deliberate intention had populated most of Polynesia. Even in the eighteenth century there was evidence of accidental voyages in which men, women, and children had survived storms and distances of over a thousand miles. They reached, if they survived at all, lands from which they could not return. In this lay the secret of man's trek to the end of the infinite waters. Each generation carried its seed farther. Whether by chance or intent the result was the same at last. If anthropological investigations have produced no traces of a lost continent they have revealed an unwritten epic of man's courage. The populating of Polynesia cost infinitely more in lives than the taking of a continent. Perhaps our star ships may someday search frantically in the great dark for planets as desperately as those lost canoemen sought for a mile-long atoll in thousands of square miles of sea.

Thus there was no lost continent after all, the scholar is forced to conclude. Whatever was carved here or inscribed hieroglyphically on wooden tablets or carried as a seed in the wind from island to island was shaped by the men whom the Pacific voyagers found. But perhaps these seeds of the wind, like those escaped from cultivated gardens, grew wild in strange places, mounted perhaps some strange and effervescing growth where conditions were ripe. Perhaps at that far spot where the stone faces lie in the great quarries horizon-searching man had come at last upon some inexpressible thought to which he gave all his labor. Indeed, at the edge of the world perhaps there was nothing further he could do. The stone was there to be worked, and the deep loneliness for contemplation.

Easter Island shows linguistic differences from the rest of Polynesia that indicate, not lack of relationship, but long isolation from the rest of the Polynesian world. The people's limited traditional memory is partly the result of the raids of South American slavers who descended upon the island in 1863 and carried the greater part of the men away into exile. Yet it is not without significance that the statues and stone platforms are not unique, even though they show local peculiarities and monstrous enhancement. Rather they appear to reflect a cultural trait which flickered sporadically across the Pacific as time and appropriate stone gave opportunities. Man has always been a builder. Perhaps he has built best in loneliness. At least this appears to be the case in the isles of Polynesia.

As sunset falls on the inscrutable stylized faces, one thinks again how appropriately, in the event of man's passing, they would symbolize the end of this age. For the faces are formless, nameless; they represent no living style. They are therefore all men and no man, and they stare indifferently upon that rolling waste which has seen man come and will see him fade once more into the primal elements from which he came. No tears are marked upon the faces, and when at last the waves close over them in the red light of some later sun than ours, the secret of mankind, if indeed man has a secret, will go with them, and all will be upon that waste as it has been before. A flight of sea birds will wind away into the west like smoke. The stars will come out. There will be no one to ask where we, or the stones on which men tried to inscribe their immortality, have gone. There will linger momentarily only a dim sense of something too tragic and too powerful to endure imprisonment in matter or long suffer itself to be reproduced in stone. This is the message from the transcendent heart of man—the seafarer and spacefarer, the figure always beckoning through the mist.

The Dance of the Frogs

H E WAS A MEMBER of the Explorers Club, and he had never been outside the state of Pennsylvania. Some of us who were world travelers used to smile a little about that, even though we knew his scientific reputation had been, at one time, great. It is always the way of youth to smile. I used to think of myself as something of an adventurer, but the time came when I realized that old Albert Dreyer, huddling with his drink in the shadows close to the fire, had journeyed farther into the Country of Terror than any of us would ever go, God willing, and emerge alive.

He was a morose and aging man, without family and without intimates. His membership in the club dated back into the decades when he was a zoologist famous for his remarkable experiments upon amphibians—he had recovered and actually produced the adult stage of the Mexican axolotl, as well as achieving remarkable tissue transplants in salamanders. The club had been flattered to have him then, travel or no travel, but the end was not fortunate. The brilliant scientist had become the misanthrope; the achievement lay all in the past, and Albert Dreyer kept to his solitary room, his solitary drink, and his accustomed spot by the fire.

The reason I came to hear his story was an odd one. I had been north that year, and the club had asked me to give a little talk on the religious beliefs of the Indians of the northern forest, the Naskapi of Labrador. I had long been a student of the strange mélange of superstition and woodland wisdom that makes up the religious life of the nature peoples. Moreover, I had come to know something of the strange similarities of the "shaking tent rite" to the phenomena of the modern medium's cabinet.

"The special tent with its entranced occupant is no different from the cabinet," I contended. "The only difference is the type of voices that emerge. Many of the physical phenomena are identical—the movement of powerful forces shaking the conical hut, objects thrown, all this is familiar to Western psychical science. What is different are the voices projected. Here they are the cries of animals, the voices from the swamp and the mountain—the solitary elementals before whom the primitive man stands in awe, and from whom he begs sustenance. Here the game lords reign supreme; man himself is voiceless."

A low, halting query reached me from the back of the room. I was startled, even in the midst of my discussion, to note that it was Dreyer.

"And the game lords, what are they?"

"Each species of animal is supposed to have gigantic leaders of more than normal size," I explained. "These beings are the immaterial controllers of that particular type of animal. Legend about them is confused. Sometimes they partake of human qualities, will and intelligence, but they are of animal shape. They control the movements of game, and thus their favor may mean life or death to man."

"Are they visible?" Again Dreyer's low, troubled voice came from the back of the room.

"Native belief has it that they can be seen on rare occasions," I answered. "In a sense they remind one of the concept of the archetypes, the originals behind the petty show of our small, transitory existence. They are the immortal renewers of substance—the force behind and above animate nature."

"Do they dance?" persisted Dreyer.

At this I grew nettled. Old Dreyer in a heckling mood was something new. "I cannot answer that question," I said acidly. "My informants failed to elaborate upon it. But they believe implicitly in these monstrous beings, talk to and propitiate them. It is their voices that emerge from the shaking tent."

"The Indians believe it," pursued old Dreyer relentlessly, "but do *you* believe it?"

"My dear fellow"—I shrugged and glanced at the smiling audience—"I have seen many strange things, many puzzling things, but I am a scientist." Dreyer made a contemptuous sound in his throat and went back to the shadow out of which he had crept in his interest. The talk was over. I headed for the bar.

II

The evening passed. Men drifted homeward or went to their rooms. I had been a year in the woods and hungered for voices and companionship. Finally, however, I sat alone with my glass, a little mellow, perhaps, enjoying the warmth of the fire and remembering the blue snowfields of the North as they should be remembered—in the comfort of warm rooms.

I think an hour must have passed. The club was silent except for the ticking of an antiquated clock on the mantel and small night noises from the street. I must have drowsed. At all events it was some time before I grew aware that a chair had been drawn up opposite me. I started.

"A damp night," I said.

"Foggy," said the man in the shadow musingly. "But not too foggy. They like it that way."

"Eh?" I said. I knew immediately it was Dreyer speaking. Maybe I had missed something; on second thought, maybe not.

"And spring," he said. "Spring. That's part of it. God knows why, of course, but we feel it, why shouldn't they? And more intensely."

"Look—" I said. "I guess—" The old man was more human than I thought. He reached out and touched my knee with the hand that he always kept a glove over—burn, we used to speculate—and smiled softly.

"You don't know what I'm talking about," he finished for me. "And, besides, I ruffled your feelings earlier in the evening. You must forgive me. You touched on an interest of mine, and I was perhaps overeager. I did not intend to give the appearance of heckling. It was only that . . ."

"Of course," I said. "Of course." Such a confession from Dreyer was astounding. The man might be ill. I rang for a drink and decided to shift the conversation to a safer topic, more appropriate to a scholar.

"Frogs," I said desperately, like any young ass in a china shop. "Always admired your experiments. Frogs. Yes."

I give the old man credit. He took the drink and held it up and looked at me across the rim. There was a faint stir of sardonic humor in his eyes.

"Frogs, no," he said, "or maybe yes. I've never been quite sure. Maybe yes. But there was no time to decide properly." The humor faded out of his eyes. "Maybe I should have let go," he said. "It was what they wanted. There's no doubting that at all, but it came too quick for me. What would you have done?"

"I don't know," I said honestly enough and pinched myself.

"You had better know," said Albert Dreyer severely, "if you're planning to become an investigator of primitive religions. Or even not. I wasn't, you know, and the things came to me just when I least suspected—But I forget, you don't believe in them."

He shrugged and half rose, and for the first time, really, I saw the black-gloved hand and the haunted face of Albert Dreyer and knew in my heart the things he had stood for in science. I got up then, as a young man in the presence of his betters should get up, and I said, and I meant it, every word: "Please, Dr. Dreyer, sit down and tell me. I'm too young to be saying what I believe or don't believe in at all. I'd be obliged if you'd tell me."

Just at that moment a strange, wonderful dignity shone out of the countenance of Albert Dreyer, and I knew the man he was. He bowed and sat down, and there were no longer the barriers of age and youthful ego between us. There were just two men under a lamp, and around them a great waiting silence. Out to the ends of the universe, I thought fleetingly, that's the way with man and his lamps. One has to huddle in, there's so little light and so much space. One——

III

"It could happen to anyone," said Albert Dreyer. "And especially in the spring. Remember that. And all I did was to skip. Just a few feet, mark you, but I skipped. Remember that, too.

"You wouldn't remember the place at all. At least not as it was then." He paused and shook the ice in his glass and spoke more easily.

"It was a road that came out finally in a marsh along the Schuylkill River. Probably all industrial now. But I had a little house out there with a laboratory thrown in. It was convenient to the marsh, and that helped me with my studies of amphibia. Moreover, it was a wild, lonely road, and I wanted solitude. It is always the demand of the naturalist. You understand that?"

"Of course," I said. I knew he had gone there, after the

death of his young wife, in grief and loneliness and despair. He was not a man to mention such things. "It is best for the naturalist," I agreed.

"Exactly. My best work was done there." He held up his black-gloved hand and glanced at it meditatively. "The work on the axolotl, newt neoteny. I worked hard. I had—" he hesitated —"things to forget. There were times when I worked all night. Or diverted myself, while waiting the result of an experiment, by midnight walks. It was a strange road. Wild all right, but paved and close enough to the city that there were occasional street lamps. All uphill and downhill, with bits of forest leaning in over it, till you walked in a tunnel of trees. Then suddenly you were in the marsh, and the road ended at an old, unused wharf.

"A place to be alone. A place to walk and think. A place for shadows to stretch ahead of you from one dim lamp to another and spring back as you reached the next. I have seen them get tall, tall, but never like that night. It was like a road into space."

"Cold?" I asked.

"No. I shouldn't have said 'space.' It gives the wrong effect. Not cold. Spring. Frog time. The first warmth, and the leaves coming. A little fog in the hollows. The way they like it then in the wet leaves and bogs. No moon, though; secretive and dark, with just those street lamps wandered out from the town. I often wondered what graft had brought them there. They shone on nothing—except my walks at midnight and the journeys of toads, but still . . ."

"Yes?" I prompted, as he paused.

"I was just thinking. The web of things. A politician in town gets a rake-off for selling useless lights on a useless road. If it hadn't been for that, I might not have seen them. I might not even have skipped. Or, if I had, the effect—How can you tell about such things afterwards? Was the effect heightened? Did it magnify their power? Who is to say?"

"The skip?" I said, trying to keep things casual. "I don't understand. You mean, just skipping? Jumping?"

Something like a twinkle came into his eyes for a moment. "Just that," he said. "No more. You are a young man. Impulsive? You should understand."

"I'm afraid—" I began to counter.

"But of course," he cried pleasantly. "I forget. You were not there. So how could I expect you to feel or know about this skipping. Look, look at me now. A sober man, eh?"

I nodded. "Dignified," I said cautiously.

"Very well. But, young man, there is a time to skip. On country roads in the spring. It is not necessary that there be girls. You will skip without them. You will skip because something within you knows the time—frog time. Then you will skip."

"Then I will skip," I repeated, hypnotized. Mad or not, there was a force in Albert Dreyer. Even there under the club lights, the night damp of an unused road began to gather.

IV

"It was a late spring," he said. "Fog and mist in those hollows in a way I had never seen before. And frogs, of course. Thousands of them, and twenty species, trilling, gurgling, and grunting in as many keys. The beautiful keen silver piping of spring peepers arousing as the last ice leaves the ponds—if you have heard that after a long winter alone, you will never forget it." He paused and leaned forward, listening with such an intent inner ear that one could almost hear that far-off silver piping from the wet meadows of the man's forgotten years.

I rattled my glass uneasily, and his eyes came back to me.

"They come out then," he said more calmly. "All amphibia have to return to the water for mating and egg laying. Even toads will hop miles across country to streams and waterways. You don't see them unless you go out at night in the right places as I did, but that night——

"Well, it was unusual, put it that way, as an understatement. It was late, and the creatures seemed to know it. You could feel the forces of mighty and archaic life welling up from the very ground. The water was pulling them—not water as we know it, but the mother, the ancient life force, the thing that made us in the days of creation, and that lurks around us still, unnoticed in our sterile cities.

"I was no different from any other young fool coming home on a spring night, except that as a student of life, and of amphibia in particular, I was, shall we say, more aware of the creatures. I had performed experiments"—the black glove gestured before my eyes. "I was, as it proved, susceptible.

"It began on that lost stretch of roadway leading to the river, and it began simply enough. All around, under the street lamps, I saw little frogs and big frogs hopping steadily toward the river. They were going in my direction.

"At that time I had my whimsies, and I was spry enough to feel the tug of that great movement. I joined them. There was no mystery about it. I simply began to skip, to skip gaily, and enjoy the great bobbing shadow I created as I passed onward with that leaping host all headed for the river.

"Now skipping along a wet pavement in spring is infectious, particularly going downhill, as we were. The impulse to take mightier leaps, to soar farther, increases progressively. The madness worked into me. I bounded till my lungs labored, and my shadow, at first my own shadow, bounded and labored with me.

"It was only midway in my flight that I began to grow conscious that I was not alone. The feeling was not strong at first. Normally a sober pedestrian, I was ecstatically preoccupied with the discovery of latent stores of energy and agility which I had not suspected in my subdued existence.

"It was only as we passed under a street lamp that I noticed, beside my own bobbing shadow, another great, leaping grotesquerie that had an uncanny suggestion of the frog world about it. The shocking aspect of the thing lay in its size, and the fact that, judging from the shadow, it was soaring higher and more gaily than myself.

"'Very well,' you will say"—and here Dreyer paused and looked at me tolerantly—"'Why didn't you turn around? That would be the scientific thing to do.'

"It would be the scientific thing to do, young man, but let me tell you it is not done—not on an empty road at midnight —not when the shadow is already beside your shadow and is joined by another, and then another.

"No, you do not pause. You look neither to left nor right, for fear of what you might see there. Instead, you dance on madly, hopelessly. Plunging higher, higher, in the hope the shadows will be left behind, or prove to be only leaves dancing, when you reach the next street light. Or that whatever had joined you in this midnight bacchanal will take some other pathway and depart.

"You do not look—you cannot look—because to do so is to destroy the universe in which we move and exist and have our transient being. You dare not look, because, beside the shadows, there now comes to your ears the loose-limbed slap of giant batrachian feet, not loud, not loud at all, but there, definitely there, behind you at your shoulder, plunging with the utter madness of spring, their rhythm entering your bones until you too are hurtling upward in some gigantic ecstasy that it is not given to mere flesh and blood to long endure.

"I was part of it, part of some mad dance of the elementals behind the show of things. Perhaps in that night of archaic and elemental passion, that festival of the wetlands, my careless hopping passage under the street lights had called them, attracted their attention, brought them leaping down some fourth-dimensional roadway into the world of time.

"Do not suppose for a single moment I thought so coherently then. My lungs were bursting, my physical self exhausted, but I sprang, I hurtled, I flung myself onward in a company I could not see, that never outpaced me, but that swept me with the mighty ecstasies of a thousand springs, and that bore me onward exultantly past my own doorstep, toward the river, toward some pathway long forgotten, toward some unforgettable destination in the wetlands and the spring.

"Even as I leaped, I was changing. It was this, I think, that stirred the last remnants of human fear and human caution that I still possessed. My will was in abeyance; I could not stop. Furthermore, certain sensations, hypnotic or otherwise, suggested to me that my own physical shape was modifying, or about to change. I was leaping with a growing ease. I was——

"It was just then that the wharf lights began to show. We were approaching the end of the road, and the road, as I have said, ended in the river. It was this, I suppose, that startled me back into some semblance of human terror. Man is a land animal. He does not willingly plunge off wharfs at midnight in the monstrous company of amphibious shadows.

"Nevertheless their power held me. We pounded madly toward the wharf, and under the light that hung above it, and the beam that made a cross. Part of me struggled to stop, and part of me hurtled on. But in that final frenzy of terror before the water below engulfed me I shrieked, '*Help! In the name of God, help me! In the name of Jesus, stop!*'"

Dreyer paused and drew in his chair a little closer under the light. Then he went on steadily.

"I was not, I suppose, a particularly religious man, and the cries merely revealed the extremity of my terror. Nevertheless this is a strange thing, and whether it involves the crossed beam, or the appeal to a Christian deity, I will not attempt to answer.

"In one electric instant, however, I was free. It was like the release from demoniac possession. One moment I was leaping in an inhuman company of elder things, and the next moment I was a badly shaken human being on a wharf. Strangest of all, perhaps, was the sudden silence of that midnight hour. I looked down in the circle of the arc light, and there by my feet hopped feebly some tiny froglets of the great migration. There was nothing impressive about them, but you will understand that I drew back in revulsion. I have never been able to handle them for research since. My work is in the past."

He paused and drank, and then, seeing perhaps some lingering doubt and confusion in my eyes, held up his black-gloved hand and deliberately pinched off the glove.

A man should not do that to another man without warning, but I suppose he felt I demanded some proof. I turned my eyes away. One does not like a webbed batrachian hand on a human being.

As I rose embarrassedly, his voice came up to me from the depths of the chair.

"It is not the hand," Dreyer said. "It is the question of choice. Perhaps I was a coward, and ill prepared. Perhaps"—his voice searched uneasily among his memories—"perhaps I should have taken them and that springtime without question. Perhaps I should have trusted them and hopped onward. Who knows? They were gay enough, at least."

He sighed and set down his glass and stared so intently into empty space that, seeing I was forgotten, I tiptoed quietly away.

The Fifth Planet

"IT ISN'T THERE any more," he said. He was the only man I ever knew who hunted for bones in the stars, and I remember we were standing out among his sheep in the clear starshine when he said it. "It isn't there any more," he repeated. And innocently enough I asked, "What isn't?"—not really thinking at all but just making conversation and watching the silver light drifting over the gray backs of the sheep.

"The fifth planet," he answered. I thought a minute and counted in my head twice over to make sure, and then I said a little soothingly, as one talks to a confused child, "But the fifth planet is Jupiter. There it is over there. All you have to do is swing the tripod around and you'll pick it up all right. Planets don't disappear that easy, thank God."

"The fifth one did all the same," he said. But he shifted the tripod and took a sight toward Jupiter, and the sheep went on munching. I decided I hadn't heard him correctly. After all there is a good deal of wind in those high valleys, and even the frost made little pinging sounds cracking the stones.

We might have been standing on the moon ourselves, it was so cold. I wanted badly to go in and sit by the fire, but it was like this night after night, and nothing for me to do about it if I wanted companionship. He was a bone hunter like myself, in a kind of way. There aren't many of us, nor enough help to stand by, so I couldn't let him down even though I knew he was crazy as hell. If I was playing a hundred-to-one shot when I looked for bones around his ranch by day, he was playing it a million to one at night. That just doesn't make sense. Your life isn't long enough to fool with odds like that. Just the same, there we were out with those sheep and pointing his little telescope looking for bones in the sky.

I suppose that was the thing that intrigued me the first season we met. I'm not a star hunter, though like most people who work with the earth and its past, I have my normal share of curiosity about the universe. It was a far-off interest, though, and besides I was nearsighted from too much peering at the ground. Radnor was different. He already had a squint when I met him, from straining over that telescope, and he was the sort of chap who generally ends up in a cistern or a cesspool while marching around after the stars.

He wasn't a professional. That kind never is. He was a pure bona fide amateur, and part of the lunatic fringe. He got into this thing by accident, and like most such people, he didn't know when to stop. It began with Williams and a flight of meteors. It ended with—well that's the story. But this is how it began.

II

Williams came up there on the hunt for a meteor. Not an ordinary one, you understand, but the real stuff. A twenty-tonner heard and seen passing over three states. I knew Williams quite well—we worked at the same institution once—and I knew there wasn't a keener eye for a starfall and where it was going.

He'd plotted it all out from the reports, triangulated, and headed up there into that high tableland. Of course he ran into Radnor. He would have to in order to operate in there at all. It was Radnor's ranch and Radnor that helped him. They were lucky, sure. A ranch hand had seen the hole where it struck. They got it, and it was quite a feather in Williams's cap. In any sane place in the world they would have packed it off posthaste to a museum, with a few drinks for the local boys, and things would have gone back to normalcy fast.

The trouble was that nothing up there was normal. In the first place, digging out that big chunk of iron took time, equipment, and all sorts of supplies. Williams stayed at Radnor's place for weeks, and knowing Williams as I did it was easy to guess what happened: Williams simply took the malleable clay of Jim Radnor, sheepman, and made a star hunter out of him. That might have been all right, in a nice amateur sort of way, if they had kept it to the front porch on a Sunday evening, but that wasn't how Williams did things.

He was a born teacher—if he wasn't an astronomer I'd have said preacher and been closer to the truth—and he set out to convert Radnor. He aimed to convince Radnor of the importance of meteorite observation, which might have been all right too—there's no real harm in it if your mind runs that way—but then he added that last devilish touch that only a fanatic like Williams would have used to corrode the soul of a good sheepman.

It was unethical, to my mind, immoral really, because it is the kind of thing which the innocent amateur isn't ready to withstand. He hasn't built up to it with the necessary preparation. You take him, addlepated and open-mouthed, and let him look into space until his brain is reeling. Then you whisper over his shoulder something about life out there in that void, and the only way we can ever learn if it exists. And you speak— oh, I knew old Williams well, you know—of the freezing dark that surrounds us and the loneliness that comes to the astronomer in that room under the slit dome. You speak of the suns going by, and the great fires roaring in the solitude of space. You speak of endless depths, great distances all cold and still and empty of the life of man. And then far off, like an insect singing, you begin to whisper the hope of life on other planets, and whether it is true or untrue, and whether there has ever been or will be things like ourselves out there to share our loneliness. And then you tell again how the secret may be found.

III

It's not my secret. I won't vouch for it, you understand. It's Williams's field, not mine, but I know what he is capable of and the kind of kinks he can tie in people's heads. I've heard that wistful insect singing too, down another dimension— time—which is why I go on bone hunting. Well, he must have done it to Radnor. When I dropped in, a year later, the man had all the symptoms.

I've never known a nicer fellow than Jim. One of those odd, big-handed, practical fellows with a streak of romance that had lain dormant till Williams touched it off. Unmarried, too. Probably a wife could have handled him, or driven Williams away before the damage was done, but he had had no protection, so there he was with the little telescope Williams had gotten for him, and nothing on his mind but stars.

He had a bunch of file cards in boxes all over his bedroom. I've never seen anything like it. There was no system, no alphabet, but all observations on meteorites and things pertaining to them filed in a colossal jumble. It made him feel scientific, I guess. He told me about the boxes and said he had "facts" in them.

"Now, Jim," I said, for I knew him well, having been in and out of that country before on a mission concerning a small mouselike animal a million years deceased, "now, Jim, what in the name of God are *you* looking for? One crank in this country is bad enough. I don't like competition."

"Bones," he said.

"What?" I exclaimed.

"Bones," he repeated.

"I don't get you," I said. "You've got a nice little telescope that Williams gave you. You've card files full of stuff on meteorites. You're mumbling to me about the Doppler effect and the red shift. What has all that to do with bones? You're in astronomy."

"I'm hunting for interplanetary bones," he said then. He looked at his shoes, and I guess he thought I wouldn't like it very well, being in bones myself. He needn't have worried. I saw where Williams had led him, and I had no intention of following.

"Oh!" I said. And then I just remarked casually I'd be off for the hills in the morning. I knew what Williams could do to a man. He might have done it to me once, if I hadn't been near-sighted. Williams was a fanatic. It wasn't that there was basically anything wrong with his theory, if he'd been content to leave it at that. You'll find it all there in the books. Where the fanaticism came in was in his utter ruthlessness, and his power over people. He knew his theory was a million-to-one shot—a twice-ten-million-to-one shot. And his idea of shortening the odds was to get twenty thousand jug-headed enthusiasts to waste their lives looking for the proof.

The proof, of course, was bones, interplanetary fossils. Nobody had ever found them and nobody ever—well, I don't know now. The older you get the less you know. At the last Radnor had his say, but he never showed me anything. That isn't science, you know.

It was meteorites that were the key. The theory runs that

they, or some of them, are the products of a smash-up in space, the fragments of a lost planetary body. The chemical composition suggests it. Some are heavy metal such as we expect at our own earth's core; others resemble rocks at the earth's surface. The variation suggests that some of them, at least, originated at different depths in another planet. A planet blown to hell and gone. Just here is where Williams got in his play. Instead of sitting quietly in a good club with his drink and the meteorites, he takes the next impossible step.

The meteorites, he says, keep coming in—big ones and little ones, raining in with the earth's attraction. Okay then, keep looking, keep watching. Some day you'll get an unconsumed mass of sedimentary rock off that vanished planet. Sedimentary rock, mind you; fossil-bearing rock. Get it and you've got the secret of the galaxies. One fossil speeding in from outer space, one bit of fossil life unknown to this planet, one skull from a meteor's heart, and space out there—I remember Williams's gesture at that point of his talk—becomes alive. Man is no longer lonely. Life is no longer a unique and terrible accident. It, too, holds its place with the spinning suns.

My God, when I was younger Williams could make the hair stand up along my neck. Think of it! To know if life was out there, or even had been out there. To know whether there were interplanetary parallelisms, perhaps even men—beings, anyhow. To know and to find it all in a whirling piece of rock coming down from the night sky.

No, Williams never told me the chances; I learned of them from other men. I learned that meteorites of sedimentary rock have never been found. I learned that the surface layers of a planet form such a small proportion of its actual bulk that, dispersed in outer space, the number of such fragments coming in, if they came at all, would be infinitesimal. The fossil-bearing rocks of a planet would be in about the same proportion as dirt on the surface of a suddenly exploded golf ball to the substance of the ball itself. The odds were too great, you see, too great for a short-lived creature like man. The odds, those infernal odds!

But I have said that Williams was a fanatic. He was—a cold madman. He tracked meteors everywhere. He organized societies of meteor enthusiasts. He lectured on meteors. He told his tale. He increased the watchers. I followed his progress for years. I saw him pile up meteors like iron junk in his rooms. All iron, no once-living stone that had felt air and known the sun. The fossils might be out there—yes, they might—but living men don't win on a million-to-one shot. Williams was getting old.

He was getting old when he came the long way up that mesa in pursuit of his last flaming disappointment. He was old when he met Radnor. And he took the last of his fury and his hunger and his eloquence and poured them into that romantic, stubborn, religious-minded man.

"It's a question of time, that's all," Radnor said to me later after I had come down from the hills for supplies. He said it with the assurance of one of those damned one-track amateurs that pull Pluto out of a hat after the professionals have gone home to bed.

I spat into the fire.

"All we need is watchers, more and more watchers. Then . . ."

It was Williams all over again. But there was a difference. I could hear it at the back of his talk. "You will see. It will prove how small we are and how great, also. There will be signs, progress, evidence eventually of how high life can go, evidence from an older planet. Evidence of the Great Plan."

"All right, Jim," I said, seeing him eye the door. I was growing used to his routine. "All right," I said meekly, "we'll go out and watch." It was one of those periods of heavy star fall, but the heavy ones don't come often, not even high up where we were.

IV

I think it was on the third night that I heard him mention that fifth planet, and again the next night. I tried tactfully to correct him once more. Actually I was afraid the stuff was going to his head.

"Look, Radnor," I said, "what do you mean by the fifth planet? I tell you it's Jupiter."

He looked at me and I could see his stubborn, intense eyes back in their shadowy hollows. "Not by Bode's law," he contradicted, and mumbled some figures at me. I remembered his piles of cards, then, and the way Williams must have hammered him.

"Jupiter is the sixth planet," he went on, more to himself than me. "There's a gap between Mars and Jupiter—something is missing there. There's a gap in the planet distances. The fifth planet isn't there."

"Pure accident," I grumbled, feeling out of my depth.

"No," said Radnor, and we both instinctively stared up at something we couldn't see. "There's something there, all right. You've forgotten the asteroids—Ceres and the rest. They're moving where there should be a planet. It's the fifth one, and they're all that's left of it. Something went wrong there."

A meteor trailed high above us, dissolved, and vanished. "Of course," I said. "Stones, aren't they? Just fragments, I remember. Part of something bigger. No air, nothing but thousands of stones on a planet's pathway."

"The fifth planet," he said again. "It's part of the fifth planet that comes down here in little pieces. Williams was sure of it. He said——"

"It was beyond Mars," I conceded to humor him. "There might have been a chance for life there. Curious, what happened . . ."

I turned away then. It was cold and our breath made little frost rings in the dark. "Come on," I urged, but he just stood there brooding as I went in. I left early the next morning. I won't say with relief, for I liked Radnor. But, after all, this was Radnor in a new phase. Later, in the smoker, as the train started to roll eastward, I thought briefly of Williams and his strange influence upon men of Radnor's type. I thought also of that wide and red-stoned plateau, and the things that came down upon it out of the dark. Then someone started a game of poker, and with the miles clicking off behind us, I forgot, a little deliberately perhaps, the whole episode. One has to live, you know, and I've always had a feeling that space was not in my line.

V

I don't think, as things turned out, that it was in Radnor's line either. Men of his religious nature get centered on a symbol, finally. With some of them it may be gold tablets buried on a mountain, or a book with cryptic inscriptions. Or it may be just a voice in the woods that they alone can hear. With Radnor I guess it was the fifth planet, and I should have known it when I left him out there in the frost that night.

I never expected to see him again. In a job like mine you go many places, have casual contacts with any number of people, and are apt never to turn up in the same spot twice. As it happens in this case, I wish I never had, but I did.

It was a small matter—a question of the proper zoning of a new fossil—but it brought me back two thousand miles, and two years later. I thought of Radnor, of course. I even went to the trouble of clipping a couple of astronomical articles out of the science section of the *New York Times* for him. I figured he'd like them for those overloaded fact boxes he kept.

Things never change in a country like that. There were the same little desert owls flitting from fence post to fence post as when I went away. The road unwound into the same red distance. It was evening when I got to the ranch, and Radnor received me hospitably enough. I must say he had aged a bit, and his eyes seemed apathetic, but after all, none of us gets any younger.

After supper we talked sheep for a while. It seemed safe, and I was not unskilled at it, but after a time a silence fell between us. It was then I drew out the clippings.

He read the first one without comment and dropped it on the white tablecloth. I handed him the second. I remember it had something to do with a recent discovery about the Martian atmosphere and gave strong support to the theory that there might be life on that planet. Knowing Radnor as I had, I expected this one to be good for an evening's conversation. He looked at the headline with an expression of mingled indifference and dislike. "They shouldn't publish that sort of thing," he said.

"I thought you might like it for those big files of yours," I countered uneasily. He raised his head and looked through me the way that little black telescope of his used to pick up holes in space. "I burned them," he said flatly.

It was awkward, and the whole thing was beginning to get on my nerves. I wanted to get up, and I did. I lit a cigar and studied his face over it. I made one more effort. "How about the scope?" I suggested. "It's a clear night and I reckon you can show me a lot now." Then I started moving for the door.

He pulled himself together with a visible effort and followed me out. For all his peculiarities he was a courteous man, and I had been his friend. On the porch he halted me. There was a high, thin starlight, I remember, and it did odd things to his face.

"I don't look any more," he said, and then repeated it. I dropped into a chair and he sat uneasily facing me on the porch railing. A sort of tension was building up steadily between us.

"I don't look any more because I know," he said. "I know about it already. 'Seek and you shall find,' the Book says. It doesn't say what you will find, it just says you will find. Up here there are ways. Williams knew them."

I looked past him into the night. There was nowhere else to look except out on that great windswept plateau. A long streamer of green light shot across the horizon. The stones are still coming in, I thought wearily, but with the other part of my mind I said to Radnor, putting my words carefully together, "I don't follow you. Do you mean you found something?"

He ignored the interruption. "I believed in the Plan," he said, "what some people call the Divine Plan. I believed in life. I believed it was advancing, rising, becoming more intelligent. I believed it might have been further along out there"—he gestured mutely. "I believed it would give us hope to know."

I heard him, but I put the question bluntly. After all, it might be a matter for science and scientist is what I called myself. "What did you find," I asked, "specifically?"

"The Plan is not what you think," he said. His eyes in that strange light were alien and as cryptic as before. "The Plan is not what you think it is. I know about it now. And life—" He made another gesture, wide, indifferent, and final. There was a greater emptiness than space within him. I could feel it grow as we sat there.

I did not ask that question again. You can call me a fool, but you did not sit there as I did in that valley out of time, while star falls whispered overhead, and a fanatic talked icy insanity at your elbow. I tell you the man frightened me—or maybe it

was space itself. I got the feeling somehow that he wanted me
to ask again what he had found. And by then I didn't want to
hear. Why? Well, that kind of experience is painful, and you try
to forget afterward, but I think he must have hit some weak
spot in my psychology, probed unaware some unexpressed
deep horror of my own. Anyhow I had a feeling that I might
believe his answer, and I knew in the same clairvoyant instant
of revulsion that I could not bear to hear him give it. "Have
you got a match?" I said.

He sighed and came a long way back from somewhere. I
thought I saw how it was with him then. Williams had finished
him as he had finished others. Well, space is not my job. I
checked my data in the hills next morning and talked a lot
about sheep, and left—maybe a little sooner than I normally
would.

I had a card once from Radnor afterward. It was five years
later, and I remember it well, because it came a week after Hi-
roshima. There was nothing on the card but one line, as
though it had been hanging in the air all that time and had just
caught up to me. "The Plan is not what you think it is," it read.
"Do you see now?"

For a time I puzzled over it, unsure that I did. But after
Nagasaki the thing began to be a tune in my head like the little
songs the wind makes under telegraph wires. It just went on
sighing, "You see? You see?" Of course I knew he was crazy,
but just the same he was right about one thing. The fifth planet
is gone. And maybe I see.

Science and the Sense of the Holy

WHEN I WAS a young man engaged in fossil hunting in the Nebraska badlands I was frequently reminded that the ravines, washes, and gullies over which we wandered resembled the fissures in a giant exposed brain. The human brain contains the fossil memories of its past—buried but not extinguished moments—just as this more formidable replica contained deep in its inner stratigraphic convolutions earth's past in the shape of horned titanotheres and stalking, dirk-toothed cats. Man's memory erodes away in the short space of a lifetime. Jutting from the coils of the earth brain over which I clambered were the buried remnants, the changing history, of the entire age of mammals—millions of years of vanished daylight with their accompanying traces of volcanic outbursts and upheavals. It may well be asked why this analogy of earth's memory should so preoccupy the mind of a scientist as to have affected his entire outlook upon nature and upon his kinship with—even his concern for—the plant and animal world about him.

Perhaps the problem can best be formulated by pointing out that there are two extreme approaches to the interpretation of the living world. One was expressed by Charles Darwin at the age of twenty-eight; one by Sigmund Freud in his mature years. Other men of science have been arrayed on opposite sides of the question, but the eminence of these two scholars will serve to point up a controversy that has been going on since science arose, sometimes quietly, sometimes marked by vitriolic behavior, as when a certain specialist wedded to his own view of the universe hurled his opponent's book into his wastebasket only to have it retrieved and cherished by a graduate student who became a lifelong advocate of the opinions reviled by his mentor. Thus it is evident that, in the supposed objective world of science, emotion and temperament may play a role in our selection of the mental tools with which we choose to investigate nature.

Charles Darwin, at a time when the majority of learned men looked upon animals as either automatons or creatures created merely for human exploitation, jotted thoughtfully into one of his early journals upon evolution the following observation:

"If we choose to let conjecture run wild, then animals, our fellow brethren in pain, disease, suffering and famine—our slaves in the most laborious works, our companions in our amusements—they may partake of our origin in one common ancestor—we may be all netted together."

What, we may now inquire, is the world view here implied, one way in which a great scientist looked upon the subject matter that was to preoccupy his entire working life? In spite of the fact that Darwin was, in his later years, an agnostic, in spite of confessing he was "in thick mud" so far as metaphysics was concerned, the remark I have quoted gives every sign of that feeling of awe, of dread of the holy playing upon nature, which characterizes the work of a number of naturalists and physicists down even to the present day. Darwin's remark reveals an intuitive sensitivity to the life of other creatures about him, an attitude quite distinct from that of the laboratory experimentalist who is hardened to the infliction of pain. In addition, Darwin's final comment that we may be all netted together in one gigantic mode of experience, that we are in a mystic sense one single diffuse animal, subject to joy and suffering beyond what we endure as individuals, reveals a youth drawn to the world of nature by far more than just the curiosity to be readily satisfied by the knife or the scalpel.

If we turn to Sigmund Freud by way of contrast we find an oddly inhibited reaction. Freud, though obviously influenced by the elegant medical experimenters of his college days, groped his way alone, and by methods not subject to quantification or absolute verification, into the dark realms of the subconscious. His reaction to the natural world, or at least his feelings and intuitions about it, are basically cold, clinical, and reserved. He of all men recognized what one poet has termed "the terrible archaeology of the brain." Freud states that "nothing once constructed has perished, and all the earlier stages of development have survived alongside the latest." But for Freud, convinced that childhood made the man, adult reactions were apt to fall under the suspicion of being childhood ghosts raised up in a disguised fashion. Thus, insightful though he could be, the very nature of his study of man tended to generate distrust of that outgoing empathy we observed in the young Darwin. "I find it very difficult to work with these intangible qualities," confessed Freud. He was suspicious of their representing some lingering monster of childhood, even if reduced in size. Since Freud regarded any type of religious feeling—even the illuminative quality of the universe—as an illusion, feelings of awe before natural phenomena such as that manifested by Darwin were to him basically remnants of childhood and to be dismissed accordingly.

In *Civilization and Its Discontents* Freud speaks with slight condescension of a friend who claimed a sensation of eternity, something limitless, unbounded—"oceanic," to use the friend's expression. The feeling had no sectarian origin, no assurance of immortality, but implied just such a sense of awe as might lie at the root of the religious impulse. "I cannot," maintained Freud, "discover this 'oceanic' impulse in myself." Instead he promptly psychoanalyzes the feeling of oneness with the universe into the child's pleasure ego which holds to itself all that is comforting; in short, the original ego, the infant's ego, included everything. Later, by experience, contended Freud, our adult ego becomes only a shrunken vestige of that far more extensive feeling which "expressed an inseparable connection . . . with the external world."

In essence, then, Freud is explaining away one of the great feelings characteristic of the best in man by relegating it to a childhood atavistic survival in adult life. The most highly developed animals, he observes, have arisen from the lowest. Although the great saurians are gone, the dwarfed crocodile remains. Presumably if Freud had completed the analogy he would have been forced to say that crocodilian adults without awe and with egos shrunken safely into their petty concerns represented a higher, more practical evolutionary level than the aberrant adult who persists in feelings of wonder before which Freud recoiled with a nineteenth-century mechanist's distaste, although not without acknowledging that this lurking childlike corruption might be widespread. He chose to regard it, however, as just another manifestation of the irrational aspect of man's divided psyche.

Over six decades before the present, a German theologian, Rudolf Otto, had chosen for his examination what he termed *The Idea of the Holy (Das Heilige)*. Appearing in 1917 in a time of bitterness and disillusionment, his book was and is still widely read. It cut across denominational divisions and spoke to all those concerned with that *mysterium tremendum*, that very awe before the universe which Freud had sighed over and dismissed as irrational. I think it safe to affirm that Freud left adult man somewhat shrunken and misjudged—misjudged because some of the world's scientists and artists have been deeply affected by the great mystery, less so the child at one's knee who frequently has to be disciplined to what in India has been called the "opening of the heavenly eye."

Ever since man first painted animals in the dark of caves he has been responding to the holy, to the numinous, to the mystery of being and becoming, to what Goethe very aptly called "the weird portentous." Something inexpressible was felt to lie behind nature. The bear cult, circumpolar in distribution and known archaeologically to extend into Neanderthal times, is a further and most ancient example. The widespread beliefs in descent from a totemic animal, guardian helpers in the shapes of animals, the concept of the game lords who released or held back game to man are all part of a variety of a sanctified, reverent experience that extends from the beautiful rock paintings of South Africa to the men of the Labradorean forests or the Plains Indian seeking by starvation and isolation to bring the sacred spirits to his assistance. All this is part of the human inheritance, the wonder of the world, and nowhere does that wonder press closer to us than in the guise of animals which, whether supernaturally as in the caves of our origins or, as in Darwin's sudden illumination, perceived to be, at heart, one form, one awe-inspiring mystery, seemingly diverse and apart but derived from the same genetic source. Thus the *mysterium* arose not by primitive campfires alone. Skins may still prickle in a modern classroom.

In the end, science as we know it has two basic types of practitioners. One is the educated man who still has a controlled sense of wonder before the universal mystery, whether it hides in a snail's eye or within the light that impinges on that delicate organ. The second kind of observer is the extreme reductionist who is so busy stripping things apart that the tremendous mystery has been reduced to a trifle, to intangibles not worth troubling one's head about. The world of the secondary qualities—color, sound, thought—is reduced to illusion. The *only* true reality becomes the chill void of ever-streaming particles.

If one is a biologist this approach can result in behavior so remarkably cruel that it ceases to be objective but rather suggests a deep grain of sadism that is not science. To list but one example, a recent newspaper article reported that a great urban museum of national reputation had spent over a half-million dollars on mutilating experiments with cats. The experiments are too revolting to chronicle here and the museum has not seen fit to enlighten the public on the knowledge gained at so frightful a cost in pain. The cost, it would appear, lies not alone in animal suffering but in the dehumanization of those willing to engage in such blind, random cruelty. The practice was defended by museum officials, who in a muted show of scientific defense maintained the right to study what they chose "without regard to its demonstrable practical value."

This is a scientific precept hard to override since the days of Galileo, as the official well knew. Nevertheless, behind its seamless façade of probity many terrible things are and will be done. Blaise Pascal, as far back as the seventeenth century, foresaw our two opposed methods. Of them he said: "There are two equally dangerous extremes, to shut reason out, and to let nothing else in." It is the reductionist who, too frequently, would claim that the end justifies the means, who would assert reason as his defense and let that *mysterium* which guards man's moral nature fall away in indifference, a phantom without reality.

"The whole of existence frightens me," protested the philosopher Søren Kierkegaard; "from the smallest fly to the mystery of the Incarnation, everything is unintelligible to me, most of all myself." By contrast, the evolutionary reductionist Ernst Haeckel, writing in 1877, commented that "the cell consists of matter . . . composed chiefly of carbon with an admixture of hydrogen, nitrogen and sulphur. These component parts, properly united, produce the soul and body of the animated world, and suitably nourished become man. With this single argument the mystery of the universe is explained, the Deity annulled and a new era of infinite knowledge ushered in." Since these remarks of Haeckel's, uttered a hundred years ago, the genetic alphabet has scarcely substantiated in its essential intricacy Haeckel's carefree dismissal of the complexity of life. If anything, it has given weight to Kierkegaard's wary statement or at least heightened the compassionate wonder with which we are led to look upon our kind.

"A conviction akin to religious feeling of the rationality or intelligibility of the world lies behind all scientific work of a high order," says Albert Einstein. Here once more the eternal dichotomy manifests itself. Thoreau, the man of literature, writes compassionately, "Shall I not have intelligence with the earth? Am I not partly leaves and vegetable mould myself?" Or Walt Whitman, the poet, protests in his *Song of Myself*: "whoever walks a furlong without sympathy walks to his own funeral drest in a shroud."

"Magnifying and applying come I"—he thunders—
"Outbidding at the start the old cautious hucksters . . .
Not objecting to special revelations, considering a curl of smoke
    or a hair
on the back of my hand just as curious as any revelation."

Strange, is it not, that so many of these voices are not those of children, but those of great men—Newton playing on the vast shores of the universe, or Whitman touched with pity or Darwin infused with wonder over the clambering tree of life. Strange, that all these many voices should be dismissed as the atavistic yearnings of an unreduced childlike ego seeking in "oceanic" fashion to absorb its entire surroundings, as though in revolt against the counting house, the laboratory, and the computer.

II

Not long ago in a Manhattan art gallery there were exhibited paintings by Irwin Fleminger, a modernist whose vast lawless Martianlike landscapes contain cryptic human artifacts. One of these paintings attracted my eye by its title: "Laws of Nature."

Here in a jumbled desert waste without visible life two thin laths had been erected a little distance apart. Strung across the top of the laths was an insubstantial string with even more insubstantial filaments depending from the connecting cord. The effect was terrifying. In the huge inhuman universe that constituted the background, man, who was even more diminished by his absence, had attempted to delineate and bring under natural law an area too big for his comprehension. His effort, his "law," whatever it was, denoted a tiny measure in the midst of an ominous landscape looming away to the horizon. The frail slats and dangling string would not have sufficed to fence a chicken run.

The message grew as one looked. With all the great powers of the human intellect we were safe, we understood, in degree, a space between some slats and string, a little gate into the world of infinitude. The effect was crushing and it brought before one that sense of the "other" of which Rudolf Otto spoke, the sense beyond our senses, unspoken awe, or, as the reductionist would have it, nothing but waste. There the slats stood and the string drooped hopelessly. It was the natural law imposed by man, but outside its compass, again to use the words of Thoreau, was something terrific, not bound to be kind to man. Not man's at all really—a star's substance totally indifferent to life or what laws life might concoct. No man would greatly extend that trifling toy. The line sagged hopelessly. Man's attempt had failed, leaving but an artifact in the wilderness. Perhaps, I thought, this is man's own measure. Perhaps he has already gone. The crepitation at my spine increased. I felt the mood of the paleolithic artists, lost in the mysteries of birth and coming, as they carved pregnant beasts in the dark of caves and tried by crayons to secure the food necessarily wrung from similar vast landscapes. Their art had the same holy quality that shows in the ivory figurines, the worship before the sacred mother who brought man mysteriously into the limited world of the cave mouth.

The numinous then is touched with superstition, the reductionist would say, but all the rituals suggest even toward hunted animals a respect and sympathy leading to ceremonial treatment of hunted souls; whereas by contrast in the modern world the degradation of animals in experiments of little, or vile, meaning, were easily turned to the experimental human torture practiced at Dachau and Buchenwald by men dignified with medical degrees. So the extremes of temperament stand today: the man with reverence and compassion in his heart whose eye ranges farther than the two slats in the wilderness, and the modern vandal totally lacking in empathy for life beyond his own, his sense of wonder reduced to a crushing series of gears and quantitative formula, the educated vandal without mercy or tolerance, the collecting man that I once tried to prevent from killing an endangered falcon, who raised his rifle, fired, and laughed as the bird tumbled at my feet. I suppose Freud might have argued that this was a man of normal ego, but I, extending my childlike mind into the composite life of the world, bled accordingly.

Perhaps Freud was right, but let us look once more at this brain that in many distinguished minds has agonized over life and the mysterious road by which it has come. Certainly, as Darwin recognized, it was not the tough-minded, logical inductionists of the early nineteenth century who in a deliberate distortion of Baconian philosophy solved the problem of evolution. Rather, it was what Darwin chose to call "speculative" men, men, in other words, with just a touch of the numinous in their eye, a sense of marvel, a glimpse of what was happening behind the visible, who saw the whole of the living world as though turning in a child's kaleidoscope.

Among the purely human marvels of the world is the way the human brain after birth, when its cranial capacity is scarcely larger than that of a gorilla or other big anthropoid, spurts onward to treble its size during the first year of life. The human infant's skull will soar to a cranial capacity of 950 cubic centimeters while the gorilla has reached only 380 cubic centimeters. In other words, the human brain grows at an exponential rate, a spurt which carries it almost to adult capacity at the age of fourteen.

This clever and specifically human adaptation enables the human offspring successfully to pass the birth canal like a reasonably small-headed animal, but in a more larval and helpless condition than its giant relatives. The brain burgeons after birth, not before, and it is this fact which enables the child, with proper care, to assimilate all that larger world which will be forever denied to its relative the gorilla. The big anthropoids enjoy no such expansion. Their brains grow without exponential quickening into maturity. Somewhere in the far past of man something strange happened in his evolutionary development. His skull has enhanced its youthful globularity; he has lost most of his body hair and what remains grows strangely. He demands, because of his immature emergence into the world, a lengthened and protected childhood. Without prolonged familial attendance he would not survive, yet in him reposes the capacity for great art, inventiveness, and his first mental tool, speech, which creates his humanity. He is without doubt the oddest and most unusual evolutionary product that this planet has yet seen.

The term applied to this condition is neoteny, or pedomorphism. Basically the evolutionary forces, and here "forces" stands for complete ignorance, seem to have taken a rough-hewn ordinary primate and softened and eliminated the adult state in order to allow for a fantastic leap in brain growth. In fact, there is a growing suspicion that some, at least, of the African fossils found and ascribed to the direct line of human ascent in eastern Africa may never, except for bipedalism and some incipient tool-using capacities, have taken the human road at all.

Some with brains that seem to have remained at the same level through long ages have what amounts quantitatively to no more than an anthropoid brain. Allowing for upright posture and free use of the hand, there is no assurance that they spoke or in any effective way were more than well-adapted bipedal apes. Collateral relatives, in short, but scarcely to be termed men. All this is the more remarkable because their history can now be traced over roughly five if not six million years—a singularly unprogressive period for a creature destined later to break upon the world with such remarkable results after so long a period of gestation.

Has something about our calculations gone wrong? Are we studying, however necessarily, some bipedal cousins but not ancestral men? The human phylogeny which we seemed well on the way to arranging satisfactorily is now confused by a multiplicity of material contended over by an almost equal number of scholars. Just as a superfluity of flying particles is embarrassing the physicist, so man's evolution, once thought to be so clearly delineated, is showing signs of similar strain. A skull from Lake Rudolf with an estimated capacity of 775 cubic centimeters or even 800 and an antiquity ranging into the three-million-year range is at the human Rubicon, yet much younger fossils are nowhere out of the anthropoid range.

Are these all parts of a single variable subhumanity from which we arose, or are some parts of this assemblage neotenous of brain and others not? The scientific exchanges are as stiff with politeness as exchanges between enemies on the floor of the Senate. "Professor so-and-so forgets the difficult task of restoring to its proper position a frontal bone trampled by cattle." A million years may be covertly jumped because there is nothing to be found in it. We must never lose sight of one fact, however: it is by neotenous brain growth that we came to be men, and certain of the South African hominids to which we have given such careful attention seem to have been remarkably slow in revealing such development. Some of them, in fact, during more years than present mankind has been alive seem to have flourished quite well as simple grassland apes.

Why indeed should they all have become men? Because they occupied the same ecological niche, contend those who would lump this variable assemblage. But surely paleontology does not always so bind its deliberations. We are here dealing with a gleam, a whisper, a thing of awe in the mind itself, that oceanic feeling which even the hardheaded Freud did not deny existed though he tried to assign it to childhood.

With animals whose precise environment through time may overlap, extinction may result among contending forms; it can and did happen among men. But with the first stirrings of the neotenous brain and its superinduced transformation of the family system a new type of ecological niche had incipiently appeared —a speaking niche, a wondering niche which need not have been first manifested in tools but in family organization, in wonder over what lay over the next hill or what became of the dead. Whether man preferred seeds or flesh, how he regarded his silent collateral relatives, may not at first have induced great competition. Only those gifted with the pedomorphic brain would in some degree have fallen out of competition with the real. It would have been their danger and at the same time their beginning triumph. They were starting to occupy, not a niche in nature, but an invisible niche carved into thought which in time would bring them suffering, superstition, and great power.

It cannot, in the beginning, be recognized clearly because it is not a matter of molar teeth and seeds, or killer instincts and ill-interpreted pebbles. Rather it was something happening in the brain, some blinding, irradiating thing. Until the quantity of that gray matter reached the threshold of human proportions no one could be sure whether the creature saw with a human eye or looked upon life with even the faint stirrings of some kind of religious compassion.

The new niche in its beginnings is invisible; it has to be inferred. It does not lie waiting to be discovered in a pebble or a massive molar. These things are important in the human dawn but so is the mystery that ordained that mind should pass the channel of birth and then grow like a fungus in the night— grow and convolute and overlap its older buried strata, while a 600-pound gorilla retains by contrast the cranial content of a very small child. When man cast off his fur and placed his trust in that remarkable brain linked by neural pathways to his tongue he had potentially abandoned niches for dreams. Henceforth the world was man's niche. All else would live by his toleration —even the earth from which he sprang. Perhaps this is the hardest, most expensive lesson the layers of the fungus brain have yet to learn: that man is not as other creatures and that without the sense of the holy, without compassion, his brain can become a gray stalking horror—the deviser of Belsen.

Its beginning is not the only curious thing about that brain. There are some finds in South Africa dating into immediately post-glacial times that reveal a face and calvaria more "modern" in appearance, more pedomorphic, than that of the average European. The skull is marked by cranial capacities in excess of 1700 cubic centimeters—big brained by any standards. The mastoids are childlike, the teeth reduced, the cranial base foreshortened. These people, variously termed Boskopoid or Strandlooper, have, in the words of one anthropologist, "the amazing cranium to face ratio of almost five to one. In Europeans it is about three to one. Face size has been modernized and subordinated to brain growth." In a culture still relying on coarse fare and primitive implements, the face and brain had been subtly altered in the direction science fiction writers like to imagine as the direction in which mankind is progressing. Yet here the curious foetalization of the human body seems to have outrun man's cultural status, though in the process giving warning that man's brain could still pass the straitened threshold of its birth.

How did these people look upon the primitive world into which they found themselves precipitated? History gives back no answer save that here there flourished striking three-dimensional art—the art of the brother animal seen in beauty. Childlike, Freud might have muttered, with childlike dreams, rushed into conflict with the strong, the adult and shrunken ego, the ego that gets what it wants. Yet how strangely we linger over this lost episode of the human story, its pathos, its possible meaning. From whence did these people come? We are not sure. We are not even sure that they derive from one of the groups among the ruck of bipedal wandering apes long ago in Kenya that reveal some relationship to ourselves. Their development was slow, if indeed some of them took that road, the strange road to the foetalized brain that was to carry man outside of the little niche that fed him his tuberous, sandy diet.

We thought we were on the verge of solving the human story, but now we hold in our hands gross jaws and delicate, and are unsure of our direction save that the trail is longer than we had imagined twenty years ago. Yet still the question haunts us, the numinous, the holy in man's mind. Early man laid gifts beside the dead, but then in the modern unbelieving world, Ernst Haeckel's world, a renowned philosopher says, "The whole of existence frightens me," or another humbler thinker says, "In the world there is nothing to explain the world" but remembers the gold eyes of the falcon thrown brutally at his feet. He shivers while Freud says, "As for me I have never had such feelings." They are a part of childhood, Freud argues, though there are some who counter that in childhood—yes, even Freud might grant it—the man is made, the awe persists or is turned off by blows or the dullness of unthinking parents. One can only assert that in science, as in religion, when one has destroyed human wonder and compassion, one has killed man, even if the man in question continues to go about his laboratory tasks.

III

Perhaps there is one great book out of all American literature which best expresses the clash between the man who has genuine perception and the one who pursues nature as ruthlessly as a hunted animal. I refer to *Moby Dick*, whose narrator, Ishmael, is the namesake of a Biblical wanderer. Every literate person knows the story of Moby Dick and his pursuit by the crazed Captain Ahab who had yielded a leg to the great albino whale. It is the whale and Ahab who dominate the story. What does the whale represent? A symbol of evil, as some critics have contended? Fate, destiny, the universe personified, as other scholars have protested?

Moby Dick is "all a magnet," remarks Ahab cryptically at one moment. "And be he agent or be he principal I will wreak my hate upon him." Here, reduced to the deck of a whaler out of Nantucket, the old immortal questions resound, the questions labeled science in our era. Nothing is to go unchallenged. Thrice, by different vessels, Ahab is warned away from his contemplated conquest. The whale does not pursue Ahab, Ahab pursues the whale. If there is evil represented in the white whale it cannot be personalized. The evils of self-murder, of megalomania, are at work in a single soul calling up its foreordained destruction. Ahab heartlessly brushes aside the supplications of a brother captain to aid in the search for his son, lost somewhere in a boat in the trail of the white whale's passing. Such a search would only impede the headlong fury of the pursuit.

In Ahab's anxiety to "strike through the mask," to confront "the principal," whether god or destiny, he is denuding himself of all humanity. He has forgotten his owners, his responsibility to his crew. His single obsession, the hidden obsession that lies at the root of much Faustian overdrive in science, totally possesses him. Like Faust he must know, if the knowing kills him, if naught lies beyond. "All my means are sane," he writes, like Haeckel and many another since. "My motive and my object mad."

So it must have been in the laboratories of the atom breakers in their first heady days of success. Yet again on the third day Starbuck, the doomed mate, cries out to Ahab, "Desist. See. Moby Dick seeks thee not. It is thou, thou, that madly seekest him." This then is not the pursuit of evil. It is man in his pride that the almighty gods will challenge soon enough. Not for nothing is Moby Dick a white snow hill rushing through Pacific nights. He carries upon his brow the inscrutability of fate. Agent or principal, Moby Dick presents to Ahab the mystery he may confront but never conquer. There is no harpoon tempered that will strike successfully the heart of the great enigma.

So much for the seeking peg-legged man without heart. We know he launched his boats and struck his blows and in the fury of returning vengeance lost his ship, his comrades, and his own life. If, indeed, he pierced momentarily the mask of the "agent," it was not long enough to tell the tale. But what of the sometimes silent narrator, the man who begins the book with the nonchalant announcement, "Call me Ishmael," the man whose Biblical namesake had every man's hand lifted against him? What did he tell? After all, Moby Dick is his book.

Ishmael, in contrast to Ahab, is the wondering man, the acceptor of all races and their gods. In contrast to the obsessed Ahab he paints a magnificent picture of the peace that reigned in the giant whale schools of the 1840s, the snuffling and grunting about the boats like dogs, loving and being loved, huge mothers gazing in bliss upon their offspring. After hours of staring in those peaceful depths, "Deep down," says Ishmael, "there I still bathe in eternal mildness of joy." The weird, the holy, hangs undisturbed over the whales' huge cradle. Ishmael knows it, others do not.

At the end, when Ahab has done his worst and the *Pequod* with the wounded whale is dragged into the depths amidst shrieking seafowl, it is Ishmael, buoyed up on the calked coffin of his cannibal friend Queequeg, who survives to tell the tale. Like Whitman, like W. H. Hudson, like Thoreau, Ishmael, the wanderer, has noted more of nature and his fellow men than has the headstrong pursuer of the white whale, whether "agent" or "principal," within the universe. The tale is not of science, but it symbolizes on a gigantic canvas the struggle between two ways of looking at the universe: the magnification by the poet's mind attempting to see all, while disturbing as little as possible, as opposed to the plunging fury of Ahab with his cry, "Strike, strike through the mask, whatever it may cost in lives and suffering." Within our generation we have seen the one view plead for endangered species and reject the despoliation of the earth; the other has left us lingering in the shadow of atomic disaster. Actually, the division is not so abrupt as this would imply, but we are conscious as never before in history that there is an invisible line of demarcation, an ethic that science must sooner or later devise for itself if mankind is to survive. Herman Melville glimpsed in his huge mythology of the white beast that was nature's agent something that only the twentieth century can fully grasp.

It may be that those childlike big-brained skulls from Africa are not of the past but of the future, man, not, in Freud's words, retaining an atavistic child's ego, but pushing onward in an evolutionary attempt to become truly at peace with the universe, to know and enjoy the sperm-whale nursery as did Ishmael, to paint in three dimensions the beauty of the world while not to harm it.

Yesterday, wandering along a railroad spur line, I glimpsed a surprising sight. All summer long, nourished by a few clods of earth on a boxcar roof, a sunflower had been growing. At last, the car had been remembered. A train was being made up. The boxcar with its swaying rooftop inhabitant was coupled in. The engine tooted and slowly, with nodding dignity, my plant began to travel.

Throughout the summer I had watched it grow but never troubled it. Now it lingered and bowed a trifle toward me as the winds began to touch it. A light not quite the sunlight of this earth was touching the flower, or perhaps it was the watering of my aging eye—who knows? The plant would not long survive its journey but the flower seeds were autumn-brown. At every jolt for miles they would drop along the embankment. They were travelers—travelers like Ishmael and myself, outlasting all fierce pursuits and destined to re-emerge into future autumns. Like Ishmael, I thought, they will speak with the voice of the one true agent: "I only am escaped to tell thee."

The Winter of Man

"WE FEAR," remarked an Eskimo shaman responding to a religious question from the explorer Knud Rasmussen some fifty years ago. "We fear the cold and the things we do not understand. But most of all we fear the doings of the heedless ones among ourselves."

Students of the earth's climate have observed that man, in spite of the disappearance of the great continental ice fields, still lives on the steep edge of winter or early spring. The pulsations of these great ice deserts, thousands of feet thick and capable of overflowing mountains and valleys, have characterized the nature of the world since man, in his thinking and speaking phase, arose. The ice which has left the marks of its passing upon the landscape of the Northern Hemisphere has also accounted, in its long, slow advances and retreats, for movements, migrations and extinctions throughout the plant and animal kingdoms.

Though man is originally tropical in his origins, the ice has played a great role in his unwritten history. At times it has constricted his movements, affecting the genetic selection that has created him. Again, ice has established conditions in which man has had to exert all his ingenuity in order to survive. By contrast, there have been other times when the ice has withdrawn farther than today and then, like a kind of sleepy dragon, has crept forth to harry man once more. For something like a million years this strange and alternating contest has continued between man and the ice.

When the dragon withdrew again some fifteen or twenty thousand years ago, man was on the verge of literacy. He already possessed great art, as the paintings in the Lascaux cavern reveal. It was an art devoted to the unseen, to the powers that control the movement of game and the magic that drives the hunter's shaft to its target. Without such magic man felt weak and helpless against the vagaries of nature. It was his first attempt at technology, at control of nature by dominating the luck element, the principle of uncertainty in the universe.

A few millennia further on in time man had forgotten the

doorway of snow through which he had emerged. He would only rediscover the traces of the ice age in the nineteenth century by means of the new science of geology. At first he would not believe his own history or the reality of the hidden ice dragon, even though Greenland and the polar world today lie shrouded beneath that same ice. He would not see that what the Eskimo said to Rasmussen was a belated modern enactment of an age-old drama in which we, too, had once participated. "We fear," the Eskimo sage had said in essence, "we fear the ice and cold. We fear nature which we do not understand and which provides us with food or brings famine."

Man, achieving literacy on the far Mediterranean shores in an instant of golden sunlight would take the world as it was, to be forever. He would explore the intricacies of thought and wisdom in Athens. He would dream the first dreams of science and record them upon scrolls of parchment. Twenty-five centuries later those dreams would culminate in vast agricultural projects, green revolutions, power pouring through great pipelines, or electric energy surging across continents. Voices would speak into the distances of space. Huge jet transports would hurtle through the skies. Radio telescopes would listen to cosmic whispers from beyond our galaxy. Enormous concentrations of people would gather and be fed in towering metropolises. Few would remember the Greek word *hubris*, the term for overweening pride, that pride which eventually causes some unseen balance to swing in the opposite direction.

Today the ice at the poles lies quiet. There have been times in the past when it has maintained that passivity scores of thousands of years—times longer, in fact, than the endurance of the whole of urban civilization since its first incipient beginnings no more than seven thousand years ago. The temperature gradient from the poles to the equator is still steeper than throughout much of the unglaciated periods of the past. The doorway through which man has come is just tentatively closing behind him.

So complex is the problem of the glacial rhythms that no living scientist can say with surety the ice will not return. If it does the swarming millions who now populate the planet may mostly perish in misery and darkness, inexorably pushed from their own lands to be rejected in desperation by their neighbors. Like the devouring locust swarms that gather in favorable summers, man may have some of that same light-winged ephemeral quality about him. One senses it occasionally in those places where the dropped, transported boulders of the ice fields still hint of formidable powers lurking somewhere behind the face of present nature.

These fractured mementoes of devastating cold need to be contemplated for another reason than themselves. They constitute exteriorly what may be contemplated interiorly. They contain a veiled warning, perhaps the greatest symbolic warning man has ever received from nature. The giant fragments whisper, in the words of Einstein, that "nature does not always play the same game." Nature is devious in spite of what we have learned of her. The greatest scholars have always sensed this. "She will tell you a direct lie if she can," Charles Darwin once warned a sympathetic listener. Even Darwin, however, alert as he was to vestigial traces of former evolutionary structures in our bodies, was not in a position to foresee the kind of strange mental archaeology by which Sigmund Freud would probe the depths of the human mind. Today we are aware of the latent and shadowy powers contained in the subconscious: the alternating winter and sunlight of the human soul.

Has the earth's glacial winter, for all our mastery of science, surely subsided? No, the geologist would answer. We merely stand in a transitory spot of sunshine that takes on the illusion of permanence only because the human generations are short.

Has the wintry bleakness in the troubled heart of humanity at least equally retreated?—that aspect of man referred to when the Eskimo, adorned with amulets to ward off evil, reiterated: "Most of all we fear the secret misdoings of the heedless ones among ourselves."

No, the social scientist would have to answer, the winter of man has not departed. The Eskimo standing in the snow, when questioned about his beliefs, said: "We do not believe. We only fear. We fear those things which are about us and of which we have no sure knowledge. . . ."

But surely we can counter that this old man was an ignorant remnant of the Ice Age, fearful of a nature he did not understand. Today we have science; we do not fear the Eskimo's malevolent ghosts. We do not wear amulets to ward off evil spirits. We have pierced to the far rim of the universe. We roam mentally through light-years of time.

Yes, this could be admitted, but we also fear. We fear more deeply than the old man in the snow. It comes to us, if we are honest, that perhaps nothing has changed the grip of winter in our hearts, that winter before which we cringed amidst the ice long ages ago.

For what is it that we do? We fear. We do not fear ghosts but we fear the ghost of ourselves. We have come now, in this time, to fear the water we drink, the air we breathe, the insecticides that are dusted over our giant fruits. Because of the substances we have poured into our contaminated rivers, we fear the food that comes to us from the sea. There are also those who tell us that by our own heedless acts the sea is dying.

We fear the awesome powers we have lifted out of nature and cannot return to her. We fear the weapons we have made, the hatreds we have engendered. We fear the crush of fanatic people to whom we readily sell these weapons. We fear for the value of the money in our pockets that stands symbolically for food and shelter. We fear the growing power of the state to take all these things from us. We fear to walk in our streets at evening. We have come to fear even our scientists and their gifts.

We fear, in short, as that self-sufficient Eskimo of the long night had never feared. Our minds, if not our clothes, are hung with invisible amulets: nostrums changed each year for our bodies whether it be chlorophyll toothpaste, the signs of astrology, or cold cures that do not cure: witchcraft nostrums for our society as it fractures into contending multitudes all crying for liberation without responsibility.

We fear, and never in this century will we cease to fear. We fear the end of man as that old shaman in the snow had never had cause to fear it. There is a winter still about us—the winter of man that has followed him relentlessly from the caverns and the ice. The old Eskimo spoke well. It is the winter of the heedless ones. We are in the winter. We have never left its breath.

Man Against the Universe

MAN AGAINST the universe. But who is man and how is the universe to be defined? Sigmund Freud, in the modern era, remarked that man's mind has suffered from the impact of three significant events. The first took place when Nicolaus Copernicus, over four centuries ago, succeeded in demonstrating that the earth revolved around the sun, thus removing man from his privileged position at the center of the cosmos. The second blow which man's religious sensitivity sustained might well be dated to Darwin's demonstration in 1859 that man was only one part of nature's living web and was akin to, indeed was descended from, the animal life of the past. Finally, Freud himself, the great conquistador of psychology, created the third trauma by revealing the subterranean irrational qualities of the human mind.

The five-hundredth anniversary of Copernicus's birth was only recently celebrated. In the 1970s science has lengthened out the period in which man was a mere wandering proto-hominid on the African savanna. Thanks to the researches of Harlow Shapley, for fifty years we have known we are not even located at the heart of our own galaxy but like a sand grain are drifting on a remote arm of a spiral nebula which contains uncounted members. In truth, we can find no center. As far as the eye can reach, the objects of our attention are fleeing outward through billions of light-years.

Yet to say that man's self-examination began with the dawn of Copernican science would be to ignore that tremendous confrontation between Job and the voice from the whirlwind, in which the humbled Job is asked where he was when the foundations of the earth were laid. Furthermore, the ancient Orient had always viewed the world as illusory and envisioned the good life as primarily a way of hastening one's escape from the suffering wheel of existence. Thus man's sense of alienation, his feelings of inadequacy and trepidation before the natural world about him, long preceded the psychical disturbances that Freud regarded as induced by modern science. As Robin Collingwood pointed out some years ago, Anicius Boethius's *Consolation of Philosophy*, written in the sixth century A.D., had the distinction of being one of the most widely read books of the Middle Ages. For a thousand years every literate individual sought solace and comfort in the *Consolation*. It was never a proscribed book. Nevertheless it touches upon the infinitesimal space occupied by man in the scheme of things. Copernicus, on the other hand, had actually opened the possibility that human power extended into the celestial realms. In no mean sense he was a necessary forerunner of the space voyagers.

Yet any great scholar or artist is likely to find his conceptions denigrated in some quarters. To say that the entirety of mankind has been overwhelmed and psychologically traumatized beyond recall is to overestimate the achievements of any single intellectual. Today there exist millions of people who are totally encapsulated in another era and to whom Darwin and Darwin's ideas mean nothing. The accretion of ideas through the centuries does change the intellectual climate. Rarely, however, is the contemporary mass conscious of the innovator in its midst. This was particularly true before the rise of the news-disseminating media, but even today the content of much of science and philosophy is confined to learned circles and only rarely reaches a wider audience. As our probes into nature become more sophisticated, the greater becomes our reliance upon the specialist, while he, in turn, appeals to a minute audience of his peers.

The truth is that no man expounds upon great ideas to a single audience. He speaks, instead, to audiences, and these in turn will be receiving his message, like the far-traveling light from a star, sometimes centuries after he has delivered it. Man is not one public; he is many and the messages he receives are likely to become garbled in transmission. Again, the ideas of the most honest and well-intentioned scholar may be distorted, reoriented, or trimmed to fit the public needs of a given epoch. In addition, it could be argued that no great act of scientific synthesis is really fixed in the public mind until that public has been prepared to receive it through anticipatory glimpses.

Darwin, for example, had the way partly prepared for his ideas by the geological and paleontological efforts of the generation before him. The fact that geological time had been vastly extended had been recognized. Animal breeders were beginning to discern a lurking dynamism, a potential for change concealed in their domestic creations. In the 1840s Robert Chambers, an enterprising journalist and amateur geologist, had written a widely circulated popular book espousing, albeit anonymously, the evolutionary cause. Even the concept of natural selection, Darwin's major claim to originality, had been anticipated, in admittedly firefly glimpses, by several previous writers. Without detracting in the least from Darwin's massive and major achievement, one may observe that the literate public was in some measure ready to receive his views. In spite of some contemporary furor, the educated world accepted him within his lifetime.

By contrast, Gregor Mendel, as significant in his own way as Darwin, never received serious recognition in scientific circles and had been dead for thirty-five years before his discoveries in genetics were appreciated. He was ahead of his century and was what today might be called a laboratory geneticist carrying out seemingly unspectacular experiments upon pea plants in a kitchen garden. He was a monk burdened with the religious duties of his monastery. He had no romantic aura of wealth, no spectacular world voyage, no eminent scientific colleagues and defenders to heighten his prestige. Mendel's discoveries, though essential to the full understanding of the evolutionary mechanism, were of a sufficiently mathematical cast to belong to the biological studies of the twentieth century, not the nineteenth.

Though his work could be regarded as leading on to a far more sophisticated understanding of the miracle of life and its interrelatedness, Mendel went unnoticed by the public. No philosopher has tried to describe what Mendel's discoveries might have done to our world view, in the way either of shock or of renewed uplift. Such horizons of thought are implicit in his work and need not be equated with a more simplistic nineteenth-century Darwinism. The simple point is, however, that Mendel never had the kind of philosophical attention that would have attracted Freud. His thought lay outside the stream of public attention and he never would have gained Freud's interest as the purveyor of psychological shock.

Freud gauged his own impact upon society, but I wonder, with all due respect to his discoveries in the basement depths of the intellect, whether his claim to having destroyed man's faith in the godlike attributes of reason is justified, or whether he has expressed merely the happy ardor of the triumphant psychiatrist. For out of the depths of unreason, the murkiness of the subconscious, have come also some of the most poignant works of great art and literature. Even scientists have, on occasion, acknowledged indebtedness to that subterranean river. Freud did not in actuality destroy man's faith in mind. He merely added to its mystery by the realization that it could create besides flawed half-idiot phantasms, a more incredible beauty than could be conjured up in daylight. Before Freud was born, Ralph Waldo Emerson, a man of no inconsiderable literary gifts, had written: "I conceive a man as always spoken to from behind, and unable to turn his head and see the speaker." From those words emerges the voice of nineteenth-century romanticism. With Emerson and Darwin as opposed yet converging forces in nineteenth-century thought I shall now concern myself.

II

George Boas, one of our most eminent intellectual historians, remarked, some thirty years ago, that there are always at least two philosophies in a country: one based upon the way people live, and the other upon the results of meditation upon the universe. In the end, one is apt to contend against the other. A perfect example of this may be observed in the rise of those doctrines labeled by critics as romanticism. They reached a peculiar intensity in the early nineteenth century, chiefly as a revolt against the formalism and social restraint of the eighteenth century.

One can venture that one of the first principles to emerge from the romantic revolt was the assertion of the self against the universe—the self, "dirty and amused" or titillated with midnight terrors but for all that having escaped forever out of the constraining formal gardens of custom into a wilder nature of crags and leaping torrents. Reason gave way for the moment to the long-restrained but impassioned reality of the heart.

People wept over the new poetry, and the new poetry—that of Byron, Shelley, Keats, Coleridge—was full of picturesque revolutionaries and moon-haunted landscapes that would have seemed absurd to Lord Chesterfield and his contemporaries— men who believed that gentlemen neither laughed nor wept, at least in public. So much was the nostalgic time-sense deepened that people of means began to replace their formal gardens with artificially constructed ruins in which to brood. The experience of nature became tinged with a belief in a higher awareness, as though in the observation of nature itself one saw into the mind of the Divinity. Something subjective, lingering behind one's casual impressions, was thus sensed in nature. An intensified empathy, a willingness to transcend the ordinary modes of thinking, became a part of the suddenly emancipated and magnified self.

This feeling for nature as a thought to be encompassed, a human ego sustained by a creative power greater than itself yet capable of assimilation by the individual, crossed the Atlantic and took on a peculiarly American tinge among Emerson and his followers, who became known as transcendentalists. The mystical aspect of this experience is described by Emerson early in his career. "Standing on bare ground," he says, "my head bathed by the blithe air and uplifted into infinite space, all mean egotism vanishes. I become a transparent eyeball; I am nothing, I see all; the currents of the Universal Being circulate through me; I am part or parcel of God."

If such remarks now seem a trifle grandiose it is because this first careless rapture, what we might call the utter intoxication with wild nature which descended upon the first romantics, has largely departed. These first innovators were dreamers, sleepwalkers upon mountain heights, who groped their way out of formal gardens to be hurried along through obscure lanes and falling leaves. One's destination mattered less than the sudden freedom from restraint.

In New England with its Puritan heritage it is not surprising that, instead of being engaged with the tale of the Ancient Mariner or the defiant acts of Byronic heroes, verse should remain spare and clipped, but that the granite hillsides should take on an unearthly light in the prose that flowered by Walden Pond. Emerson was the basic sustainer, teacher, and father of the American movement. He had traveled abroad, talked with Coleridge, visited Stonehenge with Carlyle. There he had remarked in a flash of insight that the huge broken slabs reminded him of some ancient egg out of which all the ecclesiastical structures and history of the British isles had proceeded. Unconsciously the great essayist's gift for words had forecast something of his own role in American thought. His utterance was destined to inspire the democratic embrace of Whitman, as well as the austerities of Thoreau.

Of all the Concord circle Emerson was perhaps the most widely read in science. He was familiar with Sir Charles Lyell's work in geology and was well aware that Christian chronology had become a mere "kitchen clock" compared with the vast time depths the earth sciences were beginning to reveal. "What terrible questions we are learning to ask," brooded the man sometimes accused of walking with his head in the clouds. He saw us as already divesting ourselves of the theism of our fathers.

No, it cannot be asserted that this romantic of the winds and stars did not comprehend true nature. Louis Agassiz, the exponent of the Ice Age, was his friend. Emerson could speak without reluctance of early man's chewed marrow bones, of pain and disillusion, of the exploration of dreams and their midnight revelations. Yet he remained deceptively aloof, and it is perhaps this quality which has led to much castigation and to assumptions that his puritanism could never tolerate a full evolutionary philosophy.

"I found," he once humorously remarked, "when I had finished my new lecture that it was a very good house, only the architect had unfortunately omitted the stairs." And so indeed Emerson had. For all that, however, the stairs, or the somewhat wispy and transparent ghost of them, exist. It is what he called the infinitude of the private man. But if the private man is to understand his infinitude he must be led to explore it, to clamber up any available ladder. This was Emerson's primary occupation and to it he brought not alone a truly prophetic glimpse of nature before Darwin, but also a remarkably clear perception of the fauna contained in man's own psyche.

To explain the intellectual relationship between such opposites as Darwin and Emerson is not easy. At first glance, though contemporaries in time, they appear poles apart in thought, even though both have derived much from the scientific discoveries of their elders. Darwin would never have compared himself to a transparent eyeball, neither would Emerson have ever used Darwin's words "on the clumsy . . . blundering and horribly cruel works of Nature." Darwin was also willing to confess that nature was capable of telling "a direct lie," but in spite of his occasional protestations, weariness, and doubt I cannot quite visualize him confiding, as did Emerson, that his journals were full of disjointed dreams and all manner of rambling reveries and "audacities." "I delight in telling what I think," Emerson affirms in a letter, "but if you ask how I dare say so, or why it is so, I am the most helpless of mortal men." Here Emerson is the honest romantic admitting that the voice of the speaker he was never destined to see pre-empted his thoughts. Yet he knew his own gifts well, and the powers he had "by the help of some fine words" to make "every old wagon and woodpile oscillate a little and threaten to dance."

Darwin was capable of perceiving, from the presence of vestigial organs in living creatures, the fact that they were engaged on an invisible journey. Animals slipped through the interstices between one medium and another, dragging with them evolutionary traces of the past in the shape of functionless claws or rudimentary teeth. Nevertheless, no professional biologist should be unaware of Emerson's pronouncement that "there is a crack in everything God has made." Emerson is perfectly aware that the oak glades about Concord present, at best, peripheral vistas into a nature whose "interiors are terrific, full of hydras and crocodiles." In his journals there is a hidden melancholy not always to be found in his published essays.

The Romantic Movement has been studied in many aspects —its effects on the social order, its effects upon philosophy, literature, music, and the graphic arts. I know of no adequate treatment of its impact upon the science of the nineteenth century and I shall not attempt one here. It should be recognized, however, that nature in the eyes of the early romantics was revelatory. History, the ruins of the past, partook of that revelation. So did the geological catastrophism of the early century. Men began to look behind self-evident nature, the fixed nature of the existing world, toward some *mysterium* not evident to the directly observant eye. The fixed scale of nature that satisfied the scholars of the seventeenth and eighteenth centuries showed signs of disintegration. Progress and innovation came to be regarded as ushering in a better world. The existent began to be replaced by process. As Emerson put it, "We wake and find ourselves on a stair; there are stairs below us which we seem to have ascended, there are stairs above us . . . which go out of sight."

Though Darwin, in public moments, claimed his thinking to be inductive and purely Baconian, he was already as a youth being swept along in the romantic current which included enthusiasm for Odyssean voyages, evidences of past time, and the looming shadow, not just of tomorrow, but of a different tomorrow in some manner derived from today. When in 1831 he set sail in the *Beagle* all these matters were swirling in the heads of his contemporaries, however much the first discoverers would politely deny themselves in order to avert the wrath of the orthodox. In 1818, when Darwin was a boy of nine, Keats wrote about the struggle for existence, of which men were to hear so much after the publication of the *Origin of Species*:

> I was at home
> And should have been most happy—but I saw
> Too far into the sea, where every maw
> The greater on the less feeds evermore.
> But I saw too distinct into the core
> Of an eternal fierce destruction,
> And so from happiness I was far gone.

Keats, the prescient romantic, saw what Darwin was later to see and what Emerson also glimpsed and recorded as "the virulence that still remains uncured in the universe."

As Darwin centered upon that "fierce destruction" whose creative role he sought to unravel, he appealed less to the tame logicians of his era and more and more to "speculative men," men with imagination, men who loved extremes of argument —in short, romantic men. In a burst of enthusiasm Darwin himself once cried, shedding his Baconian mask for a moment, "I am but a gambler and love a wild experiment." In those words Darwin had revealed the soul of a romantic, a man willing to follow a dancing boglight through the obscurity of forgotten ages.

Nevertheless, when he came to write the conclusion of the *Origin of Species*, a certain orthodox benignity is allowed once more to conceal the ferocity of the world whose cruelty and waste he had once exclaimed over. "As natural selection works solely by and for the good of each being," the author philosophizes, "all corporeal and mental endowments will tend to progress toward perfection." "Thus," concludes Darwin, "from the war of nature, from famine and death, the most exalted object which we are capable of conceiving, namely, the production of the higher animals, directly follows. There is grandeur in this view of life, with its several powers, having been originally breathed by the Creator into a few forms or into one; and that, whilst this planet has gone cycling on according to the fixed law of gravity, from so simple a beginning endless forms most beautiful and most wonderful have been and are being evolved." Of this type of conclusion the mystical Emerson had remarked soberly on an earlier but similar occasion, "What is so ungodly as these polite bows to God in English books?"

The man whom reality eluded, or so it was said, produced in 1841 a statement that anticipates the full flowering of process philosophy in the twentieth century but would equally and more eloquently have graced Darwin's final paragraphs in the *Origin of Species*. "The method of nature," Emerson muses, "who could ever analyze it? That rushing stream will not stop to be observed. We can never surprise nature in a corner, never find the end of a thread, never tell where to set the first stone. The bird hastens to lay her egg. The egg hastens to be a bird. [Nature's] smoothness is the smoothness of the pitch of a cataract. Its permanence is a perpetual inchoation."

Unlike Darwin's somewhat sly intimations of perfection and progress it was the idealist, not the concealed materialist, who wrote: "That no single end may be selected and nature judged thereby, appears from this, that if man himself be considered as the end, and it be assumed that the final cause of the world is to make holy or wise or beautiful men, we see that it has not succeeded."

As a result of his contemplations, even though he did not

possess the Darwinian key of natural selection, Emerson is aware of nature's infinite prodigality and wastefulness of suns and systems. He recognizes that nature can only be conceived as existing to a universal end, and not to a particular one, such as man. "To a universe of ends," Emerson adds as an afterthought, "a work of ecstasy." Nature, he maintained, is unspecific. "[It] knows neither palm nor oak, but only vegetable life, which sprouts into forests and festoons the globe." Nowhere is anything final. Nature has no private will; it will answer no private question. "The world," Emerson meditates, looking on with that far-reaching sun-struck eyeball which earned him critical derision, "leaves no track in space, and the greatest action of man no mark in the vast idea."

And yet Emerson cared—more, perhaps, than has been allowed. Crouched midway on that desperate stair whose steps pass from dark to dark, he spoke as Darwin chose not to speak in his final peroration. Emerson saw, with a terrible clairvoyance, the downward pull of the past. "The transmigration of souls is no fable," he wrote. "I would it were; but men and women are only half human. Every animal in the barnyard, the field and the forest, of the earth and of the waters . . . has contrived to get a footing and to leave the print of its features and form in someone or other of these upright heaven-facing speakers." He could sense, not Darwin's automatic trend toward perfection, but the weary slipping, the sensed entropy, the ebbing away of the human spirit into fox and weasel as it struggled upward while all its past tugged upon it from below. This is the Gethsemane of the man whom Walt Whitman called, and rightly, "transcendental of limits, a pure American for daring."

III

When I was young, in a time of boyhood marked by a world as fresh and green and utterly marvelous as on the day of its creation, I found myself attracted by a huge tropical shell which lay upon my aunt's dressing table. The twentieth century was scarcely a decade old, and people did not travel or collect as they do now. My uncle and aunt lived far inland in the central states and what wandering relative had given them the beauti-

ful iridescent shell I do not know. It was held up to my youthful ear and I was told to listen carefully and I would hear the
sea. Out of the great shell, even in that silent bedroom, I, who
had never seen the ocean, heard the whispered sibilance, the
sigh of waves upon the beach, the little murmurs of moving
water, the confused mewing of gulls in the sun-bright air. It
was my first miracle, indeed perhaps my first awareness of the
otherness of nature, of myself outside, in a sense, and listening,
as though beyond light-years, to a remote event. Perhaps, in
that Victorian bedroom with its knickknacks and curios, I had
suddenly fallen out of the nature I inhabited and turned, for
the first time, to survey her with surprise.

The sounds stayed with me through the years or I would
not be able to recall them now. Neither does it matter that in
my college days I learned that it was not the sea to which I had
listened, but the vastly magnified whispers of my blood and
the house around me. Either was marvel enough—that a shell,
a shell shaped in the seas' depths, should, without intent, so
concentrate the essence of the world as to bring its absent images before me.

The taxonomists, the classifiers, have tried with Latin appel-
lations to define man to their satisfaction. They have called
him wise and raised up justifiable doubt, as did Freud. They
have called him the tool user, the fabricator. They have, by
turns, characterized him as the only being who laughs or who
weeps. They have spoken of him as a time binder who trans-
mits thought through the generations and thus reorients and
changes the world. There are also those who have categorized
him as the sole religious animal or, finally, as *Homo duplex*, the
creature composed of flesh and spirit.

The appeal of this last definition gives me pause, even
though I am a professional anthropologist who must employ
the diction of his trade. For is it not true, as Emerson indicated
before the rise of scientific anthropology, that man, in becoming aware of nature, has entered upon a confused and endless
exploration, a transcendental search for order? Both the theologian and the scientist, each in his way, pursue that quest.

In one of the most profound and succinct analogies ever
penned by a philosopher, George Santayana once ventured:
"The universe is the true Adam, the creation the true fall." He

saw, immured in his study, that in the instant when the universe was brought out of the void of non-being its particles, achieving such powers as are present in man, would yearn for understanding of their destiny. Alienated and alone, listening to the murmur in the shell, the individual would search his mind in vain. Primitively he would seek to placate the unseen spirits in running water, or the ghost that rules in the fir tree. Divorced from the whole, he would always be intimidated by those mocking questions from the whirlwind, "Where wast thou, where wast thou? Declare if thou knowest it all."

No clearer evidence is needed to refute Freud's argument that the great traumas from which man has suffered are the products of modern science. The fall out of nature into knowledge was sustained long ago in the caverns of mankind's birth. Between the telescope and the microscope the Adamic universe has widened, that is all. If we still suffer from renewed shocks of a scientific nature, we have, at the same time, been released from the bonds of barbarous superstition. If the particle cannot rejoin the mass, it has at least achieved, in twentieth-century quantum mechanics, a creative liberty not granted under the Newtonian Mechanic God.

When I first read Emerson's *Method of Nature* I was amazed at how much of a forerunner of process philosophy he was, and how, some twenty years before the *Origin of Species*, as I have noted, he had expressed so skillfully nature's lack of any single observable objective. Not for man, not for mouse. Total nature, as he put it, was coincident with no private will, yet it was "growing like a field of maize in July."

At the time I encountered those passages I was a young man steeped in the scientific tradition, and I was struck with Emerson's articulate insight. Why then, one might ponder, if Emerson glimpsed only a universe of ends, and if every natural object is only an emanation from another, and if—I can put it in no other way—each emanation explodes into another future, did Emerson, after beclouding human hope, direct so much of his attention toward the species he had so eloquently dismissed? For, along with fox and woodchuck, we would appear as but momentary and superficial tenants of the globe.

It was not until many years later, when I had abandoned certain of the logical disciplines of my youth, that I began to sense why Emerson, with seeming inconsistency, had rounded home in that same essay to extol the nature of man. I think it was upon encountering a phrase in Whitman, and knowing there was an intellectual affinity between the two men, that I paused. The lines read:

There was a child went forth every day
And the first object he looked upon, that object he became.

It is not alone that the species is an emanation, I considered; our very thoughts transform us from minute to minute, hour to hour. How powerfully this quotation reflects Emerson's earlier statement: "The termination of the world in a man appears to be the last victory of intelligence. The universal does not attract us until housed in an individual. Who heeds the waste abyss of possibility?"

Who indeed, until the possibility is embodied, "not to be diffused," in Whitman's words, but to be realized. I had listened long ago to the impingement of secret and rumorous whispers from the air upon the coiled interior of a shell. Man it was who held and interpreted the shell, first in the romantic sea vision of youth, last as a symbol of all that the human ear might encompass from the resounding shores of the universe, as well as his own interior.

"We must admire in man," continued Emerson, who, in *The Method of Nature*, was careful not to admire him overmuch, "the form of the formless, the concentration of the vast, the house of reason, the cave of memory." The cave of memory! It is this, this echoing upheld shell, that enabled Emerson to interpose insignificant transitory man as the counterweight to stars and wasteful galaxies, to say, in fact, that man can carry the chemistry and the distance of a star inside his head. "What is a man," he jotted in his journals, "but nature's finer success in self-explication?" The explication is not eternal, any more than today is for always. The world, Emerson made clear, is in process, is departing, as men and ideas are similarly departing. The self-explication of today is approximate and will demand rearrangement as long as a critical and enlightened eye remains to examine nature. Nor does Emerson have illusions about the number of minds that can genuinely perceive nature. Furthermore, he is aware that what he terms nature's "suburbs and extremities" may contain truths that turn the world from a lumber room to an ordered creation. This observation was written in the same year that Darwin, home from probing such natural extremities as the Galápagos, conceived of the principle of natural selection.

Emerson had had, like Darwin, an illness and a voyage—that strange road taken by so many of the nineteenth-century romantics—romantics who were finally to displace the sedate white doorstone into nature by something wild and moon-haunted, whether in science or art. He may have had, as he himself once ventured, "an excess of faith"—faith in man that may cause us to stir uneasily now, but which he expressed at a time when London was truly a city of dreadful night. Above all, he seemed to sense intuitively what Alfred Russel Wallace had believed—that man possesses latent mental powers beyond what he might culturally express in a given epoch. In Ice Age caverns he had painted with an artist's eye; modern primitives can master music, writing, and machines they have never pre-viously experienced. In the words of the eminent French biol-ogist Jean Rostand, "Already at the origin of the species man was equal to what he was destined to become." A careful reading of the American transcendentalist would demonstrate that he had an intuitive grasp of this principle—so firm that neither the size of the universe nor the imperfections of our common humanity distressed him overmuch. He knew, with a surety our age is in danger of losing, that if there was ever a good man there will be more. Nature strives at better than her actual creatures. We are, Emerson maintains, "a conditional population." If atavistic reptiles still swim in the depths of man's psyche, they are not the only inhabitants of that hidden region.

Tomorrow lurks in us, the latency to be all that was not achieved before. This is what led proto-man, five million years ago, to start upon a journey, at a time when night and day were strange and miraculous, as was the trumpeting of mam-moths or the march of reindeer. It was for this that man adorned his caverns in the morning of time. It was for this that he worshiped the bear. For man had fallen out of the secure world of instinct into a place of wonder. That wonder is still expanding, changing as man's mind keeps pace with it. He stands and listens with a shell pressed to his ear. He is still a child before the infinite spaces but he is in no way frightened. It was thus that his journey began—perhaps with a message drawn from an echoing shell. Now he listens with his own giant fabricated ear to messages from beyond infinity. In the old house of nature there are monsters in every cupboard. That is why, as nature's children, we are inveterate romantics and go visiting. This is why the great American essayist antici- pated the whole of the unborn science of anthropology when he said, "The entrance of [nature] into his mind seems to be the birth of man." If it brought him fear it opened to his aroused curiosity every nook and cranny of the world. It left him, in fact, the inheritor of an echoing and ghost-ridden mansion. The shifting unseen potential that we call nature has left to man but one observable dictum, to grow. Only our un- foreseeable tomorrow can determine whether we will grow in the wisdom Emerson anticipated.

Thoreau's Vision of the Natural World

SOMEWHERE in the coverts about Concord, a lynx was killed well over a century ago and examined by Henry David Thoreau; measured, in fact, and meditated upon from nose to tail. Others called it a Canadian lynx, far strayed from the northern wilds. No, insisted Thoreau positively, it is indigenous, indigenous but rare. It is a night haunter; it is a Concord lynx. On this he was adamant. Not long ago, over in Vermont, an intelligent college girl told me that, walking in the woods, her Labrador retriever had startled and been attacked by a Canadian lynx which she had been fully competent to recognize. I was too shy, however, to raise the question of whether the creature might have been, as Thoreau defiantly asserted, a genuine New England lynx, persisting but rare since colonial times.

Thoreau himself was a genuine Concord lynx. Of that there can be no doubt. We know the place of his birth, his rarity, something of his habits, his night travels, that he had, on occasion, a snarl transferred to paper, and that he frequented swamps, abandoned cellar holes, and woodlots. His temperament has been a subject of much uncertainty, as much, in truth, as the actual shape of those human figures which he was wont to examine looming in fogs or midsummer hazes. Thoreau sometimes had difficulty in seeing men or, by contrast, sometimes saw them too well. Others had difficulty in adjusting their vision to the Concord lynx himself, with the result that a varied and contentious literature has come down to us. Even the manner of his death is uncertain, for though the cause is known, some have maintained that he benefited a weak constitution by a rugged outdoor life. Others contend that he almost deliberately stoked the fires of consumption by prolonged exposure in inclement weather.

As is the case with most wild animals at the periphery of the human vision, Thoreau's precise temperament is equally a matter of conjecture even though he left several books and a seemingly ingenuous journal which one eminent critic, at least, regards as a cunningly contrived mythology. He has been termed a stoic, a contentious moralizer, a parasite, an arsonist, a misanthrope, a supreme egotist, a father-hater who projected his animus on the state, a banal writer who somehow managed to produce a classic work of literature. He has also been described as a philosophical anarchist and small-town failure, as well as an intellectual aristocrat. Some would classify him simply as a nature writer, others as a failed scientist who did not comprehend scientific method. Others speak of his worldwide influence upon the social movements of the twentieth century, of his exquisite insight and style, of his relentless searching for something never found—a mystic in the best sense of the term. He has also been labeled a prig by a notable man of English letters. There remains from those young people who knew him the utterly distinct view that he was a friendly, congenial, and kindly man. At the time of his funeral, and long before fame attended his memory, the schools of Concord were dismissed in his honor.

In short, the Concord lynx did not go unwept to his grave. Much of the later controversy that has created a Browning-esque *Ring and the Book* atmosphere about his intentions and character is the product of a sophisticated literary world he abhorred in life. Something malevolent frequently creeps into this atmosphere even though it is an inevitable accompaniment of the transmission of great books through the ages. Basically the sensitive writer should stay away from his own kind. Jealousies, tensions, feuds, unnecessary discourtesies that are hard to bear in print are frequently augmented by close contact, even if friendships begin well. The writer's life is a lonely one and doubtless should remain so, but a prurient curiosity allows no great artist to rest easy in his grave. Critical essays upon Thoreau now number hundreds of items, very little of which, I suspect, would move the Concord lynx to do more than retire farther into whatever thickets might remain to him. Neither science nor literature was his total concern. He was a fox at the wood's edge, regarding human preoccupations with doubt. Indeed he had rejected an early invitation to join the American Association for the Advancement of Science. The man had never entertained illusions about the course of technological progress and the only message that he, like an Indian, had gotten from the telegraph was the song of the wind through its wires.

As a naturalist he possessed the kind of memory which fixes certain scenes in the mind forever—a feeling for the vastness and mystery contained in nature, a powerful aesthetic response when, in his own words, "a thousand bare twigs gleam like cobwebs in the sun." He had been imprinted, as it were, by his home landscape at an early age. Similarly, whatever I may venture upon the meaning of Thoreau must come basically from memory alone, with perhaps some examination of the tone of his first journal as contrasted with his last. My present thoughts convey only the residue of what, in the course of time, I have come to feel about his intellectual achievement—the solitary memory that Thoreau himself might claim of a journey across austere uplands.

In addition, my observations come mostly from the realm of science, whereas the preponderance of what had been written about Thoreau has come from the region he appeared to inhabit: namely, literature. His scientific interests have frequently been denigrated, and he himself is known more than once to have deplored the "inhumanity of science." Surely then, one cannot, as a representative of this alien discipline, be accused of either undue tolerance or weak sympathy for a transcendental idealism that has largely fallen out of fashion. Yet it is an aspect of this philosophy, more particularly as it is found modified in Thoreau, that I wish to consider. For Thoreau, like Emerson, is an anticipator, a forerunner of the process philosophers who have so largely dominated the twentieth century. He stands at the border between existent and potential nature.

II

Behind all religions lurks the concept of nature. It persists equally in the burial cults of Neanderthal man, in Cro-Magnon hunting art, in the questions of Job and in the answering voice from the whirlwind. In the end it is the name of man's attempt to define and delimit his world, whether seen or unseen. He knows intuitively that nature is a reality which existed before him and will survive his individual death. He may include in his definition that which is, or that which may be. Nature remains an otherness which incorporates man, but which man instinctively feels contains secrets denied to him.

A professional atheist must still account for the fleeting particles that appear and vanish in the perfected cyclotrons of modern physics. We may see behind nature a divinity which rules it, or we may regard nature itself as a somewhat nebulous and ill-defined deity. Man knows that he springs from nature and not nature from him. This is very old and primitive knowledge. Man, as the "thinking reed," the memory beast, and the anticipator of things to come, has devised hundreds of cosmogonies and interpretations of nature. More lately, with the dawn of the scientific method, he has sought to probe nature's secrets by experiment rather than unbounded speculation.

Still, of all words coming easily to the tongue, none is more mysterious, none more elusive. Behind nature is hidden the chaos as well as the regularities of the world. And behind all that is evident to our senses is veiled the insubstantial deity that only man, of all earth's creatures, has had the power either to perceive or to project into nature.

As scientific agnostics we may draw an imaginary line beyond which we deny ourselves the right to pass. We may adhere to the tangible, but we will still be forced to speak of the "unknowable" or of "final causes" even if we proclaim such phrases barren and of no concern to science. In our minds we will acknowledge a line we have drawn, a definition to which we have arbitrarily restricted ourselves, a human limit that may or may not coincide with reality. It will still be nature that concerns us as it concerned the Neanderthal. We cannot exorcise the word, refine it semantically though we may. Nature is the receptacle which contains man and into which he finally sinks to rest. It implies all, absolutely all, that man knows or can know. The word ramifies and runs through the centuries, assuming different connotations.

Sometimes it appears as ghostly as the unnamed shadow behind it; sometimes it appears harsh, prescriptive, and solid. Again matter becomes interchangeable with energy; fact becomes shadow, law becomes probability. Nature is a word that must have arisen with man. It is part of his otherness, his humanity. Other beasts live within nature. Only man has ceaselessly turned the abstraction around and around upon his tongue and found fault with every definition, found himself looking ceaselessly outside of nature toward something invisible to any eye but his own and indeed not surely to be glimpsed by him.

To propound that Henry David Thoreau is a process philosopher, it is necessary first to understand something of the concepts entertained by early nineteenth-century science and philosophy, and also to consider something of the way in which these intellectual currents were changing. New England's ties to English thought preponderated, although, as is well known, some of the ideas had their roots on the continent. An exhaustive analysis is unnecessary to the present purpose. Thoreau was a child of his time but he also reached beyond it.

In early-nineteenth-century British science there was a marked obsession with Baconian induction. The more conservative-minded, who dreaded the revelations of the new geology, sought, in their emphatic demand for facts, to drive wide-ranging and useful hypotheses out of currency. The thought of Bacon, actually one of the innovators of scientific method, was being perverted into a convenient barrier against the advance of irreligious science. Robert Chambers, an early evolutionist, had felt the weight of this criticism, and Darwin, later on, experienced it beyond the midcentury. In inexperienced hands it led to much aimless fact-gathering under the guise of proper inductive procedures for true scientists. Some, though not all, of Thoreau's compendiums and detailed observations suggest this view of science, just as do his occasional scornful remarks about museum taxonomy.

In a typical Thoreauvian paradox he could neither leave his bundles of accumulated fact alone, nor resist muttering "it is ebb tide with the scientific reports." Only toward the close of his journals does he seem to be inclining toward a more perceptive scientific use of his materials under the influence of later reading. It is necessary to remember that most of Thoreau's intellectual contacts were with literary men, though Emerson was a wide and eclectic reader who saw clearly that the new uniformitarian geology had transformed our conceptions of the world's antiquity.

The one striking exception to Thoreau's lack of direct contact with the scientific world was his meeting with Louis

Agassiz after the European glaciologist, taxonomist, and teacher had joined the staff at Harvard in the 1840s. Brilliant and distinguished naturalist though he was, Louis Agassiz was probably not the best influence upon Thoreau. He traced the structural relations of living things, he introduced America to comparative taxonomy, he taught Thoreau to observe such oddities as frozen and revived caterpillars. His eye, if briefly, was added to Thoreau's eye, not always to the latter's detriment. Typically, Agassiz warned against hasty generalizations while pursuing relentlessly his own interpretation of nature.

The European poet of the Ice Age was, in a very crucial sense, at the same time an anachronism. He did not believe in evolution; he did not grasp the significance of natural selection. "Geology," he wrote in 1857, "only shows that at different periods there have existed different species; but no transition from those of a preceding into those of a following epoch has ever been noticed anywhere." He saw a beneficent intelligence behind nature; he was a Platonist at heart, dealing with the classification of the eternal forms, seeing in the vestigial organs noted by the evolutionist only the direct evidence of divine plan carried through for symmetry's sake even when the organ was functionless.

It is this preternatural intelligence which Thoreau is led to see in the protective arrangement of moth cocoons in winter. "What kind of understanding," he writes, "was there between the mind . . . and that of the worm that fastened a few of these leaves to its cocoon in order to disguise it?" Plainly he is following in the footsteps of the great biologist, who, like many other scholars, recognized a spiritual succession of forms in the strata, but not the genuine organic transformations that had produced the living world. It may be noted in passing that this Platonic compromise with reality was one easily acceptable to most transcendentalists. They were part of a far-removed romantic movement which was to find the mechanistic aspects of nineteenth-century science increasingly intolerable. One need not align oneself totally with every aspect of the Darwinian universe, however, to see that Agassiz's particular teleological interpretation of nature could not be long sustained. Whatever might lie behind the incredible profusion of living forms would not so easily yield its secrets to man.

The mystery in nature Thoreau began to sense early. A granitic realism forced from him the recognition that the natural world is indifferent to human morality, just as the young Darwin had similarly brooded over the biological imperfections and savagery of the organic realm. "How can [man]" protested Thoreau, "perform that long journey who has not conceived whither he is bound, . . . who has no passport to the end?" This is a far cry from the expressions of some of the contemporary transcendentalists who frequently confused nature with a hypostatized divine reason which man could activate within himself. By contrast, Thoreau remarks wearily, "Is not disease the rule of existence?" He had seen the riddled leaf and the worm-infested bud.

If one meditates upon the picture of nature presented by both the transcendental thinkers such as Emerson and the evolutionary doctrine drawn from Darwin, one is struck by the contrast between what appears to be a real and an ideal nature. The transcendentalist lived in two worlds at once, in one of which he was free to transform himself; he could escape the ugly determinism of the real. In Emerson's words, "two states of thought diverge every moment in wild contrast."

The same idea is echoed in the first volume of his journals when the young Thoreau ventures, "On one side of man is the actual and on the other the ideal." These peculiar worlds are simultaneously existent. Life is bifurcated between the observational world and another more ideal but realizable set of "instructions" implanted in our minds, again a kind of Platonic blueprint. We must be taught through the proper understanding of the powers within us. The transcendentalist possessed the strong optimism of the early Republic, the belief in an earthly Eden to be created.

If we examine the Darwinian world of change we recognize that the nature of the evolutionary process is such as to deny any relationships except those that can be established on purely phyletic grounds of uninterrupted descent. The Platonic abstract blueprint of successive types has been dismissed as a hopeful fiction. Vestigial organs are really what their name implies—remnants lingering from a former state of existence. The tapeworm, the sleeping-sickness trypanosome, are as much a product of natural selection as man himself. All currently existing animals and plants have ancestral roots extending back into Archeozoic time. Every species is in some degree imperfect and scheduled to vanish by reason of the very processes which brought it into being. Even in this world of endless struggle, however, Darwin is forced to introduce a forlorn note of optimism which appears in the final pages of the *Origin of Species*. It is his own version of the ideal. Out of the war of nature, of strife unending, he declares, all things will progress toward perfection.

The remark rings somewhat hollowly upon the ear. Darwin's posited ideal world, such as it is, offers no immediate hope that man can embrace. Indeed, it is proffered on the same page where Darwin indicates that of the species now living very few will survive into the remote future. Teleological direction has been read out of the universe. It would re-emerge in the twentieth century in more sophisticated guises that would have entranced Thoreau, but for the moment the "Great Companion" was dead. The Darwinian circle had introduced process into nature but had never paused to examine nature itself. In the words of Alfred North Whitehead, "Science is concerned not with the causes but the coherence of nature." Something, in other words, held the thing called nature together, gave it duration in the midst of change and a queer kind of inhuman rationality.

It may be true enough that Thoreau in his last years never resolved his philosophical difficulties. (Indeed, what man has?) It may also be assumed that the confident and brilliant perceptions of the young writer began to give way in his middle years to a more patient search for truth. The scientist in him was taking the place of the artist. To some of the literary persuasion this may seem a great loss. Considering, however, the toils from which he freed himself, the hope that he renewed, his solitary achievement is remarkable.

One may observe that there are two reigning models involving human behavior today: a conservative and a progressive version. The first may be stated as regarding man in the mass as closely reflecting his primate origins. He is the "ape and tiger" of Huxley's writing. His origins are sufficiently bestial that they place limits upon his ethical possibilities. Latent aggression makes him an uncertain and dangerous creature.

Unconsciously, perhaps, the first evolutionists sought to link man more closely to the animal world from which he had arisen. Paradoxically this conception would literally freeze man upon his evolutionary pathway as thoroughly as though a similar argument had been projected upon him when he was a Cretaceous tree shrew. Certainly one may grant our imperfect nature. Man is in process, as is the whole of life. He may survive or he may not, but so long as he survives he will be part of the changing, onrushing future. He, too, will be subject to alteration. In fact, he may now be approaching the point of consciously inducing his own modification.

How did Thoreau, who matured under the influence of the transcendentalists and the design arguments of Louis Agassiz, react to this shifting, oncoming world of contingency and change? It is evident he was familiar with Charles Lyell's writings, that he knew about the development (that is, evolution) theories of Robert Chambers. The final volume of his journals even suggests that he may have meditated upon Darwin's views before his death in 1862.

Now, in the final years, he seems to gather himself for one last effort. "The development theory implies a greater vital force in nature," he writes, "because it is more flexible and accommodating and equivalent to a sort of constant new creation."

He recognizes the enormous waste in nature but tries carefully to understand its significance just as he grasps the struggle for existence. His eyes are open still to his tenderly cherished facts of snow and leaves and seasons. He counts tree rings and tries to understand forest succession. The world is perhaps vaster than he imagined, but, even from the first, nature was seen as lawless on occasion and capable of cherishing unimaginable potentials. That was where he chose to stand, at the very edge of the future, "to anticipate," as he says in *Walden*, "nature herself."

This is not the conservative paradigm of the neo-Darwinian circle. Instead it clearly forecasts the thought of twentieth-century Alfred North Whitehead. Thoreau strove with an unequaled intensity to observe nature in all its forms, whether in the raw shapes of mountains or the travelings of seeds and deer mice. His consciousness expanded like a sunflower. The more objects he beheld, the more immortal he became. "My senses," he wrote, "get no rest." It was as though he had foreseen White-head's dictum that "passage is a quality not only of nature, which is the thing known, but also of sense awareness, which is the procedure of knowing." Thoreau had constituted himself the Knower. "All change," he wrote, "is a miracle to contemplate, but it is a miracle that is taking place every instant." He saw man thrust up through the crust of nature "like a wedge and not till the wound heals . . . do we begin to discover where we are." He viewed us all as mere potential; shadowy, formless perhaps, but as though about to be formed. He had listened alone to the "unspeakable rain"; he had sought in his own way to lead others to a supernatural life in nature. He had succeeded. He had provided for others the passport that at first he thought did not exist; a passport, he finally noted, "earned from the elements."

III

To follow the involved saunterings of the journals is to observe a man sorting, selecting, questioning less nature than his own way into nature, to find, as Thoreau expressed it, "a patent for himself." Thoreau is never wholly a man of the transcendental camp. He is, in a sense, a double agent. He is drawn both to the spiritual life and to that of the savage. "Are we not all wreckers," he asks, "and do we not contract the habits of wreckers?" The term "*wreckers*," of course, he uses in the old evil sense of the shore scavengers who with false lights beckoned ships to their doom. And again he queries, "Is our life innocent enough?" It would appear he thinks otherwise, for he writes, "I have a murderer's experience . . ." and he adds, a trifle scornfully, "there is no record of a great success in history."

Thoreau once said disconsolately that he awaited a Visitor who never came, one whom he referred to as the "Great Looker." Some have thought that pique or disappointment shows in these words and affects his subsequent work. I do not choose to follow this line of reasoning, since, on another page of *Walden*, he actually recollects receiving his guest as the "old settler and original proprietor who is reported to have dug Walden pond . . . and fringed it with pine woods." As a somewhat heretical priest once observed, "God asks nothing of the highest soul but attention."

Supplementing that remark Thoreau had asserted, "There has been nothing but the sun and the eye since the beginning." That eye, in the instance of Thoreau, had missed nothing. Even in the depths of winter when nature's inscriptions lay all about him in the snow, he had not faltered to recognize among them the cruel marks of a farmer's whip, steadily, even monotonously, lashing his oxen down the drift-covered road. The sight wounded him. Nature, he knew, was not bound to be kind to man. In fact, there was a kind of doubleness in nature as in the writer. The inner eye was removed; its qualities were more than man, as natural man, could long sustain. The Visitor had come in human guise and looked out upon the world a few brief summers. It is remembered that when at last an acquaintance came to ask of Thoreau on his deathbed if he had made his peace with God, the Visitor in him responded simply, "We have never quarreled."

Thoreau had appropriated the snow as "the great revealer." On it were inscribed all the hieroglyphs that the softer seasons concealed. Last winter, trudging in the woods, I came to a spot where freezing, melting, and refreezing had lifted old footprints into little pinnacles of trapped oak leaves as though a shrub oak had walked upon some errand. Thoreau had recorded the phenomenon over a century ago. There was something uncanny about it, I thought, standing attentively in the snow. A visitor, perhaps the Visitor, had once more passed.

In the very first volume of his journals Thoreau had written, "There is always the possibility, the possibility, I say, of being all, or remaining a particle in the universe." He had, in the end, learned that nature was not an enlarged version of the human ego, that it was not, to use Emerson's phrase, "the immense shadow of man." Toward the close of his life he had turned from literature to the growing, formidable world of the new science—the science that in the twentieth century was destined to reduce everything to infinitesimal particles and finally these to a universal vortex of wild energies.

But the eye persisted, the unexplainable eye that gave even Darwin a cold shudder. All it experienced were the secondary qualities, the illusions that physics had rejected, but the eye remained, just as Thoreau had asserted—the sun and the eye from the beginning. Thoreau was gone, but the eye was multitudinous, ineradicable.

I advanced upon the fallen oak leaves. We were all the eye of the Visitor—the eye whose reason no physics could explain. Generation after generation the eye was among us. We were particles but we were also the recording eye that saw the sunlight —that which physics had reduced to cold waves in a cold void. Thoreau's life had been dedicated to the unexplainable eye.

I had been trained since youth against the illusions, the deceptions of that eye, against sunset as reality, against my own features as anything but a momentary midge swarm of particles. Even this momentary phantasm I saw by the mind's eye alone. I no longer resisted as I walked. I went slowly, making sure that the eye momentarily residing in me saw and recorded what was intended when Thoreau spoke of the quality of the eye as belonging more to God than man.

But there was a message Thoreau intended to transmit. "I suspect," he had informed his readers, "that if you should go to the end of the world, you would find somebody there going further." It is plain that he wanted a message carried that distance, but what was it? We are never entirely certain. He delighted in gnomic utterances such as that in which he pursued the summer on snowshoes.

"I do not think much of the actual," he had added. Is this then the only message of the great Walden traveler in the winter days when the years "came fast as snowflakes"? Perhaps it is so intended, but the words remain cryptic. Thoreau, as is evidenced by his final journals, had labored to lay the foundations of a then unnamed science—ecology. In many ways he had outlived his century. He was always concerned with the actual, but it was the unrolling reality of the process philosopher, "the universe," as he says, "that will not wait to be explained."

For this reason he tended to see men at a distance. For the same reason he saw himself as a first settler in nature, his house the oldest in the settlement. Thoreau reflected in his mind the dreamers of the westward crossing; in this he is totally American.

Yet Thoreau preferred to the end his own white winter spaces. He lingers, curvetting gracefully, like the fox he saw on the river or the falcon in the morning air. He identifies, he enters them, he widens the circumference of life to its utmost bounds. "One world at a time," he jokes playfully on his deathbed, but it is not, in actuality, the world that any of us know or could reasonably endure. It is simply Thoreau's world, "a prairie for outlaws." Each one of us must seek his own way there. This is his final message, for each man is forever the eye and the eye is the Visitor. Whatever remedy exists for life is never to be found at Walden. It exists, if at all, where the real Walden exists, somewhere in the incredible dimensions of the universal Eye.

Walden: Thoreau's Unfinished Business

THE LIFE OF Henry David Thoreau has been thoroughly explored for almost a century by critics and biographers, yet the mystery of this untraveled man who read travel literature has nowhere been better expressed than by his own old walking companion Ellery Channing, who once wrote: "I have never been able to understand what he meant by his life. Why was he so disappointed with everybody else? Why was he so interested in the river and the woods . . . ? Something peculiar here I judge."

If Channing, his personal friend, was mystified, it is only to be expected that as Thoreau's literary stature has grown, the ever-present enigma of his life and thought has grown with it. Wright Morris, the distinguished novelist and critic, has asked, almost savagely, the same question in another form. Putting Channing's question in a less personal but more formidable and timeless literary context he ventures, quoting from Thoreau who spent two years upon the Walden experiment and then abandoned it, "If we are alive let us go about our business." "But," counters Morris brutally, "what business?" Thoreau fails to inform us. In the words of Morris, Walden was the opening chapter of a life, one that enthralls us, but with the remaining chapters missing.

For more than a decade after *Walden* was composed, Thoreau continued his intensive exploration of Concord, its inhabitants and its fields, but upon the "business" for which he left Walden he is oddly cryptic. Once, it is true, he muses in his journal that "the utmost possible novelty would be the difference between me and myself a year ago." He must then have been about some business, even though the perceptive critic Morris felt he had already performed it and was at loose ends and groping. The truth is that the critic, in a timeless sense, can be right and in another way wrong, for looking is in itself the business of art.

In a studied paragraph Carl Jung, with no reference to

Thoreau, perhaps pierced closest to Thoreau's purpose without ever revealing it. He says in his alchemical studies, "Medieval alchemy prepared the greatest attack on the divine order
of the universe which mankind has ever dared. Alchemy is the
dawn of the age of natural sciences which, through the *daemo-
nium* of the scientific spirit, drove nature and her forces into
the service of mankind to a hitherto unheard of degree. . . .
Technics and science have indeed conquered the world, but
whether the soul has gained thereby is another matter."

Thoreau was indeed a spiritual wanderer through the deserts
of the modern world. Almost by instinct he rejected that beginning wave of industrialism which was later to so entrance
his century. He also rejected the peace he had found on the
shores of Walden Pond, the alternate glazing and reflection of
that great natural eye which impartially received the seasons. It
was, in the end, too great for his endurance, too timeless. He
was a restless pacer of fields, a reader who, in spite of occasional
invective directed against those who presumed to neglect their
homes for far places, nevertheless was apt with allusions drawn
from travel literature, and quick to discern in man uncharted
spaces.

"Few adults," once remarked Emerson, Thoreau's one-time
mentor and friend, "can see nature." Thoreau was one of those
who could. Moreover he saw nature as another civilization, a
thing of vaster laws and vagaries than that encompassed by the
human mind. When he visited the Maine woods he felt its
wind upon him like the closing of a dank door from some
forgotten cellar of the past.

Was it some curious midnight impulse to investigate such
matters that led Thoreau to abandon the sunny hut at Walden
for "other business"? Even at Walden he had heard, at midnight, the insistent fox, the "rudimental man," barking beyond
his lighted window in the forest. The universe was in motion,
nothing was fixed. Nature was "a prairie for outlaws," violent,
unpredictable. Alone in the environs of Walden Thoreau wandered in the midst of that greater civilization he had discovered
as surely as some monstrous edifice come suddenly upon in the
Mayan jungles. He never exclaimed about the Indian trails
seen just at dusk in a winter snowfall—neither where they
went, nor upon what prairie they vanished or in what direc-

tion. He never ventured to tell us, but he was one of those great artist-scientists who could pursue the future through its past. This is why he lives today in the heart of young and old alike, "a man of surfaces," one critic has said, but such surfaces —the arrowhead, the acorn, the oak leaf, the indestructible thought-print headed toward eternity—plowed and replowed in the same field. Truly another civilization beyond man, nature herself, a vast lawless mindprint shattering traditional conceptions.

Thoreau, in his final journals, had said that the ancients with their gorgons and sphinxes could imagine more than existed. Modern men, by contrast, could not imagine so much as exists. For more than one hundred years that statement has stood to taunt us. Every succeeding year has proved Thoreau right. The one great hieroglyph, nature, is as unreadable as it ever was and so is her equally wild and unpredictable offspring, man. Like Thoreau, the examiner of lost and fragile surfaces of flint, we are only by indirection students of man. We are, in actuality, students of that greater order known as nature. It is into nature that man vanishes. "Wildness is a civilization other than our own," Thoreau had ventured. Out of it man's trail had wandered. He had come with the great ice, drifting before its violence, scavenging the flints it had dropped. Whatever he was now, the ice had made him, the breath from the dank door, great cold, and implacable winters.

Thoreau in the final pages of *Walden* creates a myth about a despised worm who surmounts death and bursts from his hidden chamber in a wooden table. Was the writer dreaming of man, man freed at last from the manacles of the ice? No word of his intention remains, save of his diligent experiments with frozen caterpillars in his study—a man preoccupied with the persistent flame of life trapped in the murderous cold. Is not the real business of the artist to seek for man's salvation, and by understanding his ingredients to make him less of an outlaw to himself, civilize him, in fact, back into that titanic otherness, that star's substance from which he had arisen? Perhaps encamped sufficiently in the great living web we might emerge again, not into the blind snow-covered eye of Walden's winter, but into the eternal spring man dreams of everywhere and nowhere finds.

Man, himself, is Walden's eye of ice and eye of summer. What now makes man an outlaw, with the fox urgent at his heels, is the fact that one of his eyes is gray and wintry and blind, while with the other is glimpsed another world just tantalizingly visible and dismissed as an illusion. What we know with certainty is that a creature with such disparate vision cannot long survive. It was that knowledge which led Thoreau to strain his eyesight till it ached and to record all he saw. A flower might open a man's mind, a box tortoise endow him with mercy, a mist enable him to see his own shifting and uncertain configuration. But the alchemist's touchstone in Thoreau was to give him sight, not power. Only man's own mind, the artist's mind, can change the winter in man.

II

There are persons who, because of youthful associations, prefer harsh-etched things before their eyes at morning. The foot of an iron bedstead perhaps, or a weathered beam on the ceiling, an abandoned mine tipple, or even a tombstone. On July 14 of the year 1973, I awoke at dawn and saw above my head the chisel marks on an eighteenth-century beam in the Concord Inn. As I strolled up the street toward the cemetery I saw a few drifters, black and white, stirring from their illegal night's sleep among the gravestones. Later I came to the Thoreau family plot and saw the little yellow stone marked "Henry" that no one is any longer sure indicates the precise place where he lies. Perhaps there is justice in this obscurity because the critics are also unsure of the contradictions and intentions of his journal, even of the classic *Walden*. A ghost then, of shifting features, peers out from between the gravestones, unreal, perhaps uninterpreted still.

I turned away from the early morning damp for a glimpse of the famous pond which in the country of my youth would have been called a lake. I walked its whole blue circumference with an erudite citizen of Concord. It was still an unearthly reflection of the sky, even if here and there beer bottles were bobbing in the shallows. I walked along the tracks of the old railroad where Thoreau used to listen to the telegraph wires. He had an eye for the sharp-edged artifact, I thought. The bobbing bottles, the keys to beer cans, he would have transmitted into cosmic symbols just as he had sensed all past time in the odors of a swamp. "All the ages are represented still," he had said, with nostrils flaring above the vegetation-choked water, "and you can smell them out."

It was the same, he found, with the ashes of Indian campfires, with old bricks and cellarholes. As for arrowheads, he says in a memorable passage, "I landed on two spots this afternoon and picked up a dozen. You would say it had rained arrowheads for they lie all over the surface of America. They lie in the meeting house cellar, and they lie in the distant cowpasture. They are sown like grain . . . over the earth. Each one," Thoreau writes, "yields me a thought. . . . It is humanity inscribed on the face of the earth. It is a footprint—rather a mindprint—left everywhere. . . . They are not fossil bones, but, as it were, fossil thoughts forever reminding me of the mind that shaped them. I am on the trail of mind."

Some time ago in a graduate seminar met in honor of the visit of an eminent prehistorian I watched the scholar and his listeners try to grapple with the significance of an anciently shaped stone. Not one of those present, involved as they were with semantic involutions, could render up so simple an expression as "mindprint." The lonely follower of the plow at Concord had provided both art and anthropology with an expression of horizon-reaching application which it has inexplicably chosen to ignore.

Mindprints are what the first men left, mindprints will be what the last man leaves, even if it is only a beer can dropped rolling from the last living hand, or a sagging picture in a ruined house. Cans, too, have their edges, a certain harshness; they too represent a structure of the mind, perhaps even an attitude. Thoreau might have seen that, too. Indeed he had written long ago: "If the outside of a man is so variegated and extensive, what must the inside be? You are high up the Platte river," he admonished, "traversing deserts, plains covered with soda, with no deeper hollow than a prairie dog hole tenanted by owls. . . ."

Perhaps in those lines he had seen the most of man's journey through the centuries. At all events he had coined two incomparable phrases, the "mindprint" which marked man's strange passage through the millennia and which differentiated him completely from the bones of all those creatures that lay strewn in the basement rocks of the planet, and that magnificent expression "another civilization," coined to apply to nature. That "civilization" contained for Thoreau the mysterious hieroglyphs left by a deer mouse, or the preternatural winter concealment of a moth's cocoon in which leaves were made to cooperate. He saw in the dancing of a fox on snow-whipped Walden ice "the fluctuations of some mind."

Thoreau had extended his thought-prints to something beyond what we of this age would call the natural. He would read them into nature itself, see, in other words, some kind of trail through that prairie for outlaws that had always intimidated him. On mountain tops, he had realized a star's substance, sensed a nature "not bound to be kind to man." Nevertheless he confided firmly to his diary, "the earth which I have *seen* cannot bury me." He searches desperately, all senses alert, for a way to read these greater hieroglyphs in which the tiny interpretable minds of our forerunners are embedded. We, with a sharper knowledge of human limitations and a devotion to the empirical fact, may deny to ourselves the reality of this other civilization within whose laws and probabilities we exist. Thoreau reposed faith in the consistency of nature's habits, but only up to a point, for he was a student of change.

As Alfred North Whitehead was to remark long afterward, "We are in the world and the world is in us"—a phrase that all artists should contemplate. Something, some law of a greater civilization, sustains nature from moment to moment within and above the void of non-being. "I hold," maintains the process philosopher, "that these unities of existence, these occasions of experience are the really real things which in their collective unity compose the evolving universe." In spite of today's emphasis upon the erratic nature of the submicroscopic particle there is, warns Whitehead the mathematician, "no valid inference from mere possibility to matter of fact; or, in other words, from mere mathematics to concrete nature. . . . Apart from metaphysical presupposition there can be no civilization." I doubt if Whitehead had ever perused Thoreau's journals, yet both return to the word "civilization," that strange on-going otherness of interlinked connections that makes up the nature that we know, just as human society and its artistic productions represent it in miniature, even to its eternal novelty.

Now Thoreau was a stay-at-home who traveled much in his mind, both in travel literature and beside Walden Pond. I, by circumstance, directly after delivering a lecture at Concord and gazing in my turn at Walden, was forced immediately to turn and fly west to the badlands and dinosaur-haunted gulches of Montana, some of its natives wild, half-civilized, still, in the way that Thoreau had viewed one of his Indian guides in Maine: "He shall spend a sunny day, and in this century be my contemporary. Why read history, then, if the ages and the generations are now? He lives three thousand years deep into time, an age not yet described by poets." As I followed our mixed-breed Cheyenne, as ambivalent toward us as the savage blood in his veins demanded, it came to me, as it must have come to many others, that seeing is not the same thing as understanding.

One man sees with indifference a leaf fall; another with the vision of Thoreau invokes the whole of that nostalgic world which we call autumn. One man sees a red fox running through a shaft of sunlight and lifts a rifle; another lays a re-straining hand upon his companion's arm and says, "Please. There goes the last wild gaiety in the world. Let it live, let it run." This is the role of the alchemist, the true, if sometimes inarticulate artist. He transmutes the cricket's song in an autumn night to an aching void in the heart; snowflakes become the flying years. And when, as archaeologist, he lifts from the encrusting earth those forgotten objects Thoreau called "fossil thoughts," he is giving depth and tragedy and catharsis to the one great drama that concerns us most, the supreme mystery, man. Only man is capable of comprehending all he was and all that he has failed to be.

On those sun-beaten uplands over which we wandered, every chip of quartzite, every patinized flint, gleamed in our eyes as large as the monuments of other lands. Our vision in that thin air was incredibly enhanced and prolonged. Thoreau had conceived of nature as a single reflecting eye, the Walden eye of which he strove to be a solitary part, to apprehend with all his being. It was chance that had brought me in the span of a day to the dinosaur beds of Montana. Thoreau would have liked that. He had always regarded such places as endowed with the vapors of Nox, places where rules were annulled. He had called arrowheads mindprints. What then would he have termed a tooth of *Tyrannosaurus rex* held in my palm? The sign of another civilization, another order of mind? Or that tiny Cretaceous mammal which was a step on the way to ourselves? Surely it represented mind in embryo, our mind, but not of our devising. What would he have called it—that miracle of a bygone moment, the annulment of what had been, to be replaced by an eye, the artist's eye, that nature had never heretofore produced among her creatures? Would these have answered for him on this giant upland, itself sleeping like some tired dinosaur with outspread claws? Would he have simply called it "nature," as we sometimes do, scarcely knowing how to interpret the looming inchoate power out of which we have been born? Or would he have labeled nature itself a mindprint beyond our power to read or to interpret?

A man might sketch Triceratops, but the alphabet from which it was assembled had long since disappeared. As for man, how had his own alphabet been constructed? The nature in which he momentarily resided was a journal in which the script was always changing, like the dancing footprints of the fox on icy Walden Pond. Here, exposed about me, was the great journal Thoreau had striven to read, the business, in the end, that had taken him beyond Walden. He would have been too wise, too close to earth, too intimidated, to have called such a journal human. It was palpably inscribed from a star's substance. Tiny and brief in that journal were the hieroglyphs of man. Like Thoreau, we had come to the world's end, but not to the end of nature, not to the end of time. All that could be read was that we had a past; that was something no other life on the planet had learned. There was, we had also ascertained, a future.

In the meantime, Thoreau would have protested, there is the eye, the sun and the eye. "Nothing must be postponed; find eternity in each moment." But how few of us are endowed to sustain Thoreau's almost diabolical vision. Here and here alone the true alchemist of Jung's thought must come to exist in each of us. It is ours to transmute, not iron, not copper, not gold, but our tracks through nature, see them finally attended by self-knowledge, by the vision of the universal eye, that faculty possessed by the alchemist at Walden Pond.

"Miasma and infection come from within," he once wrote. It was as if he sought the cleanliness of flint patinized by the sun of ages, the artifact, the mindprint from which the mind itself had departed. It is something that perhaps only a few artists like Piranesi have understood amidst cromlechs, shards, and broken cities. It is man's final act as an alchemist to find the philosopher's stone in a desert-varnished flint and to watch himself, his mind, his species, evaporate into the air and sun that once had nourished the dinosaurs. Man alone knows the way he came; man alone is the alchemical animal who can vaporize himself in an utter cleansing, either by the powers of art alone, or, more terribly, by that dread device which began its active life at Los Alamos more than thirty years ago.

On a great hill in Montana on the day I had flown from Walden, I picked up a quartz knife that had the look of ten thousand years about it. It was as clean as the sun and I knew suddenly what Thoreau had been thinking about his arrowheads, his mindprints. They were free at last. They had aged out of human history, out of corruption. They were joined to that other civilization, evidence of some power that ran all through nature. They were a sign now beyond man, like all those other traceries of the frost that Thoreau had studied so avidly for evidence of some greater intelligence.

For just a moment I was back at Walden with a mind beyond infection by man, the mind of an alchemist who knew instinctively how laws might be annulled and great civilizations rise evanescent as toadstools on an autumn night. I too had taken on a desert varnish. I might have been a man but, if so, a man from whom centuries had been flayed away. I was being transmuted, worn down. There was flint by my hand that had not moved for millennia. It had ceased to radiate a message and whatever message I, as man, had carried there from Walden was also forgotten.

I lay among logs of petrified wood and found myself already stiffening. Nature was bound somewhere; the great mind was readying some new experiment but not, perhaps, for man. I sighed a little with the cleanliness of that release. I slept deep under the great sky. I slept sound. For a moment as I drowsed I thought of the little stone marked "Henry" in the Concord cemetery. He would have known, I thought—the great alchemist had always known—and then I slept. It was Henry who had once written "the best philosophy is all untrue," untrue, that is, for man. Across an untamed prairie one's footprints must always be altering, that was the condition of the world, the only one that mattered, the only one for art.

III

But why, some midnight questioner persisted in my brain, why had he left that sunny doorway of his hut in Walden for unknown mysterious business? Had he not written as though he had settled down forever? Why had Channing chronicled Thoreau's grievous disappointments? What had he been seeking and how had it affected him? If Walden was the opening chapter of a life, might not there still be a lurking message, a termination, a final chapter beyond his recorded death?

It was evident that he had seen the whole of American culture as copper-tinted by its antecedents, its people shadowy and gigantic as figures looming indistinctly in some Indian-summer haze. He had written of an old tree near Concord penetrated by a flying arrow with the shaft still attached. Some of the driving force of that flint projectile still persisted in his mind. Perhaps indeed those points that had once sung their message through every glade of the eastern woodland had spoken louder than the telegraph harp to which his ears had been attuned at Walden. Protest as he would, cultivate sauntering as he would, abhor as he would the rootless travelers whose works he read by lamplight, he was himself the eternal traveler. On the mountains of New Hampshire he had found "small and almost uninhabited ponds, apparently without fish, sources of rivers, still and cold, strange as condensed clouds." He had wandered without realizing it back into the time of the first continental ice recession.

"It is not worth the while to go around the world to count the cats in Zanzibar," he once castigated some luckless explorer, but why then this peering into lifeless tarns or engrossing himself with the meteoritic detritus of the Appalachians?

Did he secretly wish to come to a place of no more life, where a man might stiffen into immobility as I had found myself freezing into the agate limbs of petrified trees in Montana? A divided man, one might say with surety. The bold man of abrasive village argument, the defender of John Brown, the advocate of civil disobedience, the spokesman who supplied many of the phrases which youthful revolutionaries hurled at their elders in the sixties of this century, felt the world too large for him.

As a college graduate he wept at the thought of leaving Concord. Emerson's well-meant efforts to launch him into the intellectual life of New York had failed completely. He admitted that he would gladly fall "into some crevice along with leaves and acorns." To Emerson's dismay he captained huckleberry parties among children and was content to be a rural surveyor, wandering over the farms and woodlots he could not own, save for his all-embracing eye. "There is no more fatal blunderer," he protested, "than he who consumes the greater part of life getting a living." He had emphasized that contemplative view at Walden, lived it, in fact, to the point where the world came finally to accept him as a kind of rural Robinson Crusoe who, as the cities grew, it might prove wise to emulate.

"I sat in my sunny doorway," he ruminated, "from sunrise till noon, rapt in a revery, amidst the pines and hickories and sumachs, in undisturbed solitude and stillness, while the birds sang around or flitted noiseless through the house, until by the sun falling in at my west window, or the noise of some traveller's wagon on the distant highway, I was reminded of the lapse of time. I grew in those seasons like corn in the night."

This passage would seem to stand for the serene and timeless life of an Oriental sage, a well-adjusted man, as the psychiatrists of our day would have it. Nevertheless this benign façade is deceptive. There is no doubt that Thoreau honestly meant what he said at the time he said it, but the man was storm-driven. He would not be content with the first chapter of his life; he would, like a true artist, dredge up dreams even from the bottom of a pond.

In the year 1837 Thoreau confided abruptly to his journal: "Truth strikes us from behind, and in the dark." Thoreau's life was to be comparatively short and ill-starred. Our final question must, therefore, revolve, not about wanderings in autumn fields, not the drowsing in pleasant doorways where time stood still forever, but rather upon the leap of that lost arrow left quivering in an ancient oak. It was, in symbol, the hurtling purposeful arrow of a seemingly aimless life. It has been overlooked by Thoreau's biographers, largely because they have been men of the study or men of the forest. They have not been men of the seashore, or men gifted with the artist's eye. They have not trudged the naturalist's long miles through sea sand, where the war between two elements leaves even the smallest object magnified, as the bleached bone or broken utensil can be similarly magnified only on the dead lake beaches of the west.

Thoreau had been drawn to Cape Cod in 1849, a visit he had twice repeated. It was not the tourist resort it is today. It was still the country of men on impoverished farms, who went to sea or combed the beaches like wreckers seeking cargo. On those beaches, commented Thoreau, in a posthumous work which he was never destined to see in print, "a house was rarely visible . . . and the solitude was that of the ocean and the desert combined." The ceaseless roar of the surf, the strands of devil's-apron, the sun jellies, the stories of the drowned cast on the winter coast awoke in Thoreau what must have been memories of Emerson's shipwrecked friend Margaret Fuller. Here, recorded the chronicler, was a wilder, less human nature. Objects on the beach, he noted, were always more grotesque and dilated than upon approach they proved to be. A cast-up pair of gloves suggested the reality of hands.

Thoreau's account in *Cape Cod* of the Charity House to which his wanderings led him takes on a special meaning. I think it embodies something of a final answer to Channing's question about Thoreau's disappointment in his fellow men. Published two years after his death, it contains his formulation of the end of his business, or perhaps I should say of his quest. Hidden in what has been dismissed as a mere book of travel is an episode as potentially fabulous as Melville's great white whale.

First, however, I must tell the story of another coast because it will serve to illuminate Thoreau's final perception. A man, a shore dweller on Long Island Sound, told me of his discovery in a winter dawn. All night there had been a heavy surf and freezing wind. When he came to stroll along his beach at morning he had immediately seen a lifeboat cast upon the shingle and a still, black figure with the eastern sun behind it on the horizon. Gripped by a premonition he ran forward. The seaman in oilskins was alone and stiffly upright. A compass was clutched in his numb fingers. The man was sheeted in ice. Ice over his beard, his clothing, his hands, ice over his fixed, open eyes. Had he made the shore alive but too frozen to move? No one would ever know, just as no one would ever know his name or the sinking vessel from which he came. With desperate courage he had steered a true course through a wild night of breakers only to freeze within sight of help.

In those fishing days on Cape Cod, Thoreau came to know many such stories—vessels without weather warnings smashed in the winter seas, while a pittance of soaked men, perhaps, gained the shore. The sea, the intolerable sea, tumbled with total indifference the bodies of the dead or the living who were tossed up through the grinding surf of winter. These were common events in the days of sail.

The people who gained a scant living along that coast entertained, early in the nineteenth century, the thought that a few well-stocked sheds, or "Charity Houses," might enable lost seamen who made the shore to warm and feed themselves among the dunes till rescued. The idea was to provide straw and matches and provender, supervised and checked at intervals by some responsible person. Impressed at first by this signal beneficence of landsmen, Thoreau noted the instructions set down for the benefit of mariners. Finally, he approached one such Charity House. It appeared, he commented, "but a stage to the grave." The chimney had fallen. As he and his companion wished to gain an idea of a "humane house," they put their eyes, by turns, to a knothole in the door. "We had," Thoreau comments ironically, "some practice at looking inward —the pupil becomes enlarged. Nature is never so dark that a patient eye may not prevail over it."

So there, at last, he saw the end of his journey, of the business begun at Walden. He was peering into the Charity House of man, upon a Cape Cod beach. For frozen, shipwrecked mariners he saw a fireplace with no matches, no provisions, no straw upon the floor. "We looked," he said, "into the bowels of mercy, and for bread we found a stone." Shivering like castaways, "we looked through the knothole into that night without a star, until we concluded it was not a *humane* house at all." The arrow Thoreau had followed away from Walden had pierced as deep as Captain Ahab's lance. No wonder the demoniacal foxes leaping at Thoreau's window had urged him to begone. He had always looked for a crevice into the future. He had peered inward instead. It was ourselves who were rudimental men.

Recently I had a letter from one of my students who is working in the Arctic and who has a cat acquired somewhere in his travels. The cat, he explained to me, hunts in the barrens behind the Eskimo village. Occasionally it proudly brings in a lemming or a bird to his hut. The Eskimo were curious about the unfamiliar creature.

"Why does he do that?" my friend was asked.

"Because he is a good cat," my student explained. "He shares his game."

"So, so." The old men nodded wisely. "It is true for the man and for the beast—the good man and the good beast. They share, yes indeed. They share the game."

I think my young quick-witted friend had momentarily opened the eye of winter. Before laying aside his letter, I thought of the eye of Walden as I had seen it under the summer sun. It was the sharing that had impressed the people of the ice and it was a great sharing of things seen that Thoreau had attempted at his pond. A hundred years after his death people were still trying to understand what he was about. They were still trying to get both eyes open. They were still trying to understand that the town surveyor had brought something to share with his fellows, something that, if they partook of it, might transpose them to another world.

I had thought, staring across an angular gravestone at Concord and again as I held my wind-varnished flint in Montana, that "sharing" could be the word. It was appropriate, even though Thoreau in a final bitterness had felt sharing to be as impoverished as the Charity House for sailors—a knothole glimpse into the human condition. How then should the artist see? By an eye applied to a knothole? By a magnification of sand-filled gloves washed up on a beach? Could this be the solitary business that led Thoreau on his deathbed to mutter, whether in irony or confusion, "one world at a time"?

This is the terror of our age. How should we see? In what world are we? For we have fallen out of nature and see sometimes more and sometimes less. We see the past, the looming future, and then, so fearfully is the eye confused, that it stares inverted into a Charity House that appears to reflect a less than human heart. Is this Thoreau's final surrealist vision, his glance through the knothole into the "humane house"? It would appear at least to be a glimpse from one of those two great alternating eyes at Walden Pond from which in the end he had fled—the blind eye of winter and that innocent blue pupil beside which he had once drowsed when time seemed endless. Both are equally real, as the great poets and prophets have always known, but it was Thoreau's tragic destiny to see with eyes strained beyond endurance man subsiding into two wrinkled gloves grasping at the edge of infinity. It is his final contribution to literature, the final hidden conclusion of an unwritten life whose first chapter Morris had rightly diagnosed as Walden.

There is an old Biblical saying that our days are prolonged and every vision fails us. This I would dispute. The vision of the great artist does not fail. It sharpens and refines with age until everything extraneous is pared away. "Simplify," Thoreau had advocated. Two gloves, devoid of flesh, clutching the stones of the ebbing tide become, transmuted, the most dreadful object in the world.

"There has been nothing but the sun and eye from the beginning," Thoreau had written when his only business was looking and he grew, as he expressed it, "like corn in the night." The sun and the eye are the two aspects of nature which are irremediably linked. But the eye of man constitutes an awesome crystal whose diffractions are far greater than those of any Newtonian prism.

We see, as artists, as scientists, each in his own way, through the inexorable lens we cannot alter. In a nature which Thoreau recognized as unfixed and lawless anything might happen. The artist's endeavor is to make it happen—the unlawful, the oncoming world, whether endurable or mad, but shaped, shaped always by the harsh angles of truth, the truth as glimpsed through the terrible crystal of genius. This is the one sure rule of that other civilization which we have come to know is greater than our own. Thoreau called it, from the first, "unfinished business," when he turned and walked away from his hut at Walden Pond.

The Lethal Factor

T HE GREAT OLDUVAI GORGE in East Africa has been appropriately called the Grand Canyon of human evolution. Here a million, perhaps two million, years of human history are recorded in the shape of successive skulls and deposits of stone tools. The elusive story of the long road man has traveled is glimpsed momentarily in eroded strata and faded bone. Olduvai is now famous all over the world. Only to those who have the habit of searching beyond the obvious, however, may it have occurred that this precipitous rift through time parallels and emphasizes a similar rift in ourselves—a rift that lies like a defacing crack across our minds and consequently many of our institutions. From its depths we can hear the rumble of the torrent from which we have ascended and sense the disastrous ease with which both individual men and civilizations can topple backward and be lost.

Brooding upon the mysteries of time and change, a great and thoughtful scholar, Alfred North Whitehead, many years ago recorded his thoughts in a cryptic yet profound observation. He said, in brief, "We are . . . of infinite importance, because as we perish we are immortal." Whitehead was not speaking in ordinary theological terms. He was not concerned in this passage with the survival of the human personality after death—at least as a religious conception. He was, instead, struggling with that difficult idea which he describes as the "prehension of the past," the fact that the world we know, even as it perishes, remains an elusive, unfixed element in the oncoming future.

The organic world, as well as that superorganic state which exists in the realm of thought, is, in truth, prehensile in a way that the inorganic world is not. The individual animal or plant in the course of its development moves always in relation to an unseen future toward which its forces are directed: the egg is broken and a snake writhes away into the grass; the acorn seedling, through many seasons, contorts itself slowly into a gnarled, gigantic oak. Similarly, life moves against the future in another, an evolutionary, sense. The creature existing now—this serpent, this bird, this man—has only to leave progeny in order to stretch out a gray, invisible hand into the evolutionary future, into the nonexistent.

With time, the bony fin is transformed into a paw, a round, insectivore eye into the near-sighted gaze of a scholar. Moreover, all along this curious animal extension into time, parts of ourselves are flaking off, breaking away into unexpected and unforeseen adventures. One insectivore fragment has taken to the air and become a vampire bat, while another fragment draws pictures in a cave and creates a new prehensile realm where the shadowy fingers of lost ideas reach forward into time to affect our world view and, with it, our future destinies and happiness.

Thus, since the dawn of life on the planet, the past has been figuratively fingering the present. There is in reality no clearly separable past and future either in the case of nerve and bone or within the less tangible but equally real world of history. Even the extinct dead have plucked the great web of life in such a manner that the future still vibrates to their presence. The mammalian world was for a long time constricted and impoverished by the dominance of the now vanished reptiles. Similarly, who knows today what beautiful creature remains potential only because of man's continued existence, or what renewed manifestations of creative energy his presence inhibits or has indeed destroyed forever.

As the history of the past unrolls itself before the eyes of both paleontologists and archaeologists, however, it becomes evident, so far as the biological realm is concerned, that by far the greater proportion of once-living branches on the tree of life are dead, and to this the archaeologist and historian must add dead stone, dead letters, dead ideas, and dead civilizations. As one gropes amid all this attic dust it becomes ever more apparent that some lethal factor, some arsenical poison, seems to lurk behind the pleasant show of the natural order or even the most enticing cultural edifices that man has been able to erect.

In the organic world of evolution three facts, so far as we can perceive, today seem to determine the death of species: the irreversibility of the organic process in time; high specialization which, in the end, limits new adaptive possibilities; the sudden emergence of spectacular enemies or other environmental circumstances which overwhelm or ambush a living form so suddenly that the slow adjustive process of natural selection cannot be made to function. This third principle, one could say, is the factor which, given the other two limitations upon all forms of life, will result in extinction. As a drastic example one could point to the destruction of many of the larger creatures as man has abruptly extended his sway over both hemispheres and into many different environmental zones.

The past century has seen such great accessions of knowledge in relation to these natural events, as well as a growing consciousness of man's exposure to similar dangers, that there is an increasing tendency to speculate upon our own possibilities for survival. The great life web which man has increasingly plucked with an abruptness unusual in nature shows signs of "violence in the return," to use a phrase of Francis Bacon's. The juvenile optimism about progress which characterized our first scientific years was beginning early in this century to be replaced by doubts which the widely circulated *Decline of the West* by Oswald Spengler documents only too well. As the poet J. C. Squire says, we can turn

> the great wheel backward until Troy unburn,
> . . . and seven Troys below
> Rise out of death and dwindle . . .

We can go down through the layers of dead cities until the gold becomes stone, until the jewels become shells, until the palace is a hovel, until the hovel becomes a heap of gnawed bones.

Are the comparisons valid? The historians differ. Is there hope? A babble of conflicting voices confuses us. Are we safe? On this point I am sure that every person of cultivation and intelligence would answer with a resounding "No!" Spengler—not the optimist—was right in prophesying that this century would be one marked by the rise of dictators, great wars, and augmented racial troubles. Whether he was also right in foreseeing this century as the onsetting winter of Western civilization is a more difficult problem.

Faustian, space-loving man still hurls his missiles skyward. His tentacular space probes seem destined to palpate the farthest rim of the solar system. Yet honesty forces us to confess that this effort is primarily the product of conflict, that millions are now employed in the institutions erected to serve that conflict, that government and taxes are increasingly geared to it, that in another generation, if not now, it will have become traditional. Men who have spent their lives in the service of these institutions will be reluctant to dissolve them. A vested interest will exist on both sides of the iron curtain. The growing involution of this aspect of Western culture may well come to resemble the ingrowth and fantasies of that ritualized belief in *mana* which characterized late Polynesian society.

It is upon this anthropological note that I should like to examine the nature of the human species—the creature who at first glance appears to have escaped from the specialized cul-de-sac which has left his late-existing primate relative, the gorilla, peering sullenly from the little patch of sheltered bush that yet remains to him. I have said that some lethal factor seems to linger in man's endeavors. It is for this reason that I venture these words from a discipline which has long concerned itself with the origins, the illusions, the symbols, and the folly as well as the grandeur of civilizations whose records are lost and whose temples are fallen. Yet the way is not easy. As Herman Melville has written in one great perceptive passage:

> By vast pains we mine into the pyramid; by horrible gropings we come to the central room; with joy we espy the sarcophagus; but we lift the lid—and no body is there!—appallingly vacant, as vast, is the soul of man!

I have spent a sizable number of my adult years among the crude stones of man's Ice Age adventurings. The hard, clean flint in the mountain spring defines and immortalizes the race that preceded us better than our own erratic fabrications distinguish our time. There is as yet no sharp edge to our image. Will it be, in the end, the twisted gantrys on the rocket bases, or telephone wires winding voiceless through the high Sierras, or the glass from space-searching observatories pounded into moonstones in the surf on a sinking coast?

What makes the symbol, finally, for another age as the pyramids for theirs—writing for five thousand years man's hope against the sky? Before we pass, it is well to think of what our final image as a race may be—the image that will give us a kind of earthly immortality or represent, perhaps, our final collective visage in eternity. But it is to the seeds of death within us that we must address ourselves before we dare ask this other final question of what may stand for us when all else is fallen and gone down. We shall not begin with Western society, we shall begin with man. We shall open that symbolic sepulcher of which Melville speaks. We shall grope in the roiling, tumultuous darkness for that unplumbed vacancy which Melville termed so ironically the soul of man.

Since the days of Lyell and Hutton, who perceived, beneath the romantic geological catastrophism of their age, that the prosaic and unnoticed works of wind and sun and water were the real shapers of the planet, science has been averse to the recognition of discontinuity in natural events. (It should be said in justice to Sir Charles Lyell, however, that in combating the paroxysmal theories which preoccupied his contemporaries he maintained, nevertheless, that "minor convulsions and changes are . . . a *vera causa*, a force and mode of operation which we know to be true.") Nevertheless the rise of modern physics, with its emphasis on quantum theory in the realm of particles, and even certain aspects of Mendelian genetics serve to remind us that there are still abroad in nature hidden powers which, on occasion, manifest themselves in an unpredictable fashion. On a more dramatic scale no one to date has been quite able to account satisfactorily for that series of rhythmic and overwhelming catastrophes which we call the Ice Age. It is true that we no longer cloak such mysteries in an aura of supernaturalism, but they continue to remind us, nevertheless, of the latent forces still lurking within nature.

Another of these episodes is reflected in the origins of the human mind. It represents, in a sense, a quantum step: the emergence of genuine novelty. It does so because the brain brought into being what would have been, up until the time of its appearance, an inconceivable event—the world of culture. The *mundus alter*—this other intangible, faery world of dreams, fantasies, invention—has been flowing through the heads of men since the first ape-man succeeded in cutting out a portion of his environment and delineating it in a transmissible word.

With that word a world arose which will die only when the last man utters the last meaningful sound.

It is a world that lurks, real enough, behind the foreheads of men; it has transformed their natural environment. It has produced history, the unique act out of the natural world about us. "The foxes have their holes," the words are recorded of the apostle Matthew, "the wild birds have their nests, but the Son of Man has no where to lay his head." Two thousand years ago in the Judean desert men recognized that the instinctive world of the animals had been lost to man. Henceforward he would pass across the landscape as a wanderer who, in a sense, was outside of nature. His shadow would grow large in the night beside the glare of his red furnaces. Fickle, erratic, dangerous, he would wrest from stone and deep-veined metal powers hitherto denied to living things. His restless mind would try all paths, all horrors, all betrayals. In the strange individual talents nourished in his metropolises, great music would lift him momentarily into some pure domain of peace. Art would ennoble him; temptation and terror pluck his sleeve. He would believe all things and believe nothing. He would kill for shadowy ideas more ferociously than other creatures kill for food, then, in a generation or less, forget what bloody dream had so oppressed him.

Man stands, in other words, between the two most disparate kingdoms upon earth: the flesh and the spirit. He is lost between an instinctive mental domain he has largely abandoned and a realm of thought through which still drift ghostly shadows of his primordial past. Like all else that lingers along the borders of one world while gazing into another, he is imperfectly adapted. It is not only the sea lion from the deep waters that inches itself painfully up the shore into the unfamiliar sunlight. So does man in the deep interior of his mind occasionally clamber far up into sunlit meadows where his world is changed and where, in the case of some few, for such is the way of evolution, there is no return to lower earth. It has taken us far longer to discover the scars of evolution in our brains than to interpret the vestigial organs tucked into odd crannies of our bodies or the wounds and aches that reveal to us that we have not always walked upon two feet.

Alfred Russel Wallace perceived in 1864 that in man the rise of the most remarkable specialization in the organic world— the human brain—had to a considerable degree outmoded the evolution of specialized organs. The creature who could clothe himself in fur or take it off at will, who could, by extension of himself into machines, fly, swim, or roll at incredible speeds, had simultaneously mastered all of earth's environments with the same physical body. Paradoxically, this profound biological specialization appeared to have produced an organ devoted to the sole purpose of escaping specialization. No longer could man be trapped in a single skin, a single climate, a single continent, or even a single culture. He had become ubiquitous. The wind wafted his little craft to the ends of the earth, seeds changed their substance under his hands, the plague hesitated and drew back before his cities. Even his body appeared destined to remain relatively stable since he had become the supremely generalized animal whose only mutability lay in his intelligence as expressed upon his instruments and weapons.

A creature who sets out upon a new road in the wilderness, however, is apt to encounter unexpected dangers, particularly if he ventures into that invisible and mysterious environmental zone, that "other world" which has been conjured up by the sheer power of thought. Man, when he moved from the animal threshold into dawning intellectual consciousness, no longer could depend upon the instinctive promptings that carry a bird upon the wing.

Although it is difficult to penetrate into that half-world of the past, it is evident that order, simple order, must have rapidly become a necessity for survival in human groups across whose members the inchoate thoughts and impulses of the freed mind must have run as alarming vagaries. Man must, in fact, have walked the knife edge of extinction for untold years. As he defined his world he also fell victim to the shadows that lay behind it. He did not accept it, like the animal, as a thing given. He bowed to stone and heard sprites in running water. The entire universe was talking about him and his destiny. He knew the powers and heard the voices. He formulated their wishes for himself. He shaped out of his own drives and timidities the rules and regulations which reintroduced into his world a kind of facsimile of his lost instinctive animal order and simplicity.

By means of custom, strengthened by supernatural enforcement, the violent and impulsive were forced into conformity. There was a way—the way of the tribe. The individual conformed or perished. There was only one way, and one people, the tribe. That there were many tribes and many ways into the future no one knew, and at first in the wide emptiness of the world it scarcely mattered. It was enough that there was some kind of way or path. Even the Neanderthals had known this and had provided meat and tools that the dead might need upon their journey.

If we pause and contemplate this dim and unhistorical age for a moment it is, as I have intimated, with the thought that it reveals a rift or schism in man's endeavors that runs through his life in many aspects and throughout his history. The great historians have spoken of universal history as, in the words of Lord Acton, "an illumination of the soul." They have ventured to foresee the eventual reunion of man and the meeting of many little histories, ultimately, in the great history of man's final unification. It may be so. Yet whether we peer backward into the cloudy mirror of the past or look around us at the moment, it appears that behind every unifying effort in the life of man there is an opposite tendency to disruption, as if the force symbolized in the story of the Tower of Babel had been felt by man since the beginning. Eternally he builds, and across the smooth façade of his institutional structures there runs this ancient crack, this primordial flaw out of old time. We of this age have not escaped it.

Scarcely had man begun dimly and uncertainly to shape his newfound world of culture than it split into many facets. Its "universal" laws were unfortunately less than those of the nature out of which he had emerged. The customs were, in reality, confined to this island, that hill fort, or the little tribe by the river bank. The people's conception of themselves was similarly circumscribed. *They* were the people. Those who made strange sounds in a different tongue or believed in other gods were queer and, at best, tolerated for trading purposes if they did not encroach upon tribal territory.

In some parts of the world this remote life, untouched by

self-questioning of any sort, this ancient way of small magical dealings with animals and wood spirits and man, has persisted into modern times. It is a comparatively harmless but ensorcelled world in which man's innate capacities are shut up in a very tiny ring and held latent through the long passage of millennia. There are times when one wonders whether it is only a very rare accident that releases man from the ancient, hypnotic sleep into which he so promptly settled after triumphing in his first human endeavor—that of organizing a way of life, controlling the seasons, and, in general, setting up his own microscopic order in the vast shadow of the natural world. It is worth a passing thought upon primitive capacities that perhaps no existing society has built so much upon so little.

Nevertheless the rift persisted and ran on. The great neolithic empires arose and extended across the old tribal boundaries. The little peoples were becoming the great people. The individual inventor and artist, released from the restrictions of a low-energy society, enriched the whole culture. The arts of government increased. As wealth arose, however, so war, in a modern sense, also arose. Conquest empires, neolithic and classic, largely erased the old tribalism, but a long train of miseries followed in their wake. Slavery, merciless exploitation such as our paleolithic ancestors never imagined in their wildest dreams, disrupted the society and in the end destroyed both the individual and the state.

There began to show across the face of these new empires not alone the symptoms of a bottomless greed but what, in the light of times to come, was more alarming: the very evident fact that, as human rule passed from the village to the empire, the number of men who could successfully wield power for long-term social purposes grew less. Moreover, the long chain of bureaucracy from the ruler to the ruled made for greater inefficiency and graft. Man was beginning to be afflicted with bigness in his affairs, and with bigness there often emerges a dogmatic rigidity. The system, if bad, may defy individual strength to change it and simply run its inefficient way until it collapses. It is here that what we may call involution in the human drama becomes most apparent.

There tend to arise in human civilization institutions which

424 THE STAR THROWER

monopolize, in one direction or another, the wealth and attention of the society, frequently to its eventual detriment and increasing rigidity. These are, in a sense, cultural overgrowths, excessively ornate societal excrescences as exaggerated as some of the armor plate that adorned the gigantic bodies of the last dinosaurs. Such complications may be as relatively harmless as a hyperdeveloped caste system in which no social fluidity exists, or as dangerous as military institutions that employ increasingly a disproportionate amount of the capital and attention of the state and its citizens.

In one of those profound morality plays which C. S. Lewis is capable of tossing off lightly in the guise of science fiction, one of his characters remarks that in the modern era the good appears to be getting better and the evil more terrifying. It is as though two antipathetic elements in the universe were slowly widening the gap between them. Man, in some manner, stands at the heart of this growing rift. Perhaps he contains it within himself. Perhaps he feels the crack slowly widening in his mind and his institutions. He sees the finest intellects, which in the previous century concerned themselves with electric light and telephonic communication, devote themselves as wholeheartedly to missiles and supersonic bombers. He finds that the civilization which once assumed that only barbarians would think of attacking helpless civilian populations from the air has, by degrees, come to accept the inevitability of such barbarism.

Hope, if it is expressed by the potential candidates for mass extermination in this age of advanced destruction, is expressed, not in terms of living, but in those of survival, such hope being largely premised on confidence in one's own specialists to provide a nuclear blanket capable of exceeding that of the enemy. All else gives way before the technician and the computer specialist running his estimates as to how many million deaths it takes, and in how many minutes, before the surviving fragment of a nation—if any—sues for peace. Nor, in the scores of books analyzing these facts, is it easy to find a word spared to indicate concern for the falling sparrow, the ruined forest, the contaminated spring—all, in short, that still spells to man a life in nature.

One of these technicians wrote in another connection

involving the mere use of insecticides, which I here shorten and paraphrase: "Balance of nature? An outmoded biological concept. There is no room for sentiment in modern science. We shall learn to get along without birds if necessary. After all, the dinosaurs disappeared. Man merely makes the process go faster. Everything changes with time." And so it does. But let us be as realistic as the gentleman would wish. It may be we who go. I am just primitive enough to hope that somehow, somewhere, a cardinal may still be whistling on a green bush when the last man goes blind before his man-made sun. If it should turn out that we have mishandled our own lives as several civilizations before us have done, it seems a pity that we should involve the violet and the tree frog in our departure.

To perpetrate this final act of malice seems somehow disproportionate, beyond endurance. It is like tampering with the secret purposes of the universe itself and involving not only man but life in the final holocaust—an act of petulant, deliberate blasphemy.

It is for this reason that Lewis's remark about the widening gap between good and evil takes on such horrifying significance in our time. The evil man may do has this added significance about it: it is not merely the evil of one tribe seeking to exterminate another. It is, instead, the thought-out willingness to make the air unbreathable to neighboring innocent nations and to poison, in one's death throes, the very springs of life itself. No greater hypertrophy of the institution of war has ever been observed in the West. To make the situation more ironic, the sole desire of every fifth-rate nascent nationalism is to emulate Russia and America, to rattle rockets, and, if these are too expensive, then at least to possess planes and a parade of tanks. For the first time in history a divisive nationalism, spread like a contagion from the West, has increased in virulence and blown around the world.

A multitude of states are now swept along in a passionate hunger for arms as the only important symbol of prestige. Yearly their number increases. For the first time in human history the involutional disease of a single civilization, that of the West, shows signs of becoming the disease of all contemporary societies. Such, it would appear, is one of the less beneficial aspects of the communications network that we have flung around the world. The universal understanding which has been the ultimate goal sought by the communications people, that shining Telstar through which we were to promote the transmission of wisdom, bids fair, instead, to promote unsatisfied hunger and the enthusiastic reception of irrationalities that embed themselves all too readily in the minds of the illiterate.

Man may have ceased to teeter uncertainly upon his hind legs; his strange physical history may be almost over. But within his mind he is still hedged about by the shadows of his own fear and uncertainty; he still lingers at the borders of his dark and tree-filled world. He fears the sunlight, he fears truth, he fears himself. In the words of Thomas Beddoes, who looked long into that world of shadows:

> Nature's polluted.
> There's man in every secret corner of her
> Doing damned wicked deeds. Thou art, old world,
> A hoary, atheistic, murdering star.

This is the dark murmur that rises from the abyss beneath us and that draws us with uncanny fascination.

If I were to attempt to spell out in a sentence the single lethal factor at the root of declining or lost civilizations up to the present, I would be forced to say adaptability. I would have to remark, paradoxically, that the magnificent specialization of gray matter which has opened to us all the climates of the earth, which has given us music, surrounded us with luxury, entranced us with great poetry, has this one flaw: it is too adaptable. In breaking free from instinct and venturing naked into a universe which demands constant trial and experiment, a world whose possibilities were unexplored and are unlimited, man's hunger for experience became unlimited also. He has the capacity to veer with every wind or, stubbornly, to insert himself into some fantastically elaborated and irrational social institution only to perish with it.

It may well be that some will not call this last piece of behavior adaptation. Yet it is to be noted that only extreme, if unwise, adaptability would have allowed man to contrive and inhabit such strange structures. When men in the mass have once attached themselves to a cultural excrescence that grows until it threatens the life of the society, it is almost impossible to modify their behavior without violence. Yet along with this, as I have remarked, fervid waves of religious or military enthusiasm may sweep through a society and then vanish with scarcely a trace.

It would take volumes to chronicle the many facets of this problem. It is almost as though man had at heart no image, but only images, that his soul was truly as vacant and vast as Melville intimated in the passage I quoted earlier. Man is mercurial and shifting. He can look down briefly into the abyss and say, smiling, "We are beasts from the dark wood. We will never be anything else. We are not to be trusted. Never on this earth. We have come from down there." This view is popular in our time. We speak of the fossil ape encrusted in our hearts. This is one image of many that man entertains of himself. Another was left by a man who died 2000 years ago.

I have mentioned before the collective symbol that a civilization sometimes leaves to posterity and the difficulty one has with our own because of the rapidity with which our technology has altered and the restless flickering of our movement from one domain of life to another. A few months ago I read casually in my evening newspaper that our galaxy is dying. That great wheel of fire of which our planetary system is an infinitesimal part was, so the report ran, proceeding to its end. The detailed evidence was impressive. Probably, though I have not attempted to verify the figures, the spiral arm on which we drift is so vast that it has not made one full circle of the wheel since the first man-ape picked up and used a stone.

I saw no use in whispering behind my hand at the club, next morning, "They say the galaxy is dying." I knew well enough that man, being more perishable than stars, would be gone billions of years before the edge of the Milky Way grew dark. It was not that aspect of the human episode which moved me. Instead it was the sudden realization of what man could do on so gigantic a scale even if, as yet, his personal fate eluded him. Out there millions of light-years away from earth, man's hands were already fumbling in the coal-scuttle darkness of a future universe. The astronomer was foreshortening time—just as on a shorter scale eclipses can be foretold, or an apparently empty point in space can be shown as destined to receive an invisibly moving body. So man, the short-lived midge, is reaching into and observing events he will never witness in the flesh. In a psychological second, on this elusive point we call the present, we can watch the galaxy drift into darkness.

The materiality of the universe, Alfred North Whitehead somewhere remarks, is measured "in proportion to the restriction of memory and anticipation." With consciousness, memory, extended through the written word and the contributions of science, penetrates further and further into both aspects of time's unknown domain—that is, the past and the future. Although individual men do not live longer, we might say that the reach of mind in the universe and its potential control of the natural order is enormously magnified.

Material substance no longer dominates the spiritual life. There is not space here to explore all aspects of this fascinating subject, or the paradoxes with which our burgeoning technology have presented us. This strange capacity of the mind upon which we exercise so little thought, however, means that man both remains within the historical order, and, at the same time, passes beyond it.

We are present in history, we may see history as meaningless or purposeful, but as the heightened consciousness of time invades our thinking, our ability to free our intellects from a narrow and self-centered immediacy should be intensified. It is this toward which Whitehead was directing his thought: that all responsible decisions are acts of compassion and disinterest; they exist within time and history but they are also outside of it, unique and individual and, because individual, spiritually free. In the words of Erich Frank, "History and the world do not change, but man's attitude to the world changes."

I wonder if we understand this point, for it is the crux of this essay, and though I have mentioned modern thinkers, it leads straight back to the New Testament. A number of years ago in a troubled period of my life I chanced to take a cab from an airport outside a large eastern city. My route passed through the back streets of a run-down area of dilapidated buildings. I remember passing a pathetic little cemetery whose smudged crosses, dating from another era, were now being encroached upon and overshadowed by the huge gray tanks of an oil refinery. The shadow of giant machines now fell daily across the hill of the dead. It was almost a visible struggle of the symbols to which I have earlier referred—the cross that marks two thousand years of Western culture, shrinking, yet still holding its little acre in the midst of hulking beams and shadows where now no sunlight ever fell.

I felt an unreasoned distaste as we jounced deeper into these narrow alleyways, or roared beneath giant bridges toward a distant throughway. Finally, as we cut hastily through a slightly more open section, I caught a glimpse of a neighborhood church—a church of evident poverty, of a sect unknown, and destined surely to vanish from that unsavory spot. It was an anachronism as doomed as the cemetery. We passed, and a moment later, as though the sign had been hanging all that time in the cab before me, instead of standing neatly in the yard outside the church, my conscious mind unwillingly registered the words:

> Christ died to save mankind.
> Is it nothing to you, all ye that pass by?

I looked at that invisible hanging sign with surprise, if not annoyance. By some I have been castigated because I am an evolutionist. In one church which I had attended as the guest of a member I had been made the covert object of a sermon in which I had recognizably played the role of a sinning scientist. I cannot deny that the role may have fitted me, but I could not feel that the hospitality, under the circumstances, was Christian. I had seen fanatical sectarian signs of ignorant and contentious sects painted on rocks all over America, particularly in desert places. I had gazed unmoved on them all.

But here on a plain white board that would not remove itself from my eyes, an unknown man in the shadow of one of the ugliest neighborhoods in America had in some manner lifted that falling symbol from the shadow of the refinery tanks and thrust it relentlessly before my eyes. There was no evading it. "Is it nothing to you?" I was being asked—I who passed by, who had indeed already passed and would again ignore much more sophisticated approaches to religion.

But the symbol, one symbol of many in the wilderness of modern America, still exerted its power over me; a dozen lines of thinking, past and present, drew in upon me. Nothing eventful happened in the outside world. Whatever took place happened within myself. The cab sped on down the throughway.

But before my mind's eye, like an ineradicable mote, persisted the vision of that lost receding figure on the dreadful hill of Calvary who whispered with his last breath, "It is finished." It was not for himself he cried—it was for man against eternity, for us of every human generation who perform against the future the acts which justify creation or annul it. This is the power in the mind of man, a mindprint, if you will, an insubstantial symbol which holds like a strained cable the present from falling into the black abyss of nothingness. This is why, if we possess great fortitude, each one of us can say against the future he has not seen, "It is finished."

At that moment we will have passed beyond the reach of time into a still and hidden place where it was said, "He who loses his life will find it." And in that place we will have found an ancient and an undistorted way.

The Illusion of the Two Cultures

NOT LONG AGO an English scientist, Sir Eric Ashby, remarked that "to train young people in the dialectic between orthodoxy and dissent is the unique contribution which
universities make to society." I am sure that Sir Eric meant by
this remark that nowhere but in universities are the young
given the opportunity to absorb past tradition and at the same
time to experience the impact of new ideas—in the sense of a
constant dialogue between past and present—lived in every
hour of the student's existence. This dialogue, ideally, should
lead to a great winnowing and sifting of experience and to a
heightened consciousness of self which, in turn, should lead
on to greater sensitivity and perception on the part of the
individual.

Our lives are the creation of memory and the accompanying
power to extend ourselves outward into ideas and relive them.
The finest intellect is that which employs an invisible web of
gossamer running into the past as well as across the minds of
living men and which constantly responds to the vibrations
transmitted through these tenuous lines of sympathy. It would
be contrary to fact, however, to assume that our universities
always perform this unique function of which Sir Eric speaks,
with either grace or perfection; in fact our investment in man,
it has been justly remarked, is deteriorating even as the finan-
cial investment in science grows.

More than thirty years ago, George Santayana had already
sensed this trend. He commented, in a now-forgotten essay,
that one of the strangest consequences of modern science was
that as the visible wealth of nature was more and more transferred and abstracted, the mind seemed to lose courage and to
become ashamed of its own fertility. "The hard-pressed natural
man will not indulge his imagination," continued Santayana,
"unless it poses for truth; and being half-aware of this imposi-
tion, he is more troubled at the thought of being deceived
than at the fact of being mechanized or being bored; and he
would wish to escape imagination altogether."

"Man would wish to escape imagination altogether." I repeat

that last phrase, for it defines a peculiar aberration of the human mind found on both sides of that bipolar division between the humanities and the sciences, which C. P. Snow has popularized under the title of *The Two Cultures*. The idea is not solely a product of this age. It was already emerging with the science of the seventeenth century; one finds it in Bacon. One finds the fear of it faintly foreshadowed in Thoreau. Thomas Huxley lent it weight when he referred contemptuously to the "caterwauling of poets."

Ironically, professional scientists berated the early evolutionists such as Lamarck and Chambers for overindulgence in the imagination. Almost eighty years ago John Burroughs observed that some of the animus once directed by science toward dogmatic theology seemed in his day increasingly to be vented upon the literary naturalist. In the early 1900s a quarrel over "nature faking" raised a confused din in America and aroused W. H. Hudson to some dry and pungent comment upon the failure to distinguish the purposes of science from those of literature. I know of at least one scholar who, venturing to develop some personal ideas in an essay for the layman, was characterized by a reviewer in a leading professional journal as a worthless writer, although, as it chanced, the work under discussion had received several awards in literature, one of them international in scope. More recently, some scholars not indifferent to humanistic values have exhorted poets to leave their personal songs in order to portray the beauty and symmetry of molecular structures.

Now some very fine verse has been written on scientific subjects, but, I fear, very little under the dictate of scientists as such. Rather there is evident here precisely that restriction of imagination against which Santayana inveighed; namely, an attempt to constrain literature itself to the delineation of objective or empiric truth, and to dismiss the whole domain of value, which after all constitutes the very nature of man, as without significance and beneath contempt.

Unconsciously, the human realm is denied in favor of the world of pure technics. Man, the tool user, grows convinced that he is himself only useful as a tool, that fertility except in the use of the scientific imagination is wasteful and without purpose, even, in some indefinable way, sinful. I was reading

J. R. R. Tolkien's great symbolic trilogy, *The Fellowship of the Ring*, a few months ago, when a young scientist of my acquaintance paused and looked over my shoulder. After a little casual interchange the man departed leaving an accusing remark hovering in the air between us. "I wouldn't waste my time with a man who writes fairy stories." He might as well have added, "or with a man who reads them."

As I went back to my book I wondered vaguely in what leafless landscape one grew up without Hans Christian Andersen, or Dunsany, or even Jules Verne. There lingered about the young man's words a puritanism which seemed the more remarkable because, as nearly as I could discover, it was unmotivated by any sectarian religiosity unless a total dedication to science brings to some minds a similar authoritarian desire to shackle the human imagination. After all, it is this impossible, fertile world of our imagination which gave birth to liberty in the midst of oppression, and which persists in seeking until what is sought is seen. Against such invisible and fearful powers, there can be found in all ages and in all institutions—even the institutions of professional learning—the humorless man with the sneer, or if the sneer does not suffice, then the torch, for the bright unperishing letters of the human dream.

One can contrast this recalcitrant attitude with an 1890 reminiscence from that great Egyptologist Sir Flinders Petrie, which steals over into the realm of pure literature. It was written, in unconscious symbolism, from a tomb:

"I here live, and do not scramble to fit myself to the requirements of others. In a narrow tomb, with the figure of Néfermaat standing on each side of me—as he has stood through all that we know as human history—I have just room for my bed, and a row of good reading in which I can take pleasure after dinner. Behind me is that Great Peace, the Desert. It is an entity—a power—just as much as the sea is. No wonder men fled to it from the turmoil of the ancient world."

It may now reasonably be asked why one who has similarly, if less dramatically, spent his life among the stones and broken shards of the remote past should be writing here about matters involving literature and science. While I was considering this with humility and trepidation, my eye fell upon a stone in my office. I am sure that professional journalists must recall times when an approaching deadline has keyed all their senses and led them to glance wildly around in the hope that something might leap out at them from the most prosaic surroundings. At all events my eyes fell upon this stone.

Now the stone antedated anything that the historians would call art; it had been shaped many hundreds of thousands of years ago by men whose faces would frighten us if they sat among us today. Out of old habit, since I like the feel of worked flint, I picked it up and hefted it as I groped for words over this difficult matter of the growing rift between science and art. Certainly the stone was of no help to me; it was a utilitarian thing which had cracked marrow bones, if not heads, in the remote dim morning of the human species. It was nothing if not practical. It was, in fact, an extremely early example of the empirical tradition which has led on to modern science.

The mind which had shaped this artifact knew its precise purpose. It had found out by experimental observation that the stone was tougher, sharper, more enduring than the hand which wielded it. The creature's mind had solved the question of the best form of the implement and how it could be manipulated most effectively. In its day and time this hand ax was as grand an intellectual achievement as a rocket.

As a scientist my admiration went out to that unidentified workman. How he must have labored to understand the forces involved in the fracturing of flint, and all that involved practical survival in his world. My uncalloused twentieth-century hand caressed the yellow stone lovingly. It was then that I made a remarkable discovery.

In the mind of this gross-featured early exponent of the practical approach to nature—the technician, the no-nonsense practitioner of survival—two forces had met and merged. There had not been room in his short and desperate life for the delicate and supercilious separation of the arts from the sciences. There did not exist then the refined distinctions set up between the scholarly percipience of reality and what has sometimes been called the vaporings of the artistic imagination.

As I clasped and unclasped the stone, running my fingers down its edges, I began to perceive the ghostly emanations from a long-vanished mind, the kind of mind which, once having shaped an object of any sort, leaves an individual trace behind it which speaks to others across the barriers of time and language. It was not the practical experimental aspect of this mind that startled me, but rather that the fellow had wasted time.

In an incalculably brutish and dangerous world he had both shaped an instrument of practical application and then, with a virtuoso's elegance, proceeded to embellish his product. He had not been content to produce a plain, utilitarian implement. In some wistful, inarticulate way, in the grip of the dim aesthetic feelings which are one of the marks of man—or perhaps I should say, some men—this archaic creature had lingered over his handiwork.

One could still feel him crouching among the stones on a long-vanished river bar, turning the thing over in his hands, feeling its polished surface, striking, here and there, just one more blow that no longer had usefulness as its criterion. He had, like myself, enjoyed the texture of the stone. With skills lost to me, he had gone on flaking the implement with an eye to beauty until it had become a kind of rough jewel, equivalent in its day to the carved and gold-inlaid pommel of the iron dagger placed in Tutankhamen's tomb.

All the later history of man contains these impractical exertions expended upon a great diversity of objects, and, with literacy, breaking even into printed dreams. Today's secular disruption between the creative aspect of art and that of science is a barbarism that would have brought lifted eyebrows in a Cro-Magnon cave. It is a product of high technical specialization, the deliberate blunting of wonder, and the equally deliberate suppression of a phase of our humanity in the name of an authoritarian institution, science, which has taken on, in our time, curious puritanical overtones. Many scientists seem unaware of the historical reasons for this development or the fact that the creative aspect of art is not so remote from that of science as may seem, at first glance, to be the case.

I am not so foolish as to categorize individual scholars or scientists. I am, however, about to remark on the nature of science as an institution. Like all such structures it is apt to reveal certain behavioral rigidities and conformities which increase with age. It is no longer the domain of the amateur, though some of its greatest discoverers could be so defined. It is now a professional body, and with professionalism there tends to emerge a greater emphasis upon a coherent system of regulations. The deviant is more sharply treated, and the young tend to imitate their successful elders. In short, an "Establishment"—a trade union —has appeared.

Similar tendencies can be observed among those of the humanities concerned with the professional analysis and interpretation of the works of the creative artist. Here too, a similar rigidity and exclusiveness make their appearance. It is not that in the case of both the sciences and the humanities standards are out of place. What I am briefly cautioning against is that too frequently they afford an excuse for stifling original thought or constricting much latent creativity within traditional molds.

Such molds are always useful to the mediocre conformist who instinctively castigates and rejects what he cannot imitate. Tradition, the continuity of learning, are, it is true, enormously important to the learned disciplines. What we must realize as scientists is that the particular institution we inhabit has its own irrational accretions and authoritarian dogmas which can be as unpleasant as some of those encountered in sectarian circles— particularly so since they are frequently unconsciously held and surrounded by an impenetrable wall of self-righteousness brought about because science is regarded as totally empiric and open-minded by tradition.

This type of professionalism, as I shall label it in order to distinguish it from what is best in both the sciences and humanities, is characterized by two assumptions: that the accretions of fact are cumulative and lead to progress, whereas the insights of art are, at best, singular, and lead nowhere, or, when introduced into the realm of science, produce obscurity and confusion. The convenient label "mystic" is, in our day, readily applied to men who pause for simple wonder, or who encounter along the borders of the known that "awful power" which Wordsworth characterized as the human imagination. It can, he says, rise suddenly from the mind's abyss and enwrap the solitary traveler like a mist.

We do not like mists in this era, and the word imagination is less and less used. We like, instead, a clear road, and we abhor solitary traveling. Indeed one of our great scientific historians remarked not long ago that the literary naturalist was obsolescent if not completely outmoded. I suppose he meant that with our penetration into the biophysical realm, life, like matter, would become increasingly represented by abstract symbols. To many it must appear that the more we can dissect life into its elements, the closer we are getting to its ultimate resolution. While I have some reservations on this score, they are not important. Rather, I should like to look at the symbols which in the one case denote science and in the other constitute those vaporings and cloud wraiths that are the abomination, so it is said, of the true scientist but are the delight of the poet and literary artist.

Creation in science demands a high level of imaginative insight and intuitive perception. I believe no one would deny this, even though it exists in varying degrees, just as it does, similarly, among writers, musicians, or artists. The scientist's achievement, however, is quantitatively transmissible. From a single point his discovery is verifiable by other men who may then, on the basis of corresponding data, accept the innovation and elaborate upon it in the cumulative fashion which is one of the great triumphs of science.

Artistic creation, on the other hand, is unique. It cannot be twice discovered, as, say, natural selection was discovered. It may be imitated stylistically, in a genre, a school, but, save for a few items of technique, it is not cumulative. A successful work of art may set up reverberations and is, in this, just as transmissible as science, but there is a qualitative character about it. Each reverberation in another mind is unique. As the French novelist François Mauriac has remarked, each great novel is a separate and distinct world operating under its own laws with a flora and fauna totally its own. There is communication, or the work is a failure, but the communication releases our own visions, touches some highly personal chord in our own experience.

The symbols used by the great artist are a key releasing our humanity from the solitary tower of the self. "Man," says Lewis Mumford, "is first and foremost the self-fabricating animal." I shall merely add that the artist plays an enormous role in this act of self-creation. It is he who touches the hidden strings of pity, who searches our hearts, who makes us sensitive to beauty, who asks questions about fate and destiny. Such questions, though they lurk always around the corners of the external universe which is the peculiar province of science, the rigors of the scientific method do not enable us to pursue directly.

And yet I wonder.

It is surely possible to observe that it is the successful analogy or symbol which frequently allows the scientist to leap from a generalization in one field of thought to a triumphant achievement in another. For example, Progressionism in a spiritual sense later became the model contributing to the discovery of organic evolution. Such analogies genuinely resemble the figures and enchantments of great literature, whose meanings similarly can never be totally grasped because of their endless power to ramify in the individual mind.

John Donne gave powerful expression to a feeling applicable as much to science as to literature when he said devoutly of certain Biblical passages: "The literall sense is always to be preserved; but the literall sense is not alwayes to be discerned; for the literall sense is not alwayes that which the very letter and grammar of the place presents." A figurative sense, he argues cogently, can sometimes be the most "literall intention of the Holy Ghost."

It is here that the scientist and artist sometimes meet in uneasy opposition, or at least along lines of tension. The scientist's attitude is sometimes, I suspect, that embodied in Samuel Johnson's remark that, wherever there is mystery, roguery is not far off.

Yet surely it was not roguery when Sir Charles Lyell glimpsed in a few fossil prints of raindrops the persistence of the world's natural forces through the incredible, mysterious aeons of geologic time. The fossils were a symbol of a vast hitherto unglimpsed order. They are, in Donne's sense, both literal and symbolic. As fossils they merely denote evidence of rain in a past era. Figuratively they are more. To the perceptive intelligence they afford the hint of lengthened natural order, just as the eyes of ancient trilobites tell us similarly of the unchanging laws of light. Equally, the educated mind may discern in a scratched pebble the retreating shadow of vast ages of ice and gloom. In Donne's archaic phraseology these objects would bespeak the principal intention of the Divine Being—that is, of order beyond our power to grasp.

Such images drawn from the world of science are every bit as powerful as great literary symbolism and equally demanding upon the individual imagination of the scientist who would fully grasp the extension of meaning which is involved. It is, in fact, one and the same creative act in both domains.

Indeed evolution itself has become such a figurative symbol, as has also the hypothesis of the expanding universe. The laboratory worker may think of these concepts in a totally empirical fashion as subject to proof or disproof by the experimental method. Like Freud's doctrine of the subconscious, however, such ideas frequently escape from the professional scientist into the public domain. There they may undergo further individual transformation and embellishment. Whether the scholar approves or not, such hypotheses are now as free to evolve in the mind of the individual as are the creations of art. All the resulting enrichment and confusion will bear about it something suggestive of the world of artistic endeavor.

As figurative insights into the nature of things, such embracing conceptions may become grotesquely distorted or glow with added philosophical wisdom. As in the case of the trilobite eye or the fossil raindrop, there lurks behind the visible evidence vast shadows no longer quite of that world which we term natural. Like the words in Donne's Bible, enormous implications have transcended the literal expression of the thought. Reality itself has been superseded by a greater reality. As Donne himself asserted, "The substance of the truth is in the great images which lie behind."

It is because these two types of creation—the artistic and the scientific—have sprung from the same being and have their points of contact even in division that I have the temerity to assert that, in a sense, the "two cultures" are an illusion, that they are a product of unreasoning fear, professionalism, and misunderstanding. Because of the emphasis upon science in our society, much has been said about the necessity of educating the layman and even the professional student of the humanities upon the ways and the achievements of science. I admit that a barrier exists, but I am also concerned to express the view that there persists in the domain of science itself an occasional marked intolerance of those of its own membership who venture to pursue the way of letters. As I have remarked, this intolerance can the more successfully clothe itself in seeming objectivity because of the supposed open nature of the scientific society. It is not remarkable that this trait is sometimes more manifest in the younger and less secure disciplines.

There was a time, not too many centuries ago, when to be active in scientific investigation was to invite suspicion. Thus it may be that there now lingers among us, even in the triumph of the experimental method, a kind of vague fear of that other artistic world of deep emotion, of strange symbols, lest it seize upon us or distort the hard-won objectivity of our thinking—lest it corrupt, in other words, that crystalline and icy objectivity which, in our scientific guise, we erect as a model of conduct. This model, incidentally, if pursued to its absurd conclusion, would lead to a world in which the computer would determine all aspects of our existence; one in which the bomb would be as welcome as the discoveries of the physician.

Happily, the very great in science, or even those unique scientist-artists such as Leonardo, who foreran the emergence of science as an institution, have been singularly free from this folly. Darwin decried it even as he recognized that he had paid a certain price in concentrated specialization for his achievement. Einstein, it is well known, retained a simple sense of wonder; Newton felt like a child playing with pretty shells on a beach. All show a deep humility and an emotional hunger which is the prerogative of the artist. It is with the lesser men, with the institutionalization of method, with the appearance of dogma and mapped-out territories, that an unpleasant suggestion of fenced preserves begins to dominate the university atmosphere.

As a scientist, I can say that I have observed it in my own and others' specialties. I have had occasion, also, to observe its effects in the humanities. It is not science *per se*; it is, instead, in both regions of thought, the narrow professionalism which is also plainly evident in the trade union. There can be small men in science just as there are small men in government or business. In fact it is one of the disadvantages of big science, just as it is of big government, that the availability of huge sums attracts a swarm of elbowing and contentious men to whom great dreams are less than protected hunting preserves.

The sociology of science deserves at least equal consideration with the biographies of the great scientists, for powerful and changing forces are at work upon science, the institution, as contrasted with science as a dream and an ideal of the individual. Like other aspects of society, it is a construct of men and is subject, like other social structures, to human pressures and inescapable distortions.

Let me give an illustration. Even in learned journals, clashes occasionally occur between those who would regard biology as a separate and distinct domain of inquiry and the reductionists who, by contrast, perceive in the living organism only a vaster and more random chemistry. Understandably, the concern of the reductionists is with the immediate. Thomas Hobbes was expressing a similar point of view when he castigated poets as "working on mean minds with words and distinctions that of themselves signifie nothing, but betray (by their obscurity) that there walketh . . . another kingdome, as it were a kingdome of fayries in the dark." I myself have been similarly criticized for speaking of a nature "beyond the nature that we know."

Yet consider for a moment this dark, impossible realm of "fayrie." Man is not totally compounded of the nature we profess to understand. He contains, instead, a lurking unknown future, just as the man-apes of the Pliocene contained in embryo the future that surrounds us now. The world of human culture itself was an unpredictable fairy world until, in some pre-ice-age meadow, the first meaningful sounds in all the world broke through the jungle babble of the past, the nature, until that moment, "known."

It is fascinating to observe that, in the very dawn of science, Francis Bacon, the spokesman for the empirical approach to nature, shared with Shakespeare, the poet, a recognition of the creativeness which adds to nature, and which emerges from nature as "an art which nature makes." Neither the great scholar nor the great poet had renounced this "kingdome of fayries." Both had realized what Henri Bergson was later to express so effectively, that life inserts a vast "indetermination into matter." It is, in a sense, an intrusion from a realm which can never be completely subject to prophetic analysis by science. The novelties of evolution emerge; they cannot be predicted. They haunt, until their arrival, a world of unimaginable possibilities behind the living screen of events, as these last exist to the observer confined to a single point on the time scale.

Oddly enough, much of the confusion that surrounded my phrase, "a nature beyond the nature that we know," resolves itself into pure semantics. I might have pointed out what must be obvious even to the most dedicated scientific mind—that the nature which we know has been many times reinterpreted in human thinking, and that the hard, substantial matter of the nineteenth century has already vanished into a dark, bodiless void, a web of "events" in space-time. This is a realm, I venture to assert, as weird as any we have tried, in the past, to exorcise by the brave use of seeming solid words. Yet some minds exhibit an almost instinctive hostility toward the mere attempt to wonder or to ask what lies below that microcosmic world out of which emerge the particles which compose our bodies and which now take on this wraithlike quality.

Is there something here we fear to face, except when clothed in safely sterilized professional speech? Have we grown reluctant in this age of power to admit mystery and beauty into our thoughts, or to learn where power ceases? I referred earlier to one of our own forebears on a gravel bar, thumbing a pebble. If, after the ages of building and destroying, if after the measuring of light-years and the powers probed at the atom's heart, if after the last iron is rust-eaten and the last glass lies shattered in the streets, a man, some savage, some remnant of what once we were, pauses on his way to the tribal drinking place and feels rising from within his soul the inexplicable mist of terror and beauty that is evoked from old ruins—even the ruins of the greatest city in the world—then, I say, all will still be well with man.

And if that savage can pluck a stone from the gravel because it shone like crystal when the water rushed over it, and hold it against the sunset, he will be as we were in the beginning, whole—as we were when we were children, before we began to split the knowledge from the dream. All talk of the two cultures is an illusion; it is the pebble which tells man's story. Upon it is written man's two faces, the artistic and the practical. They are expressed upon one stone over which a hand once closed, no less firm because the mind behind it was submerged in light and shadow and deep wonder.

Today we hold a stone, the heavy stone of power. We must perceive beyond it, however, by the aid of the artistic imagination, those humane insights and understandings which alone can lighten our burden and enable us to shape ourselves, rather than the stone, into the forms which great art has anticipated.

Chronology

Born Loren Corey Eiseley in Lincoln, Nebraska, on September 3, 1907, the only child of Clyde and Daisy Corey Eiseley. Mother, thirty-two, grew up in Iowa; deaf and reclusive, she takes in sewing to help support the family. Father, thirty-eight, works as a hardware-store clerk in Fremont, Nebraska; the son of a prosperous German-immigrant hardware merchant, as a younger man he managed the small Opera House in Norfolk, Nebraska (which his father owned), acted in plays, and committed much Shakespeare to memory, but his family's fortunes collapsed after a bad investment in a sugar beet factory. Fifteen-year-old half-brother Leo, from the first of father's two previous marriages, begins high school in January 1909 but soon drops out to work as a messenger boy. Family moves from Fremont to Lincoln in 1910; Leo leaves home the following year, becoming a telegraph operator.

1913–17   Begins school in Lincoln in 1913. Four years later, family moves to Aurora, Nebraska, where father continues to work as a hardware-store clerk.

1918–21   Paternal grandfather Charles Frederick dies on February 2, 1918, and maternal grandfather Milo on December 12. Taking a job as a traveling salesman, father moves family back to Lincoln; he survives a case of influenza. Loren visits the Museum at the University of Nebraska with his maternal uncle Buck, an attorney; for his twelfth birthday, receives a birdhouse and a copy of Jules Verne's *From the Earth to the Moon*. Leo, now married to Mamie Harris and the father of a two-year-old daughter, returns to Lincoln in 1920; Loren visits them often. Writes in an eighth-grade essay: "I have selected Nature Writing for my vocation because at this time in my life it appeals to me more than any other subject."

1922–24   Leo and family relocate to Colorado Springs. Loren enters Lincoln High School in September 1922; moves in with Aunt Grace and Uncle Buck. Unhappy in school, does not return the following January and takes a menial job. Uncle arranges for his enrollment at Teachers College High

School where he becomes a junior in September 1923; joins the football team. Buck gives him a copy of Henry Fairfield Osborn's *Origin and Evolution of Life*.

1925   During his senior year in high school, captains the football team and is elected class president; acts in the class play, Walter Ben Hare's *Kicked out of College*. With three friends, buys a 1919 Model T Ford and drives to Los Angeles in search of summer employment, camping along the way. Back in Lincoln, enters the University of Nebraska in September, living alternately with his parents and his aunt and uncle, his uncle helping to pay his tuition. Begins a correspondence with Mabel Langdon, a former student teacher at his high school who had taken her first job in rural Arnold, Nebraska; they exchange poems.

1926   Studies evolution and genetics, elementary paleontology, social psychology, and applied psychology in summer school, but takes only two courses in the fall; failing one, is forced to take a semester's leave. Travels widely throughout the western states on freight trains.

1927   Reenters the University of Nebraska in the summer. Becomes associate editor of the recently established literary magazine *Prairie Schooner*; publishes poems and prose sketches.

1928   Does not reenroll at the University of Nebraska in January; takes job on the graveyard shift at a chicken hatchery. Father, diagnosed with stomach and liver cancer, dies on March 31; unable to afford mortgage payments, mother moves in with uncle and aunt. Returns to school in September.

1929   Diagnosed with influenza and a pulmonary infection, is urged by doctors to take a rest cure; spends time in Manitou Springs, Colorado, to recuperate. Returns to Lincoln in the fall; his doctor recommends further rest in a dry climate. "The Deserted Homeland," a poem, appears in *Poetry* in December.

1930   Travels to California, where he becomes caretaker of a property in the Mojave Desert, near Lancaster. Mabel Langdon visits. His health improved, makes his way back to Lincoln by freight train, living as a hobo; arrives "in a dreadful state." Re-enrolls in classes in September, taking

Introductory Anthropology and Field & Museum Techniques.

1931 During spring vacation, joins a group of about a dozen students on an archaeological dig led by Dr. William Duncan Strong; they excavate remains of sixteenth-century Pawnee village in central Nebraska. On May 3 becomes president of the Poetry Society, part of the Nebraska Writers' Guild. Travels through Colorado, New Mexico, Arizona, and Utah in June, camping with Helen Hopt, a young teacher, and another couple. Spends August and September before classes as part of the Morrill Paleontological Expedition (the "South Party"); discovers a promising fossil site in Banner County, Nebraska.

1932 Returns to Banner County with the South Party in May, June, and July, unearthing bones of extinct mastodon, rhinoceros, camel, and saber-toothed cat.

1933 Graduates from University of Nebraska in June with a B.A. in anthropology and English; joins the South Party for third summer. Visits Taos, New Mexico, introducing himself to literary heiress Mabel Dodge Luhan. In August discovers remains of a large Brontothere and the skull of a saber-toothed cat. Moves to Philadelphia in the fall, beginning graduate school in anthropology at the University of Pennsylvania. Visits Harold Vinal, editor of the poetry magazine *Voices* and a longtime correspondent, in New York City, and his half-brother Leo and family, who now live in Philadelphia.

1934 Spends time on weekends hiking in the New Jersey Pine Barrens with Frank Speck, a favorite professor. In June, travels to Carlsbad, New Mexico, with another graduate student, Joseph B. Townsend Jr., on a fellowship; over six weeks they excavate a Native American burial site in a cave in what is now Guadalupe Mountains National Park. (At the end of the summer's work, celebrating in Ciudad Juárez, Mexico, the two are arrested for "doing something ridiculous," and are released only after the intervention of the American consul.) Visits Santa Fe with Mabel Langdon and Dorothy Thomas, a young writer from Lincoln; they attend the funeral of writer Mary Austin, and tour Pecos Pueblo and Chaco Canyon. Returning to Penn, moves into International House with two friends; is awarded a scholarship. Later remembers the time as "one of the happiest in my life." Completes master's thesis, "A Review of the Paleontological Evidence Bearing upon the Age of the Scottsbluff Bison Quarry and Its Assorted Artifacts."

1935   Part of thesis is accepted for publication in *American Anthropologist*. Receives master's degree on February 9. In June, joins eight others at the Lindenmeier archaeological site in Larimer County, Colorado, seeking skeletal remains of Folsom man; finds a Folsom arrowhead in the neck vertebrae of a Pleistocene bison, and gathers mollusk specimens hoping they may help date the party's finds. On August 20, learns of the death of his uncle Buck, and returns to Lincoln for his funeral. Unable to afford tuition at Penn without uncle's support, moves in with family in Lincoln; Mabel Langdon helps pay his tuition at the University of Nebraska, where he takes graduate courses in sociology and German.

1936   Takes a W.P.A. job in February, assisting the editor of *Nebraska: A Guide to the Cornhusker State*; writes essays on Nebraska geology, paleontology, and Native American culture. Returns to Penn in the fall with the aid of a fellowship. Goes on canoeing and hiking trips with Frank Speck, now his dissertation advisor. Maternal grandmother Malvina dies on December 17.

1937   Receives Ph.D. degree in June after Speck approves his dissertation, "Three Indices of Quaternary Time and Their Bearing on the Problems of American History: A Critique." Spends three weeks excavating eighteenth-century Native American burial sites in Doniphan County, Kansas, with a party from the National Museum of Natural History. Accepts position as assistant professor in the sociology department at the University of Kansas, moving to Lawrence; subsequently teaches courses on General Anthropology, the American Indian, Primitive Society, Peoples of the Pacific, the Evolution of Culture, and the Elements of Sociology. In August is reunited with Mabel in New Mexico; they stay with Dorothy Thomas at Mary Austin's house in Santa Fe, visit Taos, and meet Frieda Von Richtofen Lawrence, D. H. Lawrence's widow, at her home nearby.

1938   On weekends, investigates potential archaeological sites near Lawrence, and does some excavating work in early August. Visits Albuquerque with Mabel; they marry there on August 29 in a small informal ceremony. Lacking the funds to set up a joint household, she returns to her family in Lincoln, working as a curator at the university art gallery, he to bachelor quarters in Lawrence.

1939   In April, with Bert Schultz of Nebraska State Museum, announces archaeological discoveries in southeastern Nebraska; with Schultz, signs contract to write book "They Hunted the Mammoth: The Story of Ice Age Man" for Macmillan (later cancelled because of the war). Helps to prepare an exhibit on the works of Robinson Jeffers, the modern writer he claims most to admire, at the University of Nebraska Library. Beginning in the summer, is joined by wife Mabel in Lawrence. Signs a contract with publisher Thomas Y. Crowell to write a textbook on general anthropology.

1940   In June, begins postdoctoral fellowship in physical anthropology at Columbia University and the American Museum of Natural History. Mabel, arriving in New York in September, works part-time as a typist for crime writer Rex Stout. Registers for military service in October. Friend Dorothy Thomas rents a vacant apartment downstairs; she takes the Eiseleys to parties and introduces them to her circle.

1941   Working with Harry L. Shapiro at Columbia, undertakes biometric studies of Basket Maker skeletons excavated in Arizona. Returns to Kansas in September.

1942   In January, as second author with Frank Speck, publishes "Montagnais-Naskapi Bands and Family Hunting Districts of the Central and Southeastern Labrador Peninsula," in the *Proceedings of the American Philosophical Society*. Application for a commission in the Army Air Corps is rejected in July due to his weak eyesight; fears he may soon be drafted as an infantryman. Publishes "What Price Glory? The Counterplaint of an Anthropologist" in *American Sociological Review* in December, along with "The Folsom Mystery," the first of a series of articles in *Scientific American*.

1943   In the spring becomes an instructor in human gross anatomy in the short-handed University of Kansas Medical School; receives a draft deferment on the basis of his teaching responsibilities.

1944   Mabel, diagnosed with breast cancer, undergoes a radical mastectomy and radiation treatment. In May, accepts an offer to serve as a professor in the department of sociology at Oberlin College in Ohio; moves to Oberlin with Mabel in October.

1945   Teaches over the summer at the University of Pennsylvania. Becomes friendly with writer and photographer Wright Morris, a downstairs neighbor, who nicknames him "Schmerzie" (a diminutive of *Weltschmerz*), making light of his tendency to melancholia. Shares work in progress with Morris. Publishes essays "Myth and Mammoth in Archaeology" in *American Antiquity* and "There Were Giants" in *Prairie Schooner*.

1946   Is awarded tenure at Oberlin. Teaches over the summer at Columbia University. In November receives a grant of $3,500 from The Viking Fund for "A Survey and Investigation of Researches on Early Man in South Africa"; plans extensive travel, intending to visit South African anthropologists, museum collections, and archaeological sites.

1947   In January is appointed Professor of Anthropology at the University of Pennsylvania and assumes chairmanship of his department; with Mabel, rents a small apartment in suburban Rose Valley, outside Philadelphia. Works with Froelich Rainey, newly appointed director of the University Museum, to restore long-strained ties between the Museum and the Department of Anthropology; is named curator of early man at the Museum. Postpones South African trip after he learns that a group from the University of California will soon be en route with similar objectives; hopes to revise his itinerary and sail in a year.

1948   Travels to Oaxaca, Mexico, in the fall after human skeletal remains are unearthed near those of an ancient elephant, a potentially significant discovery that ultimately proves less important than it appears. Abandons South African travel plans, citing "the unsettled state of both departmental and family affairs." Suffers from partial deafness after an ear infection in the fall, recovering about six months later. Comes to an informal agreement with Harper & Brothers to publish a collection of essays; works with editor Jack Fischer. Presents paper on "Providence and the Death of Species" at meeting of the American Anthropological Association in December, in Toronto.

1949   Teaches at Berkeley over the summer, afterward traveling
        to Wyoming, near Cody, where he investigates a distinc-
        tive arrowhead site. Moves with Mabel to a new apartment
        in Wynnewood, a prosperous suburb on the Philadelphia
        Main Line. In September is elected president of the newly
        formed American Institute of Human Paleontology, after
        a meeting of leading physical anthropologists in New
        York, which he attends. Publishes "The Fire Apes" in
        Harper's; tells Ray Bradbury, who writes him in praise of
        the essay, that he is "contemplating doing a book."

1950   Delivers a eulogy for Frank Speck after his death on Feb-
        ruary 7.

1951   Toward the end of the year, travels briefly to England to
        arrange for the university museum's acquisition of an im-
        portant collection of plaster casts of paleontological speci-
        mens.

1952   Accepts a commission from the American Philosophical
        Society to write a book on the reception of Darwin's ideas
        in America, to be published on the centennial of the first
        edition of *On the Origin of Species*, in 1959. Helps the Society
        build its Darwin collections, locating books and manuscripts
        for acquisition. (Wright Morris occasionally accompanies him
        on buying trips.) Wife Mabel takes part-time job as secretary
        to the director of the Pennsylvania Academy of the Fine Arts,
        in subsequent years becoming an associate director of that
        institution. Begins a year's leave with fellowship support to
        work on a book about "the philosophical implications of
        human evolution."

1953   In July, meets with Jack Fischer to discuss essay collection.

1954   Essay "Man the Firemaker" appears in *Scientific American*
        in September.

1955   After discussions at the American Philosophical Society,
        abandons plans for a book on Darwinism in America, and
        agrees in December to edit two volumes of the letters of
        Darwin and his contemporaries. (Subsequently renegoti-
        ates this commitment, and instead of the proposed vol-
        umes works intensely on a book about the history of
        evolution tentatively titled "The Time Voyagers," solicited
        by Doubleday editor Jason Epstein.)

1956   Early in the year, meets with Hiram Haydn, editor-in-chief at Random House, who persuades him to work with the firm on a book gathering his essays; proposes to call it "The Great Deeps." Begins a year's sabbatical in September, freeing him to advance his American Philosophical Society publication projects. Presents Haydn with a manuscript of his essay collection, now titled "The Crack in the Absolute," on November 13; they ultimately agree on the title *The Immense Journey*. Readers for the press recommend substantial revisions to Eiseley's initial submission.

1957   In January, meets with Haydn in Philadelphia to discuss alterations to *The Immense Journey*; rewrites and reorganizes it, dropping several essays. Submits a manuscript of "The Time Voyagers" to Doubleday in June. *The Immense Journey* is published on August 26; travels to New York for radio interviews about the book. Joins editorial board of *The American Scholar*.

1958   Proposes to edit an anthology of naturalists' writings for Random House; at a meeting in New York in February, Haydn convinces him instead to sign a contract for a second book of his own essays. "The Time Voyagers," now titled *Darwin's Century: Evolution and the Men Who Discovered It*, is published by Doubleday in July.

1959   *Darwin's Century* wins Phi Beta Kappa science prize. Is appointed provost of the University of Pennsylvania in October; quickly realizes he has overcommitted himself. Mother dies on November 29. Delivers six public lectures at the College of Medicine of the University of Cincinnati.

1960   Publishes *The Firmament of Time*, gathering his University of Cincinnati lectures.

1961   In March, delivers the Montgomery Lectures on Contemporary Civilization at the University of Nebraska, addressing the life and significance of Francis Bacon on the four hundredth anniversary of his birth. On April 3, is awarded the John Burroughs Medal for *The Firmament of Time*. Resigns as provost in April, and in the fall becomes a fellow of the Center for Advanced Study in the Behavioral Sciences, in Palo Alto.

1962   Aunt Grace dies on January 17. In February, gives the fifth annual lecture of the John Dewey Society, in Chicago; it is published eight months later by Harper & Row, titled *The*

Mind as Nature. Visits New York in June, meeting with editors and publishers. Returning to Penn in July, is named University Professor in the Life Sciences.

1963   *Francis Bacon and the Modern Dilemma*, an expanded version of his 1961 Nebraska lectures, is published on January 15. Delivers address "The Divine Animal" on May 22, at a New York meeting of the National Institute of Arts and Letters. Accepts a position as director of the new Richard Prentice Ettinger Program for Creative Writing at Rockefeller University; on June 27, at ceremonies inaugurating the program, gives address "The Illusion of Two Cultures."

1964   Is awarded a Guggenheim Fellowship in April, enabling a year's leave of absence from Penn. Hires Caroline E. Werkley as his administrative assistant, to oversee the affairs of the Ettinger Program; she remains his assistant for the rest of his life. In November is relieved of his duties as Ettinger director, charged with neglecting them. Named to President's Task Force on the Preservation of Natural Beauty, travels to Washington, D.C., for a press conference, and meets with Lyndon Johnson.

1965   Addresses the Nebraska Academy of Sciences in April, on "The Inner Galaxy: A Prelude to Space."

1966   Gives address "Man, Time, and Prophecy" at the University of Kansas Centennial Celebration in April. Serves as host and narrator of *Animal Secrets*, an NBC television series that airs for about eighteen months.

1967   Vacations on Sanibel Island in Florida in February. Spends the fall semester at the University of Wisconsin, where he has been appointed Johnson Research Professor. Tours the countryside around Madison with Walter Hamady, an assistant professor of art and the proprietor of The Perishable Press, who proposes to publish some of Eiseley's works in an illustrated limited edition.

1968   In Dallas at a December meeting of the American Association for the Advancement of Science, gives an interview to the *Dallas Morning News* in which he questions the value of the space program, given its immense cost; many criticize his remarks.

1969   Spends three winter weeks in Aruba with Mabel; later in
         the year, she retires from her position at the Pennsylvania
         Academy of the Fine Arts. Delivers a series of lectures at
         the University of Washington in Seattle in the fall. *The
         Unexpected Universe* is published in October to widely
         positive reviews. Receives copies of *The Brown Wasps*, a
         collection of three essays published in a limited edition by
         Walter Hamady at The Perishable Press; "the books are
         beautiful," he writes Hamady. Begins a novel, never com-
         pleted, titled "The Snow Wolf."

1970   On February 21, *The New Yorker* publishes a warmly favor-
         able review by W. H. Auden of *The Unexpected Universe*;
         Eiseley and Auden subsequently correspond, and meet for
         lunch. *The Unexpected Universe* is nominated for a Na-
         tional Book Award. Invitation to deliver the commence-
         ment address at Kent State is cancelled by the university
         after the shooting of four students on May 4. Receives
         honorary doctor of science degree from St. Lawrence
         University in New York, one of many such honors. *The
         Invisible Pyramid* is published in October.

1971   Elected to the National Institute of Arts and Letters. *The
         Night Country* is published November 10.

1972   Suffers from viral pneumonia during the fall. In November
         publishes *Notes of an Alchemist*, a book of poems, with
         Scribner's; receives a note of praise from Auden, who asks
         permission to dedicate a poem to Eiseley.

1973   *The Man Who Saw Through Time*, a revised and expanded
         edition of *Francis Bacon and the Modern Dilemma*, is
         published by Scribner's in April. Over the summer spends
         ten days with Froelich Rainey of the Museum of Archaeol-
         ogy and Anthropology at Penn, doing archaeological
         fieldwork in Dawson County, Montana. Visits Walden
         Pond and Thoreau's grave in Masssachusetts; considers
         writing a book about Thoreau and discusses the project
         with Scribner's. *The Innocent Assassins*, a second collection
         of poetry, appears in October.

1974   On September 18, at the Kennedy Center in Washington,
         D.C., is presented with the Distinguished Nebraskan
         Award. Toward the end of the year, begins working inten-
         sively on a long-postponed volume of autobiography; it is
         initially titled "The Other Player."

1975   Publishes his autobiography, *All the Strange Hours: The Excavation of a Life*, in October.

1976   Sits for a portrait bust by sculptor Kappy Wells. Learns after routine prostate surgery in September that doctors have discovered a malignant bile duct tumor.

1977   Undergoes a pancreatectomy on January 27, returning home at the end of March. Receives galleys for *Another Kind of Autumn*, a book of poems to be published by Scribner's. Dictates a letter to Scribner's proposing an outline of the contents of his last book, published posthumously by Times Books as *The Star Thrower*. Returns to the hospital in June, his tumor having returned. Dies on July 9.

Note on the Texts

This volume—the second of a two-volume set, *Collected Essays on Evolution, Nature, and the Cosmos*, by Loren Eiseley—presents the complete texts of *The Invisible Pyramid* (1970) and *The Night Country* (1971), along with all of the prose items gathered for the first time in *The Star Thrower* (1978), a posthumous collection Eiseley had planned before his death. The first volume in the set presents *The Immense Journey* (1957), *The Firmament of Time* (1960), *The Unexpected Universe* (1969), and a selection of uncollected prose writings. The text of *The Invisible Pyramid* has been taken from a corrected later printing of the first edition, as described below; the texts of *The Night Country* and items from *The Star Thrower* have been taken from the first printings.

The Invisible Pyramid. In January 1969, shortly after the names of the first lunar astronauts were announced, Eiseley signed an agreement with Charles Scribner's Sons to produce a book of essays that would "put the moon shot in perspective with other explorations." The characterization was that of his new editor, Kenneth Heuer, who had begun a correspondence with Eiseley some fifteen years before; the two met regularly in New York and spoke frequently on the telephone while Eiseley was writing the book, which he delivered early in 1970. Further information on the composition and publication history of individual essays in *The Invisible Pyramid* is provided in the list below:

"The Star Dragon." Incorporates material presented as an address at the Mayo Clinic and published as "Freedom of the Juggernaut," *Mayo Clinic Proceedings*, January 1965. An early draft is titled "The Great Chameleon." First published in *Natural History*, June–July 1970.

"The Cosmic Prison." Incorporates material presented at an October 16, 1965, symposium sponsored by the Division of Earth Sciences, National Academy of Sciences, National Research Council, and published as "Man and Novelty" in *Time and Stratigraphy in the Evolution of Man* (1967).

"The World Eaters." Incorporates material presented as a lecture at Occidental College in March 1970. First published in *The Invisible Pyramid.*

"The Spore Bearers." Drafts incorporate material from a manu-
script titled "Space: The First Decade," subsequently published as
"The Invisible Pyramid" in *Science Year: The World Book Science
Annual, 1967* (1967). First published in its entirety in *The Invisible
Pyramid*.

"The Time Effacers." First published in *The Invisible Pyramid*.

"Man in the Autumn Light." Draft titles included "The Autumn
Light," "The Vast Chameleon," and "Part of Autumn Light."
Presented as the 1970 Phi Beta Kappa oration at Harvard under
the title "Man in the Dark Wood." Published for the first time in
The Invisible Pyramid.

"The Last Magician." Presented under the title "Man: The Last
Magician" as the Danz Lecture at the University of Washington,
and later at Purdue University, in November 1969. First published in *Playboy*, August 1970.

Always a careful reader of proofs, Eiseley found an "almost incredible" number of typographical errors in *The Invisible Pyramid* as it
made its way through the press. "I am the victim of type set by a
computer," he explained to his friend Hal Borland: Scribner's had just
introduced "some kind of new IBM process" in its typesetting, and
his book became "the guinea pig." Having attempted to fix some 350
errors in his initial author's proofs, he found that his corrections had
in many cases precipitated new errors. Later, errors that seemed to
have been corrected successfully inexplicably reappeared. Though he
made an "enormous effort" to overcome these problems, he quickly
found misprints in the book's October 1, 1970, first printing, and began to keep a list of errors to be corrected.

Eiseley's subsequent correspondence with Heuer suggests that
Scribner's made two substantial sets of corrections and revisions in
later printings of *The Invisible Pyramid*: he refers to an "'A' edition,"
a "'B' edition," and a "paper edition" of the book, the hardcover
"'B' edition" (a printing prior to July 1971) correcting approximately
twenty typographical errors and the "paper edition," in 1972, correcting additional errors and in some instances giving Eiseley the chance
to correct his own errors of fact and to revise his prose. (An English
edition, published by Rupert Hart-Davis in London in 1971, predates
the corrections and revisions Eiseley made for the "paper edition"
and appears to have been prepared without his involvement.) The
text of *The Invisible Pyramid* in the present volume is the corrected
one of 1972, taken from a Charles Scribner's Sons paperback printing.

The Night Country. In a letter to his friend Earl W. Count, Eiseley modestly described *The Night Country* as "an interim book": it

gathered the unpublished and uncollected essays of several decades but relatively little of the new writing he hoped to produce. Still, he revised these older essays with care and sometimes extensively as he assembled the new collection—titled "The Uncompleted Man" and "The Night Tide" before he decided on *The Night Country.*

Further information on the composition and publication history of individual essays in *The Night Country* is provided in the list below:

"The Gold Wheel." Originally written as the opening chapter of an unfinished autobiography, "The Counting of the Days." Published in *Harper's,* August 1971.

"The Places Below." Draft titles included "The Green Door: A Memoir of Childhood" and "A Memoir of Terror." Published in *Harper's,* June 1948.

"Big Eyes and Small Eyes." Draft titles included "Moon Travel," "The Night Side of Nature," and "Journey in the Dark." Published in *Gentry,* Fall 1956, and as "Big Eyes and Little Eyes" in *The Brown Wasps* (1969), a limited edition private press book produced by Walter Hamady at the Perishable Press, Mount Horeb, Wisconsin.

"Instruments of Darkness." Presented as an address at a joint ceremony of the American Academy of Arts and Letters and the National Institute of Arts and Letters, May 22, 1963, and published as "The Divine Animal" in the American Academy of Arts and Letters *Proceedings* (1964). Published in *Harper's,* March 1964, as "The Uncompleted Man."

"The Chresmologue." First presented as "Man, Time, and Contemplation," an address at the University of Kansas Centennial Celebration, April 11, 1966. Published in *The American Scholar,* Winter 1966, as "Man, Time, and Prophecy," and under the same title, in 1966, as a new year's gift book by Harcourt Brace Jovanovich, to be distributed to friends of the publisher. First serial publication in *Harper's,* November 1971, as "The Scientist as Prophet."

"Paw Marks and Buried Towns." Published as a longer book review in *The American Scholar,* Spring 1958.

"Barbed Wire and Brown Skulls." Draft titles included "The Heads in Haggerty's Barn," "Headhunting in the USA," "The Head Takers," "In a Place of Skulls," "A Place in the Light," and "Anthropologists?" Published in *Harper's,* May 1951, as "People Leave Skulls with Me."

"The Relic Men." Draft titles included "Bones All Over That Valley" and "Listen, Professor." Published in *Harper's,* November 1948, as "Buzby's Petrified Woman."

"Strangeness in the Proportion." First published in *The Night Country*.

"The Creature from the Marsh." Draft titles included "The Creature from the Dark," "A Bone Hunter's Confession," "When the Time Comes," and "I Found the Missing Link." First published in *The Night Country*.

"One Night's Dying." A draft is titled "Doorway to the Dark." Published in *The Atlantic*, June 1963, and in *The Brown Wasps* (1969), a limited edition private press book produced by Walter Hamady at the Perishable Press, Mount Horeb, Wisconsin, as "Endure the Night."

"Obituary of a Bone Hunter." Published in *Harper's*, October 1947.

"The Mind as Nature." First presented as the fifth John Dewey Society Lecture, in Chicago, on February 16, 1962. Published as *The Mind as Nature* (New York: Harper & Row, 1962); revised for *The Night Country*.

"The Brown Wasps." A draft is titled "The Tree That Never Was." Published in *Gentry*, Fall–Winter 1956–57, and in *The Brown Wasps* (1969), a limited edition private press book produced by Walter Hamady at the Perishable Press, Mount Horeb, Wisconsin.

Eiseley is not known to have been involved in the preparation of an English edition of *The Night Country*, published by Garnstone Press in 1974. The text of *The Night Country* in the present volume has been taken from the November 10, 1971, Charles Scribner's Sons first printing.

Essays from *The Star Thrower*. Less than a month before his death on June 9, 1977, in a letter dictated from his hospital bed, Eiseley informed his publisher Charles Scribner that he had prepared the "rough outline" of a new book. To be titled "The Loren Eiseley Sampler," it would gather both poetry and prose, some of it previously unpublished or uncollected and some of it representing the best of his earlier books. His assistant Caroline E. Werkley subsequently assembled copies of the works he had selected and brought them to his bedside; he is reported to have remarked that they would make "a good 'last' book." Though he was too ill to revisit these works closely, he is said to have asked that his 1949 essay "The Fire Apes" be included only in part, its latter section having been rendered obsolete by more recent discoveries.

After Eiseley's death, with Werkley's assistance, Kenneth Heuer at Scribner's prepared the collection for the press. Arranging Eiseley's papers for the archives of the University of Pennsylvania, Werkley discovered two previously unpublished short stories—"The Dance of the Frogs" and "The Fifth Planet," both probably written in the 1940s—that Eiseley had not listed for inclusion in his "Sampler," and it was decided to add them. A dispute over remuneration led Mabel Eiseley to choose Times Books over Charles Scribner's Sons as the book's publisher; it appeared as *The Star Thrower* in June 1978, with an "Editor's Preface" that Heuer had written for the originally planned Scribner's edition.

The present volume includes all of those prose items from *The Star Thrower* not already published in one of Eiseley's previous collections, in the order and form in which they appear in *The Star Thrower*. Further information on the composition and publication history of individual items in *The Star Thrower* is provided in the list below:

"The Long Loneliness." A draft is titled "Man's Newest Intellectual Rival: The Bottle Nosed Porpoise." Published in *The American Scholar*, Winter 1960–61, as "The Long Loneliness: Man and the Porpoise—Two Solitary Destinies."

"Man the Firemaker." Published in *Scientific American*, September 1954.

"The Ghostly Guardian." First published in *The Star Thrower*.

"The Fire Apes." A draft is titled "The Apes That Make Fires." Published in *Harper's*, September 1949. Before his death, Eiseley noted that the second section of the *Harper's* text had been rendered obsolete by new scientific discoveries, and he recommended that only the first section be included in his "Sampler."

"Easter: The Isle of Faces." Published in *Holiday*, March 1962, as "The Island of Great Stone Faces."

"The Dance of the Frogs." Posthumously published in *Audubon*, May 1978.

"The Fifth Planet." First published in *The Star Thrower*.

"Science and the Sense of the Holy." Presented as an address at the Boston Museum of Science on October 18, 1976, on the occasion of Eiseley's receipt of the Bradford Washburn Award, and four days later as the keynote speech of the annual conference of the Humane Society of the United States, under the title "Darwin, Freud, and the Concept of the Animal Universe." Published in *Quest*, March–April 1978.

"The Winter of Man." Published in *The New York Times*, January 16, 1972.

"Man Against the Universe." Presented as a lecture at several institutions from 1974 to 1976. Published for the first time in *The Star Thrower*.

"Thoreau's Vision of the Natural World." Presented as an address to the Thoreau Society during the summer of 1973, and subsequently published as an afterword to *The Illustrated World of Thoreau*, edited by Howard Chapnick (Grosset & Dunlap, 1974).

"Walden: Thoreau's Unfinished Business." Presented in the President's Lecture Series at Rice University on March 9, 1975, and at other institutions. First published in *The Star Thrower*.

"The Lethal Factor." Presented as the Phi Beta Kappa–Sigma Xi Lecture at a meeting of the American Association for the Advancement of Science, near the end of December 1962. Published in *American Scientist*, March 1963, as "Man: The Lethal Factor."

"The Illusion of the Two Cultures." Presented as an address at ceremonies inaugurating the Ettinger Program for Creative Writing, Rockefeller University, June 27, 1963. Subsequently published in *The Rockefeller Institute Review*, March–April 1964, *The American Scholar*, Summer 1964, and other periodicals.

In the present volume, the texts of these items from *The Star Thrower* have been taken from the June 1978 Times Books first printing.

This volume presents the texts of the essays and other writings chosen for inclusion here, but it does not attempt to reproduce features of their typographic design, such as the display capitalization of chapter openings. The texts are reprinted without change, except for the correction of typographical errors. Spelling, punctuation, and capitalization are often expressive features, and they are not altered, even when inconsistent or irregular. The following is a list of typographical errors corrected, cited by page and line number: 8.28, sun ward; 37.4, BAITAILLON; 42.24, the the tracks; 48.6, experiment; 52.2, minature,; 57.7–8, candelight; 62.29, suported; 63.1, that; 76.9, to to; 113.31–32, University of of; 154.28, said.; 159.28–29, settler; 166.34, again a crow's; 176.1, in a; 178.3, enobling; 182.14, noncommital; 184.21, youngster, she; 186.39, life."; 204.29, us,; 209.39, stateman; 225.6, with the the; 251.10, thousands; 259.6, Sydney; 266.3, Subsconsciously; 266.19, Carl Schneer; 279.26, begins,; 284.1, Pneumatic; 285.15, growning; 330.31, fellow—I; 332.33, Schuykill; 363.25, box car; 368.28, chlorophyl; 369.20, proto-homonid; 381.14, possibility.; 419.15, adverse; 442.23–24, heart if.

Notes

In the notes below, the reference numbers denote page and line of this volume (the line count includes chapter headings but not blank lines). No note is generally made for material included in standard desk-reference works. Quotations from Shakespeare are keyed to *The Riverside Shakespeare*, ed. G. Blakemore Evans (Boston: Houghton Mifflin, 1974), and biblical references to the King James Version. For further information about Eiseley's life and works, and references to other studies, see: Gale E. Christianson, *Fox at the Wood's Edge: A Biography of Loren Eiseley* (New York: Henry Holt, 1990); Peter Heidtmann, *Loren Eiseley: A Modern Ishmael* (Hamden, CT: Archon Books, 1991); Tom Lynch and Susan N. Maher, eds., *Artifacts and Illuminations: Critical Essays on Loren Eiseley* (Lincoln: University of Nebraska, 2012); and Mary Ellen Pitts, *Toward a Dialogue of Understandings: Loren Eiseley and the Critique of Science* (Bethlehem, PA: Lehigh University Press, 1995).

THE INVISIBLE PYRAMID

2.2   Frank G. Speck] Speck (1881–1950) was chairman of the anthropology department at the University of Pennsylvania when Eiseley began his graduate studies there in 1933, and later supervised Eiseley's doctoral dissertation.

4.1–5   Once in a cycle . . . NEIHARDT] See "The Poet's Town" (1910) by Nebraska poet and ethnographer John G. Neihardt (1881–1973).

5.5–8   Like John Donne . . . beyond the body.] See the fourth meditation of Donne's *Devotions upon Emergent Occasions* (1624), the relevant text of which is quoted by Eiseley on page 36 of the present volume.

5.34–35   A. E. Housman . . . home and afar.] See *A Shropshire Lad* (1896), poem 40, by A. E. Housman (1859–1936).

6.9   Palomar Mountain] Site in San Diego County, California, of Palomar Observatory, which until 1992 possessed the world's largest telescope.

7.3–5   *Already at the origin* . . . ROSTAND] See Rostand's *The Substance of Man* (1962).

10.20–23   Pindar . . . to the end?"] See Pindar's sixth Nemean Ode (c. 465 B.C.E.).

11.19   Frémont's] John C. Frémont (1813–1890) conducted five western exploring expeditions between 1842 and 1854.

11.21–22   James Hutton] See Hutton's *Theory of the Earth; or an Investigation of the Laws Observable in the Composition, Dissolution, and Restoration of Land upon the Globe* (1788): "The result, therefore, of our present enquiry is, that we find no vestige of a beginning,—no prospect of an end."

11.38–12.6   Alexis de Tocqueville . . . touching the earth."] See Tocqueville's account of his travels in Michigan in 1831, published posthumously as *Quinze Jours au Désert* (1860) and subsequently translated as *A Fortnight in the Wilderness*.

17.9   A. L. Kroeber . . . superorganic] See Kroeber's essay "The Superorganic," first published in the April–June 1917 issue of *American Anthropologist*.

17.29–30   Hughlings Jackson] English neurologist John Hughlings Jackson (1835–1911).

18.17–18   "There never was . . . Glenn Jepsen] See "Time, Strata, and Fossils: Comments and Recommendations," in *Time and Stratigraphy in the Evolution of Man* (1967), the proceedings of a symposium held on October 16, 1965.

23.3–4   *Not till we are lost* . . . THOREAU] See Thoreau's *Walden* (1854), chapter 2: "Not till we are lost, in other words, not till we have lost the world, do we begin to find ourselves, and realize where we are and the infinite extent of our relations."

23.5–6   "A name . . . Kazantzakis.] See Kazantzakis's semi-autobiographical *Report to Greco* (1961).

30.18–25   a wise remark of George Santayana's . . . sequel."] See Santayana's *Realms of Being* (1942).

30.33–36   this statement . . . changeless."] See "Man's Place in Space and Time" by Thornton Page (1913–1996), in Page, ed., *Stars and Galaxies: Birth, Ageing, and Death in the Universe* (1962).

34.7–8   a Greek . . . vessel itself"] See Lucretius's *On the Nature of Things* (c. 50 B.C.E.), book 6.

34.20–23   Henry Vaughan . . . dismembered."] See Vaughan's poem "Vanity of Spirit" (1650).

34.27–33   Thomas Traherne . . . infinite."] See meditation 80 of Traherne's *Centuries*, first published posthumously in 1908.

36.13–19   John Donne . . . one steppe, everywhere."] See note 5.5–8.

37.3–4   *Really we create* . . . BATAILLON] French biologist Jean-Eugène Bataillon (1864–1953), quoted in Jean Rostand's *Can Man Be Modified?* (1956).

38.7–9   observers like myself . . . toward space?] In December 1968, in an interview with a reporter from the *Dallas Morning News*, Eiseley questioned the value of the space program; his remarks were widely criticized.

38.11–13 the words that Herman Melville . . . mad."] See *Moby-Dick; or, The Whale* (1851), chapter 41.

38.29–30 Arthur Clarke . . . end."] See *Childhood's End* (1953), a science fiction novel by Arthur C. Clarke (1917–2008).

39.30–32 "a tinkerer . . . R. J. Forbes.] See Forbes's *The Conquest of Nature: Technology and Its Consequences* (1968).

41.13–19 H. C. Conklin . . . culturally significant."] See Conklin's 1954 Yale Ph.D. dissertation, "The Relation of Hanunoo Culture to the Plant World," as quoted in *The Savage Mind* (1962) by Claude Lévi-Strauss.

43.8–14 Samuel Taylor Coleridge . . . human insight."] From an entry in *Specimens of the Table Talk of the Late Samuel Taylor Coleridge* (1835) dated May 1, 1830.

43.24–25 Joseph Glanvill . . . to America."] From *The Vanity of Dogmatizing, or Confidence in Opinions Manifested in a Discourse of the Shortness and Uncertainty of Our Knowledge, and Its Causes* (1661), revised as *Scepsis Scientifica* (1665).

44.32–35 Glanvill . . . a fiction."] See note 43.24–25.

44.39–40 Lewis Mumford . . . record.] See *The City in History* (1961): "Living *by* the record and *for* the record became one of the great stigmata of urban existence: indeed life as recorded—with all its temptations to over-dramatization, illusory inflation, and deliberate falsification—tended often to become more important than life as lived."

45.28–29 "We live in an epoch . . . Lovering] See "Mineral Resources from the Emerged Lands," a paper presented on June 16–17, 1968, at a joint meeting of the State Department's Policy Planning Council and the National Academy of Science's Special Panel of the Committee on Science and Public Policy, and published in *The Potential Impact of Science and Technology on Future U.S. Foreign Policy* (1968).

47.32–35 "If we must . . . place of honor."] See Whewell's *On the Philosophy of Discovery, Chapters Historical and Critical* (1860), chapter 15.

50.6–9 "The living memory . . . continuum."] See *The City in History: Its Origins, Its Transformations, and Its Prospects* (1961), chapter 17.

51.3–6 *Either the machine . . .* GARRETT] See chapter 2 of *Ouroboros; or, the Mechanical Extension of Mankind.*

52.39–40 "Thou canst not . . . Path itself."] See Helena Blavatsky, *The Voice of the Silence, Being Chosen Fragments from the "Book of the Golden Precepts"* (1889).

59.37–38 "Traveling . . . seventeenth-century writer.] See the Preface to *Villare Anglicum, or a View of the Townes of England* (1656), by Sir Henry Spelman (c. 1562–1641).

67.3–5 *The savage mind* . . . LÉVI-STRAUSS] The quotation, from *The Savage Mind* (1962), continues: "It builds mental structures which facilitate an understanding of the world in as much as they resemble it. In this sense savage thought can be defined as analogical thought."

71.33–34 "Every man . . . through life."] From an entry in Thoreau's journals dated January 5, 1860.

72.5–8 In Ruth Benedict's . . . time span.] See "Configurations of Culture in North America," first published in the January–March 1932 issue of *American Anthropologist.*

75.13–23 Henry Phillips . . . is known."] See Phillips's essay "On the Nature of Progress," published in *American Scientist* in October 1945.

77.2–4 as Joseph Campbell . . . no death."] See *The Masks of God: Primitive Mythology* (1959).

79.7–8 Margaret Mead . . . lark song."] See Mead's poem "Absolute Benison," first published in *The New Republic* on October 19, 1932: "Turn with nostalgia to that darkened garden / Where all eternal replicas are kept, / And the first rose and the first lark son / Since the first springtime have slept."

81.3–6 *From a* . . . LINDSAY] From Lindsay's poem "To Marie Delcourt-Curvers," collected in *The Origins of Alchemy in Graeco-Roman Egypt* (1970).

81.6–11 Jean Cocteau . . . the audience."] See Cocteau's essay "Le Numéro Barbette," first published in *Nouvelle Revue Française* in July 1926.

81.24–25 Blake . . . the natural.] Eiseley probably refers to an untitled poem William Blake (1757–1827) copied into a letter of November 22, 1802, to Thomas Butts, the relevant lines of which read: "For double the vision my Eyes do see / And a double vision is always with me / With my inward Eye 'tis an old Man grey / With my outward a Thistle across my way."

83.6–7 Jacques Maritain . . . thing for God.] See Maritain's *God and the Permission of Evil* (1966).

84.21–24 Thomas Love Peacock . . . crab, backward."] See Peacock's "The Four Ages of Poetry" (1820).

85.10–11 Ralph Waldo Emerson . . . no traveller."] See "Self-Reliance," from *Essays: First Series* (1841).

86.5–6 Don Stuart . . . "Twilight."] John W. Campbell (1910–1971) published "Twilight" under the pseudonym Don A. Stuart, in the November 1934 issue of *Astounding Stories.*

92.5–9 a young poet . . . and a man.] See Marlowe's *The Tragical History of Doctor Faustus* (first performed in 1592), I.i.23.

93.3–7 The human heart . . . SANTAYANA] See "The Philosophy of Travel," written around 1912 and first published posthumously in 1964.

98.29–30   Ortega y Gasset . . . only history."] See Ortega's essay "History as a System" (1935): "Man, in a word, has no nature; what he has is—history."

100.36–39   "Other sheep . . . follow me."] See John 10:16, 27.

106.39–107.2   "The poet . . . neither is of use."] See "Europe and European Books," first published in *The Dial* in April 1843.

107.16–17   "to pronounce . . . clear voice"] See *The Anatomy of Melancholy* (1621), I.2.1.2, by Robert Burton (1577–1640).

THE NIGHT COUNTRY

137.3–5   "This is your house . . . light."] See section 6 of "The House of Dust" (1920), by Conrad Aiken (1889–1973).

140.23   T. K. Oesterreich's *Demoniacal Possession*] *Possession, Demoniacal and Other: Among Primitive Races, in Antiquity, the Middle Ages, and Modern Times* by German psychologist and philosopher T. K. Oesterreich (1880–1949), first published in German in 1921.

149.16–26   "'Tis strange . . . our hope."] See *Macbeth*, I.iii.123–26 and V.viii.19–22.

151.15–20   George Santayana . . . what they are."] See "The Philanthropist," one of Santayana's *Dialogues in Limbo* (1925).

153.24–25   "Trust the divine . . . Emerson.] See "The Poet," collected in *Essays: Second Series* (1844): "As the traveller who has lost his way, throws his reins on his horse's neck, and trusts to the instinct of the animal to find his road, so we must do with the divine animal who carries us through the world."

153.35   "man crystallized,"] See Emerson's essay *Nature* (1836).

155.3–5   "Man's unhappiness . . . the Finite."] See "The Everlasting Yea," book 2, chapter 9, of *Sartor Resartus* (1836).

155.12–14   As Shakespeare . . . until he were."] See *Antony and Cleopatra*, I.iv.41–42.

157.4–10   "Former men . . . mutable cloud."] The first two sentences are quoted from Emerson's essay "Illusions" (*The Conduct of Life*, 1860), the third from "History" (*Essays: First Series*, 1841).

161.13–14   Frank Manuel's phrase . . . Ixion] See "Ixion's Wheel: The Renaissance Ponders the Vicissitudes," chapter 3 of Frank E. Manuel's *Shapes of Philosophical History* (1965). In Greek mythology, Ixion was cast out of Olympus and spun across the heavens on a burning wheel.

162.21–22   in the words . . . the sun."] See Ecclesiastes 1:9.

163.36–38   "God forbid . . . no more."] See Augustine's *The City of God* (c. 1470), book 12.14.

164.17–23   Jean Baptiste Lamarck . . . we see . . . ?"] See Lamarck's *Zoological Philosophy: An Exposition with Regard to the Natural History of Animals* (1809), chapter 3.

164.37–165.5   Arthur Weigall . . . drop of water."] See Weigall's *Travels in the Upper Egyptian Deserts* (1909), chapter 2 ("To the Quarries of Wady Hammamât").

167.13–15   the words of Karl Heim . . . fought for."] See *Christian Faith and Natural Science* (1953), chapter 2.

168.12–13   Philip Rieff . . . illness.] See *Freud: The Mind of a Moralist* (1959): "Aware at last that he is chronically ill, psychological man may nevertheless end the ancient quest of his predecessors for a healing doctrine. His experience with the latest one, Freud's, may finally teach him that every cure must expose him to new illness."

168.15–17   the greatest injunction . . . *staying*."] See John 8:14.

175.16–29   Leonard Cottrell's . . . eighty feet."] See Cottrell's *Lost Cities* (1957).

183.36   Charles Addams] American cartoonist (1912–1988) noted for his macabre characters.

186.9   Cochise, Victorio, Nana, and Geronimo] Apache leaders of the nineteenth century.

207.3–7   "I may truly . . . I live."] See Bacon's undated, unendorsed letter beginning "Cum ex literis, quas ad dominum Carey misisti," elsewhere dated c. 1609–10 and supposed to have been written to Isaac Casaubon.

207.9–12   Shakespeare . . . sleepe."] See *Hamlet*, V.ii.4–5.

208.36–40   As Francis Bacon . . . adversity."] See Bacon's *De Sapientia Veterum* ("Wisdom of the Ancients," 1609), section 6.

209.2–5   The beacon light . . . toward candlelight.] See "The Old Manse," first published in *Mosses from an Old Manse* (1846): "The light revealed objects unseen before . . . but also, as was unavoidable, it attracted bats and owls, and the whole host of night-bird, which flapped their dusky wings against the gazer's eyes, and sometimes were mistaken for fowls of angelic feather. Such delusions always hover nigh, whenever a beacon fire of truth is kindled."

209.27–28   as he put it . . . continent."] See Bacon's *Novum Organum* (1620), 114.

210.16–20   "There are . . . their career.'] From Emerson's lecture "Natural History of Intellect," as published posthumously in 1893. (He had given the lecture in various forms on multiple occasions between 1849 and 1871.)

213.15–22   "Strike a new . . . begins to operate."] From a letter dated April 27, 1897, and published for the first time in *D'Arcy Wentworth Thompson: The Scholar-Naturalist, 1860–1948* (1958) by Ruth D'Arcy Thompson.

213.28–34 "There is no nature . . . a lower."] See note 208.36–40.

213.40–214.6 Patrick Cruttwell . . . skie and earth."] See Cruttwell's *The Shakespearean Moment and Its Place in the Poetry of the 17th Century* (1954), and *Troilus and Cressida*, V.ii.147–49.

214.38–215.1 that ever-creative polity . . . citizenry.] Eiseley may refer to "The Strange Case of Dr. Jekyll and Mr. Hyde" (1886): "Others will follow, others will outstrip me on the same lines; and I hazard the guess that man will ultimately be known for a mere polity of multifarious, incongruous and independent denizens."

215.7–11 "A steep and unaccountable . . . describe them."] See the chapter titled "Friday" in Thoreau's *A Week on the Concord and Merrimack Rivers* (1849).

216.15–19 the words of G. K. Chesterton . . . our way."] From Chesterton's essay "The Ballade of a Strange Town," first collected in *Tremendous Trifles* (1909).

216.20–22 a book . . . and time] See Eiseley's essay collection *The Unexpected Universe* (1969).

218.17–25 this statement . . . overwhelmingly large."] From "The Meaning of Fitness," the second in a series of six BBC radio lectures titled and subsequently published as *The Future of Man* (1959).

219.24–25 The grim portrait . . . Jefferies' window] See "Hours of Spring" by English nature writer Richard Jefferies (1848–1883), first published in May 1886 and collected posthumously in *Field and Hedgerow* (1889).

220.12–13 Nature . . . shadow of man."] From Emerson's essay "Poetry and Imagination," published in *Letters and Social Aims* (1875).

221.11–35 what he calls a "cloud factory" . . . Where are we?"] From "Ktaadn," the first section of Thoreau's *The Maine Woods* (1864).

221.39–40 the spirit Ariel in a cloven pine.] See *The Tempest*, I.ii.277.

224.4–7 Recently it has been said . . . individual domain.] Eiseley probably refers to British novelist and chemist C. P. Snow (1905–1980) and his 1959 Rede Lecture, published the same year as *The Two Cultures and the Scientific Revolution*.

224.17–18 "There is no Excellent . . . Bacon.] See "Of Beauty," first published in the 1612 second edition of Bacon's *Essays*.

230.7–8 "Oh Lord . . . in himself."] See Jeremiah 10:23.

230.13–14 "The greatest prize . . . primitive man."] See Ward's *Modern Exploration* (1945), part 6.

235.16 Friday] A character in the novel *Robinson Crusoe* (1719) by Daniel

Defoe (c. 1660–1731), the first trace of whom comes in the form of "the Print of a Man's naked Foot on the Shore."

237.22–23 In the words . . . murder sleep."] See *Macbeth*, II.ii.33.

242.36–38 St. Paul . . . endureth all things."] See 1 Corinthians 13:7.

245.4–6 the stories . . . Tepexpan] American archaeologists joined an Italian team searching for Sybaris, in southern Italy, in 1962; the Tepexpan skeleton was discovered in 1947.

257.26 *East Lynne*] Any of several, widely popular theatrical adaptations of Ellen Wood's 1861 sensation novel *East Lynne*; the first to be produced under this title opened in 1863.

257.31–34 Elizabethan passages . . . longings in me."] See *Antony and Cleopatra*, V.ii.280–81.

258.28–35 "Uncertainty . . . immediate action."] From *The Quest for Certainty: A Study of the Relation of Knowledge and Action* (1929), lecture 9 ("The Supremacy of Method").

259.6–7 the sculptors . . . described us as being.] See Hook's *Education for Modern Man: A New Perspective* (revised edition, 1963).

259.26–33 Dewey . . . and instrumental."] See note 258.28–35.

261.4–13 Dewey . . . series of operations."] See note 258.28–35.

261.28–31 "A person too early . . . mammals."] From Rostand's *The Substance of Man* (1962).

262.7–9 "The whole of my pleasure . . . my mind."] From a letter of Charles Darwin to his fiancée Emma Wedgwood, dated January 20, 1839.

262.16–17 John Bunyan . . . of Thee"] See "The Author's Apology for His Book," preceding *The Pilgrim's Progress* (1628).

264.15–17 "A man . . . to be lost."] See *Walden* (1854), chapter 8.

264.30–39 Nathaniel Hawthorne . . . fragments of a world,"] See Hawthorne's story "The Devil in Manuscript" (1835).

265.10–13 Herman Melville . . . reveries."] See *Moby-Dick* (1851), chapter III.

265.29–33 "Humanity is not . . . importance."] From *The Public and Its Problems* (1927), chapter 5.

266.19–27 Cecil Schneer . . . labored for years."] See Schneer's *The Search for Order: The Development of the Major Ideas in the Physical Sciences from the Earliest Times to the Present* (1960).

268.19–27 Melville's words . . . here on earth."] See "Hawthorne and His

Mosses," first published in the *New York Literary World* on August 17 and 24, 1850.

268.36–269.6   Thoreau wrote . . . know them!"] From an entry in Thoreau's journals dated July 7, 1851.

269.14–16   As Thoreau comments . . . his tread."] From Thoreau's journals, July 21, 1851.

270.1–8   Thoreau's advice . . . things to her?"] From Thoreau's journals, August 19, 1851.

270.32–35   Robert Hofstadter . . . end of the trail."] Remarks at a Stanford, California, press conference in November 1961 following the news that Hofstadter would share the Nobel Prize in Physics.

272.17–20   "Who placed us . . . constant strain."] From Thoreau's journals, February 19, 1854.

272.34–38   William Butler Yeats . . . reality."] From Yeats's sonnet "Meru," first published in December 1934 and subsequently collected as the twelfth and final poem in the sequence "Supernatural Songs."

273.10–21   "I catch myself philosophizing . . . makes my sea."] The first quotation is taken from an entry in Thoreau's journals dated March 17, 1852, the second from an entry dated February 27, 1841.

273.22–26   Lawrence Kubie . . . equally gifted."] See Kubie's *Neurotic Distortion of the Creative Process* (1958).

275.16–18   Saint Paul . . . signification?] See 1 Corinthians 14:10.

276.17–18   as Emerson long ago . . . making him."] From Emerson's "Natural History of Intellect" (1893).

276.26–28   that great book . . . through it,"] Melville used the phrase to describe *Moby-Dick* in a letter to Sarah Morewood on September 12, 1851.

276.37–277.1   "As the dead man . . . significant."] See Thoreau's journals, December 8, 1859.

277.10–13   "What shall be . . . unbodied air."] See *Moby-Dick* (1851), chapter 33.

277.18–19   "The more an organism . . . itself going."] From *Experience and Nature* (1925), chapter 7.

from THE STAR THROWER

291.37–292.2   Dr. Lilly . . . to be questioned."] See "Science Notes," *New York Times*, June 26, 1960.

293.6   Three-Two-Three] In an incident described to the press by John C. Lilly (1915–2001), dolphins at the Communications Research Institute in the

Virgin Islands were observed for the first time to be mimicking human speech when one repeated the phrase "three-two-three," just spoken by a researcher, "in a 'Donald Duckish' but discernible voice." (See John W. Finney, "Navy Gives Dolphin High I.Q., Perhaps Equal to Human Beings," *New York Times*, June 21, 1960.)

297.14–19   as Melville wrote . . . pyramidical silence."] See *Moby-Dick*, chapter 79.

300.20–25   In 1865 . . . Australia."] See Lubbock's *Pre-historic Times, as Illustrated by Ancient Remains, and the Manners and Customs of Modern Savages* (1865).

302.16–26   Stewart says . . . fire vegetation."] See Stewart's essay "Why the Great Plains Are Treeless," *Colorado Quarterly*, Summer 1953.

302.34–39   Sauer argues . . . the mammoth.] See Carl O. Sauer, "A Geographic Sketch of Early Man in America," *Geographic Review*, 1944.

305.10–11   what Leslie White . . . "Fuel Revolution"] See White's *The Science of Culture: A Study of Man and Civilization* (1949).

323.1–4   "No simple . . . animals."] See Rudolf Schmid, *The Theories of Darwin and their Relation to Philosophy, Religion, and Morality* (1876), chapter 3.

344.37   Bode's law] Also termed the Titius-Bode law, a rule approximately predictive of the spacing of planets in the solar system.

347.14   'Seek . . . Book says.] See Matthew 7:7.

350.35–36   what one poet . . . brain."] See Eiseley's poem "The Face of the Lion," collected in *Notes of an Alchemist* (1972).

351.4–5   "I find it . . . confessed Freud.] See *Civilization and Its Discontents* (1930), chapter 1.

352.16–17   what Goethe . . . portentous."] See *Faust* (1808), part 2, 1.5.

353.24–27   Blaise Pascal . . . nothing else in."] See Pascal's *Pensées* (1670), 253.

353.32–35   "The whole . . . most of all myself."] From an entry in Kierkegaard's journal dated May 12, 1839.

353.35–354.2   Ernst Haeckel . . . ushered in."] See Haeckel's keynote speech at the 1877 convention of the Association of German Natural Scientists.

354.9–11   "A conviction . . . Einstein.] See Einstein's "On Scientific Truth," first published in German in 1929 and collected in English in *Ideas and Opinions* (1954).

354.12–14   Thoreau . . . mould myself?"] See *Walden; or, Life in the Woods* (1854), chapter 5.

355.21–23 the words of Thoreau . . . kind to man.] See *The Maine Woods*, chapter 1: "It was Matter, vast, terrific,—not his Mother Earth that we have heard of, not for him to tread on, or be buried in,—no, it were being too familiar even to let his bones lie there,—the home, this, of Necessity and Fate. There was there felt the presence of a force not bound to be kind to man."

360.31–34 another humbler thinker . . . his feet.] See "Days of a Doubter" in Eiseley's memoir *All the Strange Hours* (1975).

361.15–17 "all a magnet . . . upon him."] See *Moby-Dick*, chapters 100 and 36.

361.30 "strike through the mask,"] See *Moby-Dick*, chapter 36.

361.36–38 "All my means . . . object mad."] See *Moby-Dick*, chapter 41.

362.27–28 "Deep down . . . joy."] See *Moby-Dick*, chapter 87.

367.15–16 "She will tell . . . once warned] As related by Raphael Meldola in "The Speculative Method in Entomology," a presidential address to the Entomological Society of London delivered on January 15, 1896, and published in the *Transactions of the Entomological Society of London for the Year 1895* (1896).

370.1–5 As Robin Collingwood . . . Middle Ages.] See Collingwood's *The Idea of Nature* (1945), part 2, chapter 1, section 3.

371.6–7 a widely circulated . . . evolutionary cause.] See Chambers's *Vestiges of the Natural History of Creation* (1844).

372.14–16 Ralph Waldo Emerson . . . the speaker."] See Emerson's "The Method of Nature: An Oration" (1841).

372.34 "dirty and amused"] See Thomas Gray's letter of June 24, 1769, to the Rev. Norton Nicholls.

373.21–26 described by Emerson . . . parcel of God."] See Emerson's "Nature" (1836).

374.13 "kitchen clock"] From an undated entry in Emerson's journals of 1856: "The old six thousand years of chronology is a kitchen clock, no more a measure of time than an hour-glass or an egg-glass since the durations of geologic periods have come in use."

374.14–15 "What terrible questions . . . ask."] See Emerson's essay "Illusions," first collected in *The Conduct of Life* (1860).

374.28–30 "I found . . . the stairs."] From an undated essay in Emerson's journals of 1850–51.

375.5–6 Darwin's words . . . Nature."] From Darwin's letter of July 13, 1856, to Joseph Dalton Hooker.

375.11–13 "I delight . . . mortal men."] From a letter to Unitarian theologian Henry Ware Jr. dated October 8, 1838.

375.17–18 "by the help . . . to dance."] See Emerson's letter of March 14, 1841, to Margaret Fuller.

375.25–26 Emerson's pronouncement . . . has made."] See Emerson's "Compensation," from *Essays: First Series* (1841).

375.28–29 "interiors . . . crocodiles."] See Emerson's essay "The Sovereignty of Ethics," first published in the *North American Review* in May 1878.

376.6–9 As Emerson . . . out of sight."] See the essay "Experience," collected in *Essays: Second Series* (1844).

376.19–28 In 1818 . . . far gone.] From a verse letter to J. H. Reynolds dated March 25, 1818.

376.30–31 what Emerson also . . . the universe."] From an undated entry in Emerson's journals of 1854: "The existence of evil & malignant men does not depend on themselves or on men, it indicates the virulence that still remains uncured in the universe, uncured & corrupting & hurling out these pestilent rats & tigers, and men rat-like & wolf-like."

376.36–38 Darwin himself . . . wild experiment."] See Darwin's letter of March 26, 1863, to Joseph Dalton Hooker.

377.18–21 Emerson . . . English books?"] From an undated entry in Emerson's journals of 1845.

377.34–39 who wrote . . . not succeeded."] See "The Method of Nature" (1841).

378.5–8 "To a universe . . . globe."] See "The Method of Nature" (1841).

378.10–13 "The world . . . vast idea."] See Emerson's "Lecture on the Times," delivered at the Masonic Temple in Boston on December 2, 1841.

378.18–24 "The transmigration . . . heaven-facing speakers."] See "History," collected in *Essays: First Series* (1841).

378.28–30 whom Walt Whitman called . . . for daring."] From a private notation of 1863: "The most exquisite taste & caution are in him, always saving him his feet from passing beyond the limits; for his is transcendental of limits, & you see underneath the rest a secret proclivity America may be, to dare & violate & make escapades."

379.39–40 George Santayana . . . true fall."] See "Piety," first collected in *Reason and Religion* (1905).

381.3–7 a phrase in Whitman . . . he became.] The opening lines of "There Was a Child Went Forth," first published in untitled form in the first edition of *Leaves of Grass* (1855).

381.10–14 Emerson's earlier . . . of possibility?"] See "The Method of Nature" (1841).

381.15–16   "not to be . . . Whitman's words] See "To Think of Time" (1855), section 7: "It is not to diffuse you that you were born."

381.30–32   "What is a man . . . self-explication?"] From an undated entry in Emerson's journals for 1840; also subsequently included in his essay "Art" (1841).

381.39–382.1   What he terms . . . extremities"] See "The American Scholar," addressed to the Phi Beta Kappa Society in Cambridge, Massachusetts, on August 31, 1837.

382.10–11   as he himself . . . faith"] See Emerson's 1842 lecture "The Transcendentalist," collected in *Nature; Addresses and Lectures* (1849).

382.19–21   In the words . . . destined to become."] See *Can Man Be Modified?* (1959).

382.28–29   Emerson maintains . . . population."] See "Fate," collected in *The Conduct of Life* (1860).

383.10–11   "The entrance . . . birth of man."] See the 1841 lecture "Man the Reformer," collected in *Nature; Addresses and Lectures* (1849).

386.13–14   labeled a prig . . . English letters.] See "Henry David Thoreau: His Character and Opinions" by Robert Louis Stevenson, first published in *Cornhill Magazine*, June 1880.

387.6–7   in his own . . . in the sun."] From Thoreau's journals, November 22, 1860.

387.21   "inhumanity of science."] From Thoreau's journals, May 28, 1854.

389.30–31   "it is ebb . . . reports."] From Thoreau's journals, March 5, 1858.

390.14–17   "Geology . . . noticed anywhere."] See *Contributions to the Natural History of the United States of America* (1857), chapter 1, section 15.

390.25–27   "What kind . . . disguise it?"] From Thoreau's journals, January 14, 1857.

391.5–8   "How can [man] . . . the end?"] From Thoreau's journals, April 3, 1842.

391.11–12   Thoreau remarks . . . rule of existence?"] From Thoreau's journals, September 1, 1851.

391.20–21   In Emerson's . . . wild contrast."] See "The Transcendentalist" (1842).

391.22–24   in the first volume . . . the ideal."] From Thoreau's journals, April 3, 1842.

392.20–21   Alfred North Whitehead . . . nature."] See *The Concept of Nature* (1920), chapter 2.

392.37–38 the "ape . . . Huxley's writing.] See Huxley's 1893 Romanes Lecture "Evolution and Ethics."

393.21–23 "The development theory . . . new creation."] From Thoreau's journals, October 18, 1860.

393.40–394.1 "My senses . . . rest."] From Thoreau's journals, September 13, 1852.

394.1–4 Whitehead's dictum . . . knowing."] See *The Concept of Nature* (1920), chapter 3.

394.4–15 Thoreau . . . the elements."] See the chapter "Thursday" in *A Week on the Concord and Merrimack Rivers* (1849).

394.19–20 "a patent for himself."] From Thoreau's journals, November 29, 1860.

394.22–24 "Are we not . . . wreckers?"] See *Cape Cod* (1865), chapter 6.

394.26–30 "Is our life . . . success in history."] From Thoreau's journals, May 28, 1854, and August 18, 1854.

394.38–395.2 As a somewhat . . . attention."] See William Butler Yeats's "Samhain: 1905" (1905).

395.19 Thoreau . . . revealer."] From Thoreau's journals, February 14, 1854.

395.32–33 Emerson's phrase . . . of man."] See "Poetry and Imagination," first collected in *Letters and Social Aims* (1875).

396.18–21 "I suspect . . . further."] See *The Maine Woods* (1864), section 1.

396.25–33 "I do not think . . . explained."] From Thoreau's journals, July 16, 1850, and April 20, 1840.

397.2–3 "One world . . . deathbed] In an anecdote recounted by Robert Collyer ("Henry Thoreau," *Unity*, August 1, 1879), to his friend Parker Pillsbury's questions about what he can see of "the waiting world," Thoreau responds "one world at a time, Parker."

397.5 "a prairie for outlaws."] From Thoreau's journals, January 2, 1853.

399.7–11 Ellery Channing . . . I judge."] Channing wrote these lines in his notebook after Thoreau's death on May 6, 1862.

399.15–21 Wright Morris . . . business?"] See "The Question of Privacy," first published in *The American Magazine of Art*, February 1951; Morris quotes *Walden*, chapter 2.

399.28–30 he muses . . . year ago."] From Thoreau's journals, November 1, 1858.

400.22–23 "Few adults . . . nature."] See "Nature" (1836): "To speak truly, few adult persons can see nature."

401.20–21   "Wildness . . . ventured.] From Thoreau's journals, February 16, 1859.

402.34   an erudite citizen of Concord.] Eiseley visited Walden Pond in the summer of 1973 with Malcolm Ferguson (1919–2011), president of the Concord Lyceum.

403.3–5   "All the ages . . . smell them out."] From Thoreau's journals, September 24, 1859.

403.8–17   "I landed . . . trail of mind."] From Thoreau's journals, March 28, 1859.

403.33–37   "If the outside . . . tenanted by owls."] From Thoreau's journals, March 10, 1859.

404.9   "the fluctuations . . . mind."] From Thoreau's journals, January 30, 1841.

404.15   "not bound . . . man."] See *The Maine Woods* (1864), part I.

404.16–17   "the earth . . . bury me."] From Thoreau's journals, November 1, 1858.

404.25–38   As Alfred North Whitehead . . . no civilization."] See Whitehead's *Nature and Life* (1934).

405.11–14   "He shall spend . . . poets."] See *The Maine Woods* (1864), part I.

405.29–30   "fossil thoughts,"] From Thoreau' journals, March 28, 1859.

406.36–37   "Nothing must . . . moment."] From Thoreau's journals, April 24, 1859: "Nothing must be postponed. Take time by the forelock. Now or never! You must live in the present, launch yourself on every wave, find your eternity in each moment."

407.4   "Miasma . . . wrote.] From Thoreau's journals, August 23, 1853.

408.31–32   "small and almost . . . clouds."] From Thoreau's journals, July 19, 1858.

408.35–36   "It is not . . . Zanzibar,"] See *Walden*, chapter 17.

409.13–14   "into some . . . acorns."] From Thoreau's journals, December 21, 1840.

409.17–19   "There is no more . . . living."] See the essay "Life without Principle" (1862).

409.23–29   "I sat in my . . . night."] See *Walden*, chapter 4.

409.39   "Truth strikes . . . the dark."] From Thoreau's journals, November 5, 1837.

410.17–20 a posthumous work . . . combined."] Thoreau's *Cape Cod*, serialized in *Putnam's Monthly Magazine* during his lifetime (the quoted text in August 1855), but not published in book form until 1865, after his death.

413.25–26 "Simplify . . . advocated.] See *Walden*, chapter 2.

415.18–26 Alfred North Whitehead . . . prehension of the past,"] For both quotations, see "Process and Reality," published in the *Symposium in Honor of the Seventieth Birthday of Alfred North Whitehead* (1932).

417.17 "violence . . . Francis Bacon's.] See Bacon's essay "Of Nature in Men" (1612).

417.21–25 the poet J. C. Squire . . . dwindle"] From "The Birds," first collected in *The Birds and Other Poems* (1920).

418.22–28 Herman Melville . . . soul of man!] See Melville's novel *Pierre: or, The Ambiguities* (1852), book 21.

419.17–21 Sir Charles Lyell . . . true."] See Lyell's letter of March 7, 1837, to the Rev. W. Whewell.

420.6–8 "The foxes . . . lay his head."] See Matthew 8:20.

422.18 Lord Acton . . . the soul."] See Acton's letter to contributors to *The Cambridge Modern History*, published in *Lectures on Modern History* (1906), appendix I.

426.14–19 Thomas Beddoes . . . murdering star.] See Beddoes's play *Death's Jest-Book; or, The Fool's Tragedy*, published posthumously in 1850.

428.8–10 Alfred North Whitehead . . . anticipation."] See Whitehead's "Immortality," first published in *The Philosophy of Alfred North Whitehead* (Library of Living Philosophers series) in 1941.

428.32–33 Erich Frank . . . changes."] See Frank's "The Role of History in Christian Thought," first published in *The Duke Divinity School Bulletin*, November 1949.

431.2–5 Sir Eric Ashby . . . to society."] See Ashby's 1963 presidential address to the British Association for the Advancement of Science.

431.31–36 "The hard-pressed . . . altogether."] See "Locke and the Frontiers of Common Sense," collected in *Some Turns of Thought in Modern Philosophy* (1933).

432.8–9 Thomas Huxley . . . poets."] See Huxley's "Scientific Education: Notes of an After-Dinner Speech," originally presented before the Liverpool Philomathic Society on April 17, 1869, and subsequently published in *Lay Sermons, Addresses, and Reviews* (1870): "[T]here is no sight in the whole world more saddening and revolting than is offered by men sunk in ignorance of everything but what other men have written; seemingly devoid of moral belief or guidance; but with the sense of beauty so keen, and the power of expression so cultivated, that their sensual caterwauling may almost be taken for the music of the spheres."

433.27–34 "I here live . . . ancient world."] See Petrie's *Seventy Years in Archaeology* (1931).

436.33–34 "awful power . . . imagination.] See Wordsworth's *Prelude* (1850), book 6.

437.29–31 François Mauriac . . . its own.] See "Notre raison d'être," *La Table ronde*, August–September 1949.

437.36–37 "Man . . . self-fabricating animal."] See Mumford's *The Conduct of Life* (1951).

438.16–23 John Donne . . . Holy Ghost."] See Donne's Easter 1624 sermon at St. Paul's, on Revelation 20:6.

439.28–29 As Donne himself . . . lie behind."] The quoted text is from Austin Farrer's *The Glass of Vision* (1948), lecture 8.

441.13–18 Thomas Hobbes . . . the dark."] See Hobbes's *Leviathan* (1651), part 2.

441.18–19 I myself . . . we know."] See Eiseley's *The Firmament of Time* (1960), chapter 6.

441.33 "an art which nature makes."] See *The Winter's Tale*, IV.iv.91–92.

441.35–37 Henri Bergson . . . into matter."] See chapter 2 of Bergson's *Creative Evolution*, first published in French in 1907.

Index

*This book is set in 10 point ITC Galliard, a face
designed for digital composition by Matthew Carter and based
on the sixteenth-century face Granjon. The paper is acid-free
lightweight opaque that will not turn yellow or brittle with age.
The binding is sewn, which allows the book to open easily and lie flat.
The binding board is covered in Brillianta, a woven rayon cloth
made by Van Heek–Scholco Textielfabrieken, Holland.
Composition by Dedicated Book Services.
Printing and binding by Edwards Brothers Malloy, Ann Arbor.
Designed by Bruce Campbell.*